Wave Scattering Theory

Springer

Berlin
Heidelberg
New York
Barcelona
Hongkong
London
Milan
Paris
Singapore
Tokyo

Hyo J. Eom

Wave Scattering Theory

A Series Approach Based
on the Fourier Transformation

With 62 Figures

 Springer

Professor Hyo J. Eom

Korea Advanced Institute
of Science and Technology
Department of Electrical Engineering
373-1, Kusong-dong, Yusong-gu
305-701 Teajon / Korea

ISBN 3-540-41860-1 Springer-Verlag Berlin Heidelberg New York

CIP data aaplied for

Die Deutsche Bibliothek - CIP-Einheitsaufnahme
Eom, Hyo J.: Wave scattering theory : a series approach based on the Fourier
transformation / Hyo J. Eom. - Berlin ; Heidelberg ; New York ; Barcelona ; Hongkong ;
London ; Milan ; Paris ; Singapore ; Tokyo : Springer, 2001
  ISBN 3-540-41860-1

Springer-Verlag Berlin Heidelberg New York
a member of BertelsmannSpringer Science+Business Media GmbH

http://www.springer.de

© Springer-Verlag Berlin Heidelberg 2001
Printed in Germany

Typesetting: Dataconversion by author
Cover-design: Medio Technologies AG, Berlin
Printed on acid-free paper   SPIN: 10834045   62 / 3020 hu - 5 4 3 2 1 0 -

Preface

The Fourier transform technique has been widely used in electrical engineering, which covers signal processing, communication, system control, electromagnetics, and optics. The Fourier transform technique is particularly useful in electromagnetics and optics since it provides a convenient mathematical representation for wave scattering, diffraction, and propagation. Thus the Fourier transform technique has been long applied to the wave scattering problems that are often encountered in microwave antenna, radiation, diffraction, and electromagnetic interference. In order to understand wave scattering in general, it is necessary to solve the wave equation subject to the prescribed boundary conditions. The purpose of this monograph is to present rigorous solutions to the boundary-value problems by solving the wave equation based on the Fourier transform. In this monograph the technique of separation of variables is used to solve the wave equation for canonical scattering geometries such as conducting waveguide structures and rectangular/circular apertures. The Fourier transform, mode-matching, and residue calculus techniques are applied to obtain simple, analytic, and rapidly-convergent series solutions. The residue calculus technique is particularly instrumental in converting the solutions into series representations that are efficient and amenable to numerical analysis. We next summarize the steps of analysis method for the scattering problems considered in this book.

1. Divide the scattering domain into closed and open regions.
2. Represent the scattered fields in the closed and open regions in terms of the Fourier series and transform, respectively.
3. Enforce the boundary conditions on the field continuities between the open and closed regions.
4. Apply the mode-matching technique to obtain the simultaneous equations for the Fourier series modal coefficients.
5. Utilize the residue calculus to represent the scattered field in fast convergent series.

This monograph discusses time-harmonic wave scattering problems and a time factor $\exp(-i\omega t)$ is suppressed throughout the analysis. In each section, a set of simultaneous equations for the Fourier series coefficients is boxed. A set of the boxed simultaneous equations is the rigorous final formulation and must be numerically evaluated to further investigate the wave scattering behaviors. This book contains 9 chapters. In Chapter 1 electromagnetic scattering from rectangular grooves in a conducting plane is considered. In Chapter 2 electromagnetic wave radiation from multiple parallel-plate waveguide with an infinite flange is analyzed. In Chapter 3 electromagnetic, electrostatic, and magnetostatic penetrations into slits in a conducting plane are considered. In Chapter 4 electromagnetic wave guidance by a certain class of waveguides and couplers is considered. In Chapter 5 electromagnetic and acoustic wave scattering from junctions in rectangular waveguides is analyzed. In Chapter 6 wave scattering from rectangular apertures in a plane is studied. In Chapter 7 wave scattering from circular apertures in a plane is examined using the Hankel transform. In Chapter 8 wave scattering from an annular aperture in a conducting plane is considered. In Chapter 9 electromagnetic wave radiation from circumferential apertures on a circular cylinder is analyzed. All the work presented in this monograph was performed from 1992 through 2000 at the Korea Advanced Institute of Science and Technology (KAIST). My sincere thanks go to my former graduate students at KAIST (T. J. Park, K. H. Park, S. H. Kang, J. H. Lee, J. W. Lee, Y. C. Noh, K. H. Jun, Y. S. Kim, H. H. Park, S. B. Park, K.W. Lee, J. S. Seo, J. G. Lee, S. H. Min, J. K. Park, H. S. Lee, J. Y. Kwon, and Y. H. Cho), who carried out the tedious problem formulations under my guidance. My thanks also go to my wife and son for their patience with me while I was working on this monograph. Any comments and suggestions from readers to improve the monograph would be gratefully received.

Taejon, Korea   Hyo J. Eom

Contents

Notations

EM : electromagnetic

$$F_m(\eta) \;=\; \frac{e^{i\eta}(-1)^m - e^{-i\eta}}{\eta^2 - (m\pi/2)^2}$$

$H_n^{(1)}(\eta), H_n^{(2)}(\eta)$: nth order Hankel functions of the first and second kinds

$i = \sqrt{-1}$

$Im\{\cdots\}$: imaginary part of $\{\cdots\}$

$J_n(\eta), N_n(\eta)$: nth order Bessel functions of the first and second kinds

PEC : perfect electric conductor

$(\hat{r}, \hat{\phi}, \hat{z})$: unit vectors in cylindrical coordinates

$Re\{\cdots\}$: real part of $\{\cdots\}$

$(\hat{x}, \hat{y}, \hat{z})$: unit vectors in rectangular coordinates

δ_{mn} : Kronecker delta

$\varepsilon_0 = 2$

$\varepsilon_m = 1 \; (m = 1, 2, 3, \ldots)$

ω : angular frequency

$(\cdots)^*$: complex conjugate of $(\cdots)$

Transform Definitions

Note that $\widetilde{f}(\zeta)$ is called a transform of $f(x)$ and $\widetilde{f}(\zeta)$ is called an inverse transform of $f(x)$. The respective transform pairs are as follows:

Fourier transform

$$\widetilde{f}(\zeta) = \int_{-\infty}^{\infty} f(x)e^{i\zeta x}\,\mathrm{d}x$$

$$f(x) = \frac{1}{2\pi}\int_{-\infty}^{\infty} \widetilde{f}(\zeta)e^{-i\zeta x}\,\mathrm{d}\zeta$$

$$\widetilde{f}(\zeta,\eta) = \int_{-\infty}^{\infty}\int_{-\infty}^{\infty} f(x,y)\exp(i\zeta x + i\eta y)\,\mathrm{d}x\mathrm{d}y$$

$$f(x,y) = \frac{1}{4\pi^2}\int_{-\infty}^{\infty}\int_{-\infty}^{\infty} \widetilde{f}(\zeta,\eta)\exp(-i\zeta x - i\eta y)\,\mathrm{d}\zeta\mathrm{d}\eta$$

Fourier cosine transform

$$\widetilde{f}(\zeta) = \int_{0}^{\infty} f(x)\cos\zeta x\,\mathrm{d}x$$

$$f(x) = \frac{2}{\pi}\int_{0}^{\infty} \widetilde{f}(\zeta)\cos\zeta x\,\mathrm{d}\zeta$$

Fourier sine transform

$$\widetilde{f}(\zeta) = \int_{0}^{\infty} f(x)\sin\zeta x\,\mathrm{d}x$$

$$f(x) = \frac{2}{\pi}\int_{0}^{\infty} \widetilde{f}(\zeta)\sin\zeta x\,\mathrm{d}\zeta$$

1. Rectangular Grooves in a Plane

1.1 EM Scattering from a Rectangular Groove in a Conducting Plane

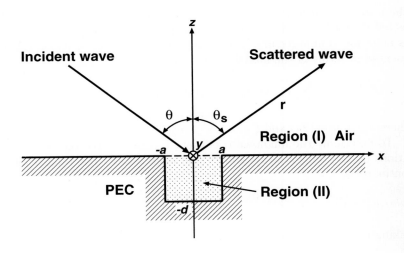

Fig. 1.1. A rectangular groove in a conducting plane

A rectangular groove in a conducting plane is a canonical structure in electromagnetic scattering study. Electromagnetic wave scattering from a groove in a conducting plane was considered in [1-5] due to its practical applications in optics and microwaves. Practical applications, for instance, include a design of optical-video disks and a radar-cross-section estimation. In the next two subsections we will consider TE (transverse electric to the y-axis) and TM (transverse magnetic to the y-axis) scattering from a two-dimensional rectangular groove in a perfectly-conducting plane.

1.1.1 TE Scattering [6]

Consider TE scattering from a rectangular groove in a perfectly-conducting plane. In region (I) $(z > 0)$ a uniform plane wave $E_y^i(x, z)$ is assumed to be incident on a rectangular groove in a perfectly-conducting plane. Region (II) $(-d < z < 0$ and $-a < x < a)$ is an infinitely-long rectangular groove engraved in the y-direction. The wavenumbers in regions (I) and (II) are k_0 $(= \omega\sqrt{\mu_0\epsilon_0})$ and k $(= \omega\sqrt{\mu\epsilon})$, respectively, where $\mu = \mu_r\mu_0$ and $\epsilon = \epsilon_r\epsilon_0$. The total E-field in region (I) is a sum of the incident, specularly-reflected, and scattered fields

$$E_y^i(x, z) = \exp(ik_x x - ik_z z) \tag{1}$$

$$E_y^r(x, z) = -\exp(ik_x x + ik_z z) \tag{2}$$

$$E_y^s(x, z) = \frac{1}{2\pi}\int_{-\infty}^{\infty} \widetilde{E}_y^s(\zeta)\exp(-i\zeta x + i\kappa_0 z)\,d\zeta \tag{3}$$

where $k_x = k_0\sin\theta$, $k_z = k_0\cos\theta$, and $\kappa_0 = \sqrt{k_0^2 - \zeta^2}$. We include the reflected field $E_y^r(x, z)$ in the total field expression, for convenience, although its inclusion is unnecessary. The total transmitted field in region (II) is represented in terms of the modal coefficient c_m

$$E_y^t(x, z) = \sum_{m=1}^{\infty} c_m \sin a_m(x + a)\sin\xi_m(z + d) \tag{4}$$

where $a_m = m\pi/(2a)$ and $\xi_m = \sqrt{k^2 - a_m^2}$.

To determine the unknown coefficient c_m, we enforce the boundary conditions on the tangential E- and H-field continuities. The tangential E-field continuity at $z = 0$ is

$$E_y^s(x, 0) = \begin{cases} E_y^t(x, 0), & |x| < a \\ 0, & |x| > a. \end{cases} \tag{5}$$

Taking the Fourier transform of (5) yields

$$\widetilde{E}_y^s(\zeta) = \sum_{m=1}^{\infty} c_m \sin(\xi_m d)a_m a^2 F_m(\zeta a). \tag{6}$$

The tangential H-field continuity along $(-a < x < a)$ at $z = 0$ is written as

$$2ik_z e^{ik_x x} - \int_{-\infty}^{\infty} \frac{i\kappa_0}{2\pi}\widetilde{E}_y^s(\zeta)e^{-i\zeta x}\,d\zeta$$

$$= \sum_{m=1}^{\infty} -\frac{c_m\xi_m}{\mu_r}\sin a_m(x + a)\cos(\xi_m d). \tag{7}$$

We multiply (7) by $\sin a_n(x + a)$ and integrate from $-a$ to a to obtain

$$2\mathrm{i}k_z a_n a^2 F_n(k_x a)$$
$$= \frac{\mathrm{i}a_n}{2\pi} \sum_{m=1}^{\infty} c_m a_m a^2 \sin(\xi_m d) \Lambda_1(k_0) - \frac{c_n \xi_n}{\mu_r} a \cos(\xi_n d) \qquad (8)$$

where

$$\Lambda_1(k_0) = \int_{-\infty}^{\infty} a^2 F_m(\zeta a) F_n(-\zeta a) \kappa_0 \,\mathrm{d}\zeta \ . \qquad (9)$$

It is convenient to transform $\Lambda_1(k_0)$ into a numerically-efficient form by performing a contour integration. The result is available in Subsect. 1.1.3 to give

$$\Lambda_1(k_0) = 2\pi \frac{\sqrt{k_0^2 - a_m^2}}{a a_m^2} \delta_{mn} - \bar{\Lambda}_1(k_0) \ . \qquad (10)$$

Note that $\bar{\Lambda}_1(k_0)$ is a numerically-efficient integral, which vanishes in high-frequency limit; thus for $k_0 a \to \infty$,

$$\Lambda_1(k_0) \to 2\pi \frac{\sqrt{k_0^2 - a_m^2}}{a a_m^2} \delta_{mn} \ . \qquad (11)$$

Substituting (11) into (8) yields an approximate high-frequency solution, which agrees with that in the Kirchhoff approximation.

The far-zone scattered field at distance r from the origin is shown to be

$$E_y^s(\theta_s, \theta) = \sqrt{\frac{k_0}{2\pi r}} \exp(\mathrm{i}k_0 r - \mathrm{i}\pi/4) \cos\theta_s \tilde{E}_y^s(-k_0 \sin\theta_s) \ . \qquad (12)$$

1.1.2 TM Scattering [7]

Consider a TM wave impinging on a rectangular groove in a conducting plane. In region (I) the total H-field is a sum of the incident, reflected, and scattered fields as

$$H_y^i(x, z) = \exp(\mathrm{i}k_x x - \mathrm{i}k_z z) \qquad (13)$$
$$H_y^r(x, z) = \exp(\mathrm{i}k_x x + \mathrm{i}k_z z) \qquad (14)$$
$$H_y^s(x, z) = \frac{1}{2\pi} \int_{-\infty}^{\infty} \tilde{H}_y^s(\zeta) \exp(-\mathrm{i}\zeta x + \mathrm{i}\kappa_0 z) \,\mathrm{d}\zeta \ . \qquad (15)$$

In region (II) the total transmitted field is

$$H_y^t(x, z) = \sum_{m=0}^{\infty} c_m \cos a_m(x + a) \cos \xi_m(z + d) \ . \qquad (16)$$

Applying the Fourier transform to the tangential E-field, $E_x(x, 0)$, continuity along the x-axis yields

$$\widetilde{H}_y^s(\zeta) = \sum_{m=0}^{\infty} c_m \xi_m \sin(\xi_m d) \frac{\zeta}{\kappa_0 \epsilon_r} a^2 F_m(\zeta a) \ . \tag{17}$$

The tangential H-field continuity along the boundary $(-a < x < a$ and $z = 0)$ requires

$$2\mathrm{e}^{\mathrm{i}k_x x} + \int_{-\infty}^{\infty} \frac{1}{2\pi} \widetilde{H}_y^s(\zeta) \mathrm{e}^{-\mathrm{i}\zeta x} \, \mathrm{d}\zeta$$

$$= \sum_{m=0}^{\infty} c_m \cos a_m (x + a) \cos(\xi_m d) \ . \tag{18}$$

Multiplying (18) by $\cos a_n (x + a)$ and integrating from $-a$ to a, we obtain

$$\boxed{\begin{aligned} &2\mathrm{i}k_x a^2 F_n(k_x a) \\ &= \frac{\mathrm{i}}{2\pi\epsilon_r} \sum_{m=0}^{\infty} c_m \xi_m a^2 \sin(\xi_m d) \Omega_1(k_0) - c_n a \varepsilon_n \cos(\xi_n d) \end{aligned}} \tag{19}$$

where

$$\Omega_1(k_0) = \int_{-\infty}^{\infty} a^2 F_m(\zeta a) F_n(-\zeta a) \zeta^2 \kappa_0^{-1} \, \mathrm{d}\zeta \ . \tag{20}$$

Performing the residue calculus, we get

$$\Omega_1(k_0) = \frac{2\pi\varepsilon_n}{a\sqrt{k_0^2 - a_m^2}} \delta_{mn} - \bar{\Omega}_1(k_0) \tag{21}$$

where $\Omega_1(k_0) \to 2\pi\varepsilon_n/(a\sqrt{k_0^2 - a_m^2})\delta_{mn}$ in high-frequency limit $(k_0 a \to \infty)$. The far-zone scattered field at distance r is

$$H_y^s(\theta_s, \theta) = \sqrt{\frac{k_0}{2\pi r}} \exp(\mathrm{i}k_0 r - \mathrm{i}\pi/4) \cos\theta_s \widetilde{H}_y^s(-k_0 \sin\theta_s) \ . \tag{22}$$

1.1.3 Appendix

Consider

$$\Lambda_1(k_0) = \int_{-\infty}^{\infty} a^2 F_m(\zeta a) F_n(-\zeta a) \kappa_0 \, \mathrm{d}\zeta \ . \tag{23}$$

When $m + n$ is odd, $\Lambda_1(k_0) = 0$. When $m + n$ is even, $\Lambda_1(k_0)$ is rewritten as

$$\Lambda_1(k_0) = \int_{-\infty}^{\infty} 2\frac{1 - (-1)^n \exp(\mathrm{i}2\zeta a)\kappa_0}{(\zeta^2 - a_m^2)(\zeta^2 - a_n^2)a^2} \, \mathrm{d}\zeta \ . \tag{24}$$

Integrating along the deformed contour $\Gamma_1, \Gamma_2, \Gamma_3$ and Γ_4 in the upper-half plane, we get

$$\Lambda_1(k_0) = \frac{2\pi\sqrt{k_0^2 - a_m^2}}{a a_m^2} \delta_{mn} - \bar{\Lambda}_1(k_0) \tag{25}$$

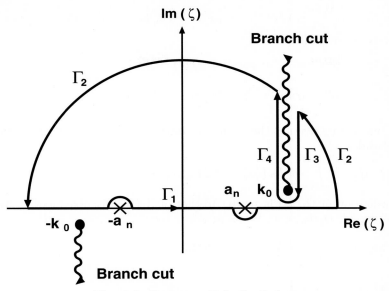

Fig. 1.2. Contour path in the ζ-plane

where the first term is a residue contribution at $\zeta = \pm a_m$ when $m = n$ and the second term results from integration along the branch cut Γ_3 and Γ_4 associated with the branch point k_0. Assuming $\zeta = k_0 + ik_0 v$, we obtain a branch cut contribution

$$\bar{\Lambda}_1(k_0) = I_1 + I_2 \tag{26}$$

where I_1 and I_2 are in numerically-efficient integral forms

$$I_1 = \int_0^\infty \frac{-4i(-1)^n \exp[2k_0 a(i-v)]\sqrt{v(-2i+v)}}{(k_0 a)^2[(1+iv)^2 - \alpha^2][(1+iv)^2 - \beta^2]}\, dv \tag{27}$$

$$I_2 = \int_0^\infty \frac{4i\sqrt{v(-2i+v)}}{(k_0 a)^2[(1+iv)^2 - \alpha^2][(1+iv)^2 - \beta^2]}\, dv$$

$$= \frac{-4i}{(k_0 a)^2(\alpha^2 - \beta^2)}\left(-\frac{\sqrt{1-\alpha^2}}{\alpha}\sin^{-1}\alpha + \frac{\sqrt{1-\beta^2}}{\beta}\sin^{-1}\beta\right) \tag{28}$$

$\alpha = a_m/k_0$ and $\beta = a_n/k_0$. It is also possible to transform I_1 into asymptotic series [8]

$$I_1 = -\frac{2\exp(2ik_0 a)(-1)^n}{(k_0 a)^2(\alpha^2 - \beta^2)}$$

$$\sum_{l=1}^\infty S_l\left\{[A(t_1) - A(t_2)]/\alpha - [A(t_3) - A(t_4)]/\beta\right\} \tag{29}$$

where $S_l = \begin{pmatrix} 0.5 \\ l-1 \end{pmatrix} (0.5\mathrm{i})^{l-1.5}$, $t_1 = (\alpha - 1)\mathrm{i}$, $t_2 = (-\alpha - 1)\mathrm{i}$, $t_3 = (\beta - 1)\mathrm{i}$, $t_4 = (-\beta - 1)\mathrm{i}$, and

$$A(t) = (-1)^l \pi t^{l-0.5} \exp(pt)\mathrm{erfc}(\sqrt{pt})$$

$$+ 2^{1-l}\sqrt{\pi}p^{0.5-l} \sum_{r=0}^{l-1} (2l - 2r - 3)!!(-2pt)^r . \tag{30}$$

Note that $\mathrm{erfc}(\cdot)$ denotes the complementary error function and $p = 2k_0 a$.

Consider

$$\Omega_1(k_0) = \int_{-\infty}^{\infty} a^2 F_m(\zeta a) F_n(-\zeta a)\zeta^2 \kappa_0^{-1} \, \mathrm{d}\zeta . \tag{31}$$

An evaluation of $\Omega_1(k_0)$ is similar to that of $\Lambda_1(k_0)$ discussed earlier. When $m + n$ is odd, $\Omega_1(k_0) = 0$. When $m + n$ is even,

$$\Omega_1(k_0) = \frac{2\pi\varepsilon_n}{a\sqrt{k_0^2 - a_m^2}}\delta_{mn} - \bar{\Omega}_1(k_0) \tag{32}$$

$$\bar{\Omega}_1(k_0) = J_1 + J_2 \tag{33}$$

where

$$J_1 = \int_0^{\infty} \frac{-4\mathrm{i}(-1)^n \exp[2k_0 a(\mathrm{i} - v)](1 + \mathrm{i}v)^2}{(k_0 a)^2[(1 + \mathrm{i}v)^2 - \alpha^2][(1 + \mathrm{i}v)^2 - \beta^2]\sqrt{v(-2\mathrm{i} + v)}} \, \mathrm{d}v \tag{34}$$

$$J_2 = \int_0^{\infty} \frac{4\mathrm{i}(1 + \mathrm{i}v)^2}{(k_0 a)^2[(1 + \mathrm{i}v)^2 - \alpha^2][(1 + \mathrm{i}v)^2 - \beta^2]\sqrt{v(-2\mathrm{i} + v)}} \, \mathrm{d}v$$

$$= \frac{-4\mathrm{i}}{(k_0 a)^2(\alpha^2 - \beta^2)}\left(-\frac{\alpha}{\sqrt{1-\alpha^2}}\sin^{-1}\alpha + \frac{\beta}{\sqrt{1-\beta^2}}\sin^{-1}\beta\right) \tag{35}$$

$\alpha = a_m/k_0$ and $\beta = a_n/k_0$.

1.2 EM Scattering from Multiple Grooves in a Conducting Plane

Electromagnetic wave scattering from multiple rectangular grooves in a conducting plane has been considered extensively in [9-11] due to optical diffraction grating and polarizer applications. Most previous studies have dealt with electromagnetic wave scattering from infinite periodic rectangular grooves by utilizing Floquet's theorem. In the next two subsections we will study TE and TM scattering from finite rectangular grooves engraved in a perfectly-conducting plane without recourse to Floquet's theorem. The present section is an extension of Sect. 1.1, which discusses scattering from a single groove in a conducting plane.

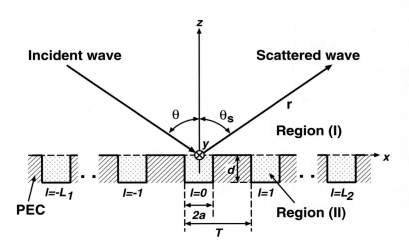

Fig. 1.3. Multiple rectangular grooves in a conducting plane

1.2.1 TE Scattering [12]

Consider a uniform plane wave $E_y^i(x, z)$ incident upon a perfectly conducting plane with multiple rectangular grooves (width $2a$, depth d, and period T). Regions (I) and (II) denote the air and a groove medium whose wavenumbers are $k_0 = \omega\sqrt{\mu_0\epsilon_0} = 2\pi/\lambda$ and $k = \omega\sqrt{\mu\epsilon}$ ($\mu = \mu_0\mu_r$ and $\epsilon = \epsilon_0\epsilon_r$), respectively. The total E-field in region (I) is composed of the incident, specularly-reflected, and scattered fields as

$$E_y^i(x, z) = \exp(ik_x x - ik_z z) \tag{1}$$

$$E_y^r(x, z) = -\exp(ik_x x + ik_z z) \tag{2}$$

$$E_y^s(x, z) = \frac{1}{2\pi}\int_{-\infty}^{\infty}\widetilde{E}_y^s(\zeta)\exp(-i\zeta x + i\kappa_0 z)\,\mathrm{d}\zeta \tag{3}$$

where $k_x = k_0\sin\theta$, $k_z = k_0\cos\theta$, and $\kappa_0 = \sqrt{k_0^2 - \zeta^2}$. The total transmitted field inside the lth groove of region (II) is

$$E_y^t(x, z) = \sum_{m=1}^{\infty} c_m^l \sin a_m(x + a - lT)\sin\xi_m(z + d) \tag{4}$$

where $a_m = m\pi/(2a)$ and $\xi_m = \sqrt{k^2 - a_m^2}$.

The tangential E-field continuity at $z = 0$ for integer l ($-L_1 \le l \le L_2$) is given by

$$E_y^s(x, 0) = \begin{cases} E_y^t(x, 0), & |x - lT| < a \\ 0, & |x - lT| > a\,. \end{cases} \tag{5}$$

Applying the Fourier transform to (5) gives

$$\widetilde{E}_y^s(\zeta) = \sum_{l=-L_1}^{L_2} \sum_{m=1}^{\infty} c_m^l e^{i\zeta lT} \sin(\xi_m d) a_m a^2 F_m(\zeta a) \ . \tag{6}$$

The tangential H-field continuity along $(lT - a) < x < (lT + a)$ at $z = 0$ is given by

$$2ik_z e^{ik_x x} - \int_{-\infty}^{\infty} \frac{i\kappa_0}{2\pi} \widetilde{E}_y^s(\zeta) e^{-i\zeta x} \, d\zeta$$

$$= -\sum_{m=1}^{\infty} \frac{c_m^l \xi_m}{\mu_r} \sin a_m (x + a - lT) \cos(\xi_m d) \ . \tag{7}$$

We substitute (6) into (7), multiply (7) by $\sin a_n(x + a - rT)$, and integrate from $(rT - a)$ to $(rT + a)$ to get

$$\boxed{\begin{aligned} &2ik_z a_n e^{ik_x rT} a^2 F_n(k_x a) \\ &= \frac{ia_n}{2\pi} \sum_{l=-L_1}^{L_2} \sum_{m=1}^{\infty} c_m^l a_m a^2 \sin(\xi_m d) \Lambda_2(k_0) - \frac{c_n^r \xi_n a}{\mu_r} \cos(\xi_n d) \end{aligned}} \tag{8}$$

where

$$\Lambda_2(k_0) = \int_{-\infty}^{\infty} a^2 F_m(\zeta a) F_n(-\zeta a) \kappa_0 \exp\left[i(l - r)\zeta T\right] \, d\zeta \ . \tag{9}$$

Using the residue calculus, we evaluate $\Lambda_2(k_0)$ in Subsect. 1.2.3

$$\Lambda_2(k_0) = 2\pi \frac{\sqrt{k_0^2 - a_m^2}}{a a_m^2} \delta_{mn} \delta_{lr} - \bar{\Lambda}_2(k_0) \tag{10}$$

where $\bar{\Lambda}_2(k_0)$ is a branch-cut integral in numerically-efficient form. It is also possible to transform $\bar{\Lambda}_2(k_0)$ into a series whose nth term is on the order of $(k_0 a)^{0.5-n}$; hence, $\bar{\Lambda}_2(k_0)$ vanishes for large $k_0 a$.

The far-zone scattered field at r is

$$E_y^s(\theta_s, \theta) = \sqrt{\frac{k_0}{2\pi r}} \exp(ik_0 r - i\pi/4) \cos\theta_s \widetilde{E}_y^s(-k_0 \sin\theta_s) \ . \tag{11}$$

1.2.2 TM Scattering

Consider scattering of a uniform plane wave $H_y^i(x, z)$ incident on a perfect conducting plane with a finite number of rectangular grooves (width $2a$, depth d, and period T). Regions (I) and (II), respectively, denote the air and rectangular grooves, which lie in parallel with the y-direction. In region (I) the total H-field consists of the incident, reflected, and scattered components

$$H_y^i(x,z) = \exp(\mathrm{i}k_x x - \mathrm{i}k_z z) \tag{12}$$

$$H_y^r(x,z) = \exp(\mathrm{i}k_x x + \mathrm{i}k_z z) \tag{13}$$

$$H_y^s(x,z) = \frac{1}{2\pi} \int_{-\infty}^{\infty} \widetilde{H}_y^s(\zeta) \exp(-\mathrm{i}\zeta x + \mathrm{i}\kappa_0 z)\,\mathrm{d}\zeta \tag{14}$$

where $k_x = k_0 \sin\theta$, $k_z = k_0 \cos\theta$, and $\kappa_0 = \sqrt{k_0^2 - \zeta^2}$. Inside the lth groove of region (II) the total transmitted field is

$$H_y^t(x,z) = \sum_{m=0}^{\infty} c_m^l \cos a_m(x + a - lT) \cos \xi_m(z + d) \tag{15}$$

where $a_m = m\pi/(2a)$ and $\xi_m = \sqrt{k^2 - a_m^2}$.

Applying the Fourier transform to the tangential E-field continuity along the x-axis ($z = 0$) yields for every integer l

$$\widetilde{H}_y^s(\zeta) = \sum_{l=-L_1}^{L_2} \sum_{m=0}^{\infty} c_m^l \xi_m \sin(\xi_m d) \frac{\zeta}{\kappa_0 \epsilon_r} a^2 F_m(\zeta a) e^{\mathrm{i}\zeta lT} . \tag{16}$$

The tangential H-field continuity along $(lT - a) < x < (lT + a)$ at $z = 0$ gives

$$2\mathrm{e}^{\mathrm{i}k_x x} + \sum_{l=-L_1}^{L_2} \sum_{m=0}^{\infty} \frac{c_m^l \xi_m}{2\pi} \sin(\xi_m d) \int_{-\infty}^{\infty} \frac{a^2 F_m(\zeta a)}{\kappa_0 \epsilon_r} \zeta \exp[-\mathrm{i}\zeta(x - lT)]\,\mathrm{d}\zeta$$

$$= \sum_{m=0}^{\infty} c_m^l \cos a_m(x + a - lT) \cos(\xi_m d) . \tag{17}$$

We multiply (17) by $\cos a_n(x + a - rT)$ and integrate with respect to x from $(rT - a)$ to $(rT + a)$ to obtain

$$\boxed{\begin{aligned} &-2\mathrm{i}k_x \mathrm{e}^{\mathrm{i}k_x rT} a^2 F_n(k_x a) \\ &= -\frac{\mathrm{i}}{2\pi\epsilon_r} \sum_{l=-L_1}^{L_2} \sum_{m=0}^{\infty} c_m^l \xi_m a^2 \sin(\xi_m d)\Omega_2(k_0) + c_n^r a\varepsilon_n \cos(\xi_n d) \end{aligned}} \tag{18}$$

where

$$\Omega_2(k_0) = \int_{-\infty}^{\infty} a^2 F_m(\zeta a) F_n(-\zeta a)\zeta^2 \exp[\mathrm{i}(l - r)\zeta T]\kappa_0^{-1}\,\mathrm{d}\zeta . \tag{19}$$

It is possible to transform $\Omega_2(k_0)$ into a numerically efficient form based on a residue integral technique as shown in Subsect. 1.2.3. The result is

$$\Omega_2(k_0) = \frac{2\pi\varepsilon_n}{a\sqrt{k_0^2 - a_m^2}}\delta_{mn}\delta_{lr} - \bar{\Omega}_2(k_0) . \tag{20}$$

The far-zone scattered field at r is

$$H_y^s(\theta_s, \theta) = \exp(\mathrm{i}k_0 r - \mathrm{i}\pi/4)\sqrt{\frac{k_0}{2\pi r}}\cos\theta_s \widetilde{H}_y^s(-k_0 \sin\theta_s) . \tag{21}$$

1.2.3 Appendix

Consider

$$\Lambda_2(k_0) = \int_{-\infty}^{\infty} a^2 F_m(\zeta a) F_n(-\zeta a) \kappa_0 \exp[i(l-r)\zeta T]\,d\zeta \tag{22}$$

in the complex ζ-plane as shown in Fig. 1.2 of Subsect. 1.1.3. Integrating along the deformed contour $\Gamma_1, \Gamma_2, \Gamma_3$, and Γ_4 in the upper-half plane, we obtain

$$\Lambda_2(k_0) = \frac{2\pi}{aa_m^2} \sqrt{k_0^2 - a_m^2}\,\delta_{mn}\delta_{lr} - \bar{\Lambda}_2(k_0) \tag{23}$$

where the first term is a residue contribution at $\zeta = \pm a_m$ when $m = n$ and $l = r$, and the second term $\bar{\Lambda}_2(k_0)$ is due to integration along the branch cut Γ_3 and Γ_4.

When $l = r$, $\bar{\Lambda}_2(k_0)$ degenerates into $\bar{\Lambda}_1(k_0)$ as considered in Subsect. 1.1.3. When $l \neq r$, we get

$$\bar{\Lambda}_2(k_0) = 2\big\{[(-1)^{m+n} + 1]I_3(qT) - (-1)^m I_3(qT + 2a)$$
$$-(-1)^n I_3(qT - 2a)\big\} \tag{24}$$

where $q = l - r$ and

$$I_3(c) = \int_{\Gamma_4} \frac{\exp(i\zeta|c|)\sqrt{k_0^2 - \zeta^2}}{a^2(\zeta^2 - a_m^2)(\zeta^2 - a_n^2)}\,d\zeta \tag{25}$$

$\alpha = a_m/k_0$ and $\beta = a_n/k_0$. By letting $\zeta = k_0 + ik_0 v$, we obtain a numerically-efficient integral

$$I_3(c) = \int_0^{\infty} \frac{i\exp[k_0|c|(i-v)]\sqrt{v(-2i+v)}}{(k_0 a)^2[(1+iv)^2 - \alpha^2][(1+iv)^2 - \beta^2]}\,dv\;. \tag{26}$$

The evaluation of $\Omega_2(k_0)$ is similar to that of $\Lambda_2(k_0)$, leading to

$$\Omega_2(k_0) = \frac{2\pi\varepsilon_n}{a\sqrt{k_0^2 - a_m^2}}\,\delta_{mn}\delta_{lr} - \bar{\Omega}_2(k_0)\;. \tag{27}$$

When $l = r$, $\bar{\Omega}_2(k_0)$ degenerates into $\bar{\Omega}_1(k_0)$ given in Subsect. 1.1.3. When $l \neq r$,

$$\bar{\Omega}_2(k_0) = 2\big\{[(-1)^{m+n} + 1]J_3(qT) - (-1)^m J_3(qT + 2a)$$
$$-(-1)^n J_3(qT - 2a)\big\} \tag{28}$$

where $q = l - r$ and

$$J_3(c) =$$
$$\int_0^{\infty} \frac{i(1+iv)^2 \exp[k_0|c|(i-v)]}{(k_0 a)^2 \sqrt{v(-2i+v)}[(1+iv)^2 - \alpha^2][(1+iv)^2 - \beta^2]}\,dv \tag{29}$$

$\alpha = a_m/k_0$ and $\beta = a_n/k_0$.

1.3 EM Scattering from Grooves in a Dielectric-Covered Ground Plane

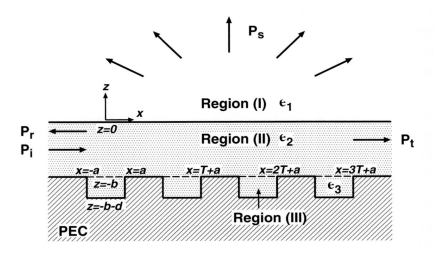

Fig. 1.4. Rectangular grooves in a dielectric-covered ground plane

Surface wave scattering from multiple grooves in a dielectric-covered ground plane was studied for Bragg reflector and leaky wave antenna applications [13]. A use of multiple grooves in the leaky wave antenna can improve the antenna radiation characteristics in terms of bandwidth, gain, and beamwidth. In the next two subsections we will analyze a problem of TE and TM surface wave scattering from grooves in a dielectric-covered ground plane.

1.3.1 TE Scattering [14]

A surface wave, which is transverse electric (TE) to the x-axis, is incident on periodic rectangular grooves. Regions (I), (II), and (III), respectively, denote the air (wavenumber; $k_1 = \omega\sqrt{\mu\epsilon_1}$), a dielectric slab (wavenumber; $k_2 = \omega\sqrt{\mu\epsilon_2} = 2\pi/\lambda$), and an N number of grooves (wavenumber; $k_3 = \omega\sqrt{\mu\epsilon_3}$). In region (I) the total E-field has the incident and scattered fields

$$E_y^{iI}(x,z) = \sin(k_{z2}b)\,\exp(\mathrm{i}k_x x - k_{z1}z) \tag{1}$$

$$E_y^{I}(x,z) = \frac{1}{2\pi}\int_{-\infty}^{\infty} \widetilde{E}_y^{I}(\zeta)\,\exp(-\mathrm{i}\zeta x + \mathrm{i}\kappa_1 z)\,\mathrm{d}\zeta \tag{2}$$

where $\kappa_1 = \sqrt{k_1^2 - \zeta^2}$ and $k_{z1} = \sqrt{k_x^2 - k_1^2}$. The E-field in region (II) similarly has the incident and scattered components

$$E_y^{iII}(x, z) = \sin k_{z2}(z + b)e^{ik_x x} \tag{3}$$

$$E_y^{II}(x, z) = \frac{1}{2\pi} \int_{-\infty}^{\infty} \left[\tilde{E}_+^{II}(\zeta)e^{i\kappa_2 z} + \tilde{E}_-^{II}(\zeta)e^{-i\kappa_2 z} \right] e^{-i\zeta x} \, d\zeta \tag{4}$$

where $\kappa_2 = \sqrt{k_2^2 - \zeta^2}, k_{z2} = \sqrt{k_2^2 - k_x^2}$, and the characteristic equation $\tan(k_{z2}b) = -\dfrac{k_{z2}}{k_{z1}}$ determines k_x. In region (III) ($lT - a < x < lT + a$ and $-d - b < z < -b$: $l = 0, 1, \ldots, N - 1$), the total E-field is a sum of discrete modes

$$E_y^{III}(x, z) = \sum_{m=1}^{\infty} c_m^l \, \sin a_m(x + a - lT) \sin \xi_m(z + b + d) \tag{5}$$

where $a_m = m\pi/(2a), (m = 1, 2, 3, \ldots)$, and $\xi_m = \sqrt{k_3^2 - a_m^2}$.

The tangential E-field and H-field continuities at $z = 0$ yield

$$\tilde{E}_-^{II}(\zeta) = \left(\frac{\kappa_2 - \kappa_1}{\kappa_2 + \kappa_1} \right) \tilde{E}_+^{II}(\zeta) . \tag{6}$$

The tangential E-field continuity along ($lT - a < x < lT + a$, $z = -b$: $l = 0, 1, \ldots, N - 1$) is

$$E_y^{II}(x, -b) = \begin{cases} E_y^{III}(x, -b), & |x - lT| < a \\ 0, & |x - lT| > a . \end{cases} \tag{7}$$

Applying the Fourier transform to (7) yields

$$\tilde{E}_+^{II}(\zeta) = \sum_{l=0}^{N-1} \sum_{m=1}^{\infty} c_m^l \frac{a_m \sin(\xi_m d)(\kappa_2 + \kappa_1)e^{i\zeta lT} a^2 F_m(\zeta a)}{2\left[\kappa_2 \cos(\kappa_2 b) - i\kappa_1 \sin(\kappa_2 b) \right]} . \tag{8}$$

The tangential H-field continuity along ($rT - a < x < rT + a$ and $z = -b$: $r = 0, 1, \ldots, N - 1$) requires

$$H_x^{iII}(x, -b) + H_x^{II}(x, -b) = H_x^{III}(x, -b) . \tag{9}$$

We substitute (6) and (8) into (9), multiply (9) by $\sin a_n(x + a - rT)$, and integrate from ($rT - a$) to ($rT + a$) to obtain

$$\boxed{\begin{aligned} & k_{z2}a_n a^2 F_n(k_x a)\exp(ik_x rT) \\ & = \sum_{l=0}^{N-1} \sum_{m=1}^{\infty} c_m^l \frac{a_m a_n \sin(\xi_m d)}{2\pi i} a^2 \Lambda_4(k_1) + c_n^r \xi_n a \cos(\xi_n d) \end{aligned}} \tag{10}$$

where

$$\Lambda_4(k_1) = \int_{-\infty}^{\infty} \kappa_2 \left[\frac{\kappa_1 - i\kappa_2 \tan(\kappa_2 b)}{\kappa_2 - i\kappa_1 \tan(\kappa_2 b)} \right]$$
$$\cdot a^2 F_m(\zeta a) F_n(-\zeta a) \exp[i\zeta(l - r)T] \, d\zeta . \tag{11}$$

Using a contour integral technique as shown in Subsect. 1.1.3, it is possible to transform (11) into a numerically-efficient form.

We represent the transmitted and reflected fields at $x = \pm\infty$ in regions (II) and (I) as

$$E_y^{II}(\pm\infty, z) = K_\pm \sin k_{z2}(z + b)e^{\pm ik_x x} \tag{12}$$

$$E_y^{I}(\pm\infty, z) = K_\pm \sin(k_{z2}b) \exp(\pm ik_x x - k_{z1}z) \tag{13}$$

where

$$K_\pm = \sum_{l=0}^{N-1} \sum_{m=1}^{\infty} c_m \frac{ia_m \sin(\xi_m d)k_{z1}k_{z2}}{k_x(1 + k_{z1}b)} \exp(\mp ik_x lT)a^2 F_m(\mp k_x a) . \tag{14}$$

The time-averaged incident, transmitted, and reflected powers (P_i, P_t, and P_r) are $P_t/P_i = |1 + K_+|^2$, $P_r/P_i = |K_-|^2$, and $P_i = \dfrac{k_x}{4\omega\mu}\left(\dfrac{1 + k_{z1}b}{k_{z1}}\right)$. The far-zone scattered field at $x = r\sin\theta_s$ and $z = r\cos\theta_s$ is

$$E_y^s(r, \theta_s) = \sqrt{\frac{k_1}{2\pi r}}\,\cos\theta_s\,\exp(ik_1 r - i\pi/4)$$

$$\cdot \sum_{l=0}^{N-1} \sum_{m=1}^{\infty} c_m^l \frac{a_m \sin(\xi_m d)\kappa_2 e^{i\zeta lT}a^2 F_m(\zeta a)}{[\kappa_2\cos(\kappa_2 b) - i\kappa_1 \sin(\kappa_2 b)]}\Bigg|_{\zeta = -k_1 \sin\theta_s} \tag{15}$$

1.3.2 TM Scattering

We will consider TM wave scattering from grooves that are periodically engraved in a dielectric-covered ground plane. In region (I) the total H-field consists of the incident and scattered fields

$$H_y^{iI}(x, z) = \exp(ik_x x - k_{z1}z) \tag{16}$$

$$H_y^{I}(x, z) = \frac{1}{2\pi} \int_{-\infty}^{\infty} \widetilde{H}_y^{I}(\zeta) \exp(-i\zeta x + i\kappa_1 z)\,d\zeta \tag{17}$$

where $\kappa_1 = \sqrt{k_1^2 - \zeta^2}$ and $k_{z1} = \sqrt{k_x^2 - k_1^2}$. In region (II) the total H-field has the incident and the scattered fields

$$H_y^{iII}(x, z) = \frac{\cos k_{z2}(z + b)}{\cos(k_{z2}b)}\,e^{ik_x x} \tag{18}$$

$$H_y^{II}(x, z) = \frac{1}{2\pi} \int_{-\infty}^{\infty} \left[\widetilde{H}_+^{II}(\zeta)e^{i\kappa_2 z} + \widetilde{H}_-^{II}(\zeta)e^{-i\kappa_2 z}\right] e^{-i\zeta x}\,d\zeta \tag{19}$$

where $\kappa_2 = \sqrt{k_2^2 - \zeta^2}$, $k_{z2} = \sqrt{k_2^2 - k_x^2}$, and $\tan(k_{z2}b) = \dfrac{k_{z1}\epsilon_2}{k_{z2}\epsilon_1}$. In region (III) the total transmitted H-field is

$$H_y^{III}(x, z) = \sum_{m=0}^{\infty} c_m^l \cos a_m(x + a - lT) \cos \xi_m(z + b + d) \tag{20}$$

where $a_m = m\pi/(2a), (m = 0, 1, 2, \dots)$, and $\xi_m = \sqrt{k_3^2 - a_m^2}$.

The tangential E-field and H-field continuities at $z = 0$ give

$$\tilde{H}_-^{II}(\zeta) = \left(\frac{\epsilon_1 \kappa_2 - \epsilon_2 \kappa_1}{\epsilon_1 \kappa_2 + \epsilon_2 \kappa_1} \right) \tilde{H}_+^{II}(\zeta) . \tag{21}$$

The tangential E-field continuity at $z = -b$ is given by

$$E_x^{II}(x, -b) = \begin{cases} E_x^{III}(x, -b), & |x - lT| < a \\ 0, & |x - lT| > a . \end{cases} \tag{22}$$

Taking the Fourier transform of (22) gives

$$\tilde{H}_+^{II}(\zeta) = \sum_{l=0}^{N-1} \sum_{m=0}^{\infty} c_m^l \frac{\epsilon_2 \xi_m \sin(\xi_m d)\zeta(\epsilon_1 \kappa_2 + \epsilon_2 \kappa_1)e^{i\zeta lT} a^2 F_m(\zeta a)}{2\epsilon_3 \kappa_2 [\epsilon_2 \kappa_1 \cos(\kappa_2 b) - i\epsilon_1 \kappa_2 \sin(\kappa_2 b)]} . \tag{23}$$

We multiply the tangential H-field continuity along ($rT - a < x < rT + a$ and $z = -b$: $l = 0, 1, \dots, N - 1$) by $\cos a_n(x + a - rT)$, and integrate from $(rT - a)$ to $(rT + a)$ to get

$$\boxed{\begin{aligned} &\frac{k_x a^2 F_n(k_x a) \exp(ik_x rT)}{\cos(k_{z2} b)} \\ &= \sum_{l=0}^{N-1} \sum_{m=0}^{\infty} c_m^l \left[\frac{\epsilon_2 \xi_m \sin(\xi_m d)}{2\pi\epsilon_3} \right] a^2 \Omega_4(k_1) + ic_n^r \epsilon_n a \cos(\xi_n d) \end{aligned}} \tag{24}$$

where

$$\Omega_4(k_1) = \int_{-\infty}^{\infty} \frac{\zeta^2 [\epsilon_1 \kappa_2 - i\epsilon_2 \kappa_1 \tan(\kappa_2 b)] a^2 F_m(\zeta a) F_n(-\zeta a)}{\kappa_2 [\epsilon_2 \kappa_1 - i\epsilon_1 \kappa_2 \tan(\kappa_2 b)] \exp[i\zeta(r - l)T]} \, d\zeta . \tag{25}$$

It is expedient to transform $\Omega_4(k_1)$ into a numerically-efficient integral based on the contour integral technique. An analytic evaluation of $\Omega_4(k_1)$ is summarized in Subsect. 1.3.3.

The transmitted and reflected fields at $x = \pm\infty$ in regions (II) and (I) are

$$H_y^{II}(\pm\infty, z) = K_\pm \frac{\cos k_{z2}(z + b)}{\cos(k_{z2}b)} e^{\pm ik_x x} \tag{26}$$

$$H_y^I(\pm\infty, z) = K_\pm \exp(\pm ik_x x - k_{z1}z) \tag{27}$$

where

$$K_\pm = \sum_{l=0}^{N-1} \sum_{m=0}^{\infty} c_m^l$$
$$\cdot \left[\frac{\mp \epsilon_1^2 \epsilon_2 k_{z1} k_{z2}^2 \xi_m \sin(\xi_m d)a^2 F_m(\mp k_x a) \exp(\mp ik_x lT)}{\epsilon_3 \cos(k_{z2}b) [\epsilon_1 \epsilon_2(k_{z1}^2 + k_{z2}^2) + k_{z1}b(\epsilon_1^2 k_{z2}^2 + \epsilon_2^2 k_{z1}^2)]} \right] . \tag{28}$$

The transmission and reflection coefficients are τ $(= P_t/P_i) = |1 + K_+|^2$ and ϱ $(= P_r/P_i) = |K_-|^2$. The far-zone scattered field at $x = r\sin\theta_s$ and $z = r\cos\theta_s$ is

$$H_y^s(r,\theta_s) = \sqrt{\frac{k_1}{2\pi r}}\cos\theta_s \exp(ik_1 r - i\pi/4)\sum_{l=0}^{N-1}\sum_{m=0}^{\infty} c_m^l$$

$$\cdot \frac{\epsilon_1\epsilon_2\xi_m \sin(\xi_m d)\zeta e^{i\zeta lT}a^2 F_m(\zeta a)}{\epsilon_3\left[\epsilon_2\kappa_1\cos(\kappa_2 b) - i\epsilon_1\kappa_2\sin(\kappa_2 b)\right]}\Bigg|_{\zeta = -k_1\sin\theta_s}. \qquad (29)$$

1.3.3 Appendix

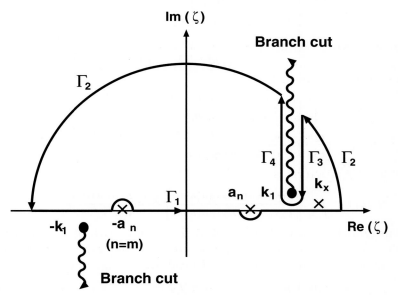

Fig. 1.5. Deformed contour path in the ζ-plane

When $l = r$, $\Omega_4(k_1)$ is rewritten as

$$\Omega_4(k_1) = \int_{-\infty}^{\infty}\left[\frac{\epsilon_1\kappa_2 - i\epsilon_2\kappa_1\tan(\kappa_2 b)}{\epsilon_2\kappa_1 - i\epsilon_1\kappa_2\tan(\kappa_2 b)}\right]\frac{\zeta^2}{\kappa_2}a^2 F_m(\zeta a)F_n(-\zeta a)\mathrm{d}\zeta . \qquad (30)$$

When $m + n$ is odd, $\Omega_4(k_1) = 0$ by odd symmetry.
When $m + n$ is even, $\Omega_4(k_1)$ is

$$\Omega_4(k_1) =$$
$$\int_{-\infty}^{\infty}\left[\frac{\epsilon_1\kappa_2 - i\epsilon_2\kappa_1\tan(\kappa_2 b)}{\epsilon_2\kappa_1 - i\epsilon_1\kappa_2\tan(\kappa_2 b)}\right]\frac{2\zeta^2\left[1 - (-1)^m e^{i2\zeta a}\right]}{a^2\kappa_2(\zeta^2 - a_m^2)(\zeta^2 - a_n^2)}\mathrm{d}\zeta. \qquad (31)$$

In view of Fig. 1.5, the integrand has a pair of branch points at the zeros of κ_1 and two single poles at $\zeta = \pm a_m$ when $m = n$. Also it has a finite number of single poles at $\zeta = \pm k_x$ (surface wave poles) that are solutions to the equation $\epsilon_2\kappa_1 = i\epsilon_1\kappa_2\tan(\kappa_2 b)$. The integral contour can be deformed in the upper-half plane by using the residue calculus, yielding

$$\Omega_4(k_1) =$$

$$\frac{\delta_{rl}}{a^2}\left\{ \frac{2\pi\epsilon_n a}{\kappa_{2m}}\left[\frac{\epsilon_1\kappa_{2m} - i\epsilon_2\kappa_{1m}\tan(\kappa_{2m}b)}{\epsilon_2\kappa_{1m} - i\epsilon_1\kappa_{2m}\tan(\kappa_{2m}b)}\right]\delta_{nm}\right.$$

$$-\int_{\Gamma_3} f_a(\zeta)\mathrm{d}\zeta - \int_{\Gamma_4} f_b(\zeta)\mathrm{d}\zeta + \sum_{k_x}$$

$$\left.\frac{4\pi k_x k_{z1}(\epsilon_1^2 k_{z2}^2 + \epsilon_2^2 k_{z1}^2)\left[1 - (-1)^m e^{2ik_x a}\right]}{[\epsilon_1\epsilon_2(k_{z1}^2 + k_{z2}^2) + k_{z1}b(\epsilon_1^2 k_{z2}^2 + \epsilon_2^2 k_{z1}^2)](k_x^2 - a_m^2)(k_x^2 - a_n^2)}\right\} \tag{32}$$

where δ_{nm}: Kronecker delta, $\kappa_{1m} = \sqrt{k_1^2 - a_m^2}$, $\kappa_{2m} = \sqrt{k_2^2 - a_m^2}$, k_x: zeros of $[\epsilon_2\kappa_1 - i\epsilon_1\kappa_2\tan(\kappa_2 b)]$, $k_{z1} = \sqrt{k_x^2 - k_1^2}$, and $k_{z2} = \sqrt{k_2^2 - k_x^2}$. Furthermore we note that

$$f_a(\zeta) = -\left[\frac{\epsilon_1\kappa_2 + i\epsilon_2\kappa_1\tan(\kappa_2 b)}{\epsilon_2\kappa_1 + i\epsilon_1\kappa_2\tan(\kappa_2 b)}\right]\frac{2\zeta^2\left[1 - (-1)^m e^{i2\zeta a}\right]}{\kappa_2(\zeta^2 - a_m^2)(\zeta^2 - a_n^2)} \tag{33}$$

$$f_b(\zeta) = \left[\frac{\epsilon_1\kappa_2 - i\epsilon_2\kappa_1\tan(\kappa_2 b)}{\epsilon_2\kappa_1 - i\epsilon_1\kappa_2\tan(\kappa_2 b)}\right]\frac{2\zeta^2\left[1 - (-1)^m e^{i2\zeta a}\right]}{\kappa_2(\zeta^2 - a_m^2)(\zeta^2 - a_n^2)}. \tag{34}$$

When $l \neq r$, $\Omega_4(k_1)$ is

$$\Omega_4(k_1) = \int_{-\infty}^{\infty}\left[\frac{\epsilon_1\kappa_2 - i\epsilon_2\kappa_1\tan(\kappa_2 b)}{\epsilon_2\kappa_1 - i\epsilon_1\kappa_2\tan(\kappa_2 b)}\right]\frac{\zeta^2 a^2 F_m(\zeta a)F_n(-\zeta a)}{\kappa_2 e^{i\zeta(r-l)T}}\mathrm{d}\zeta. \tag{35}$$

The integrand has a pair of branch points at the zeros of κ_1 and a finite number of single poles at $\zeta = \pm k_x$ that are solutions to the equation $\epsilon_2\kappa_1 = i\epsilon_1\kappa_2\tan(\kappa_2 b)$. The integral contour depends on the sign of $(l - r)$ as

$$\Omega_4(k_1)\begin{cases}\text{upper half plane} & \text{when } l > r \\ \text{lower half plane} & \text{when } l < r.\end{cases} \tag{36}$$

Using the residue calculus, we obtain

$$\Omega_4(k_1) =$$

$$\frac{1}{a^2}\begin{cases}\displaystyle\sum_{k_x}\frac{2\pi k_x k_{z1}(\epsilon_1^2 k_{z2}^2 + \epsilon_2^2 k_{z1}^2)a^4 F_m(k_x a)F_n(-k_x a)}{[\epsilon_1\epsilon_2(k_{z1}^2 + k_{z2}^2) + k_{z1}b(\epsilon_1^2 k_{z2}^2 + \epsilon_2^2 k_{z1}^2)]\,e^{ik_x(r-l)T}} \\ \displaystyle -\int_{\Gamma_3} f_a^L(\zeta)\mathrm{d}\zeta - \int_{\Gamma_4} f_b^L(\zeta)\mathrm{d}\zeta \qquad \text{when } l > r \\[2ex] \displaystyle\sum_{k_x}\frac{2\pi k_x k_{z1}(\epsilon_1^2 k_{z2}^2 + \epsilon_2^2 k_{z1}^2)a^4 F_m(-k_x a)F_n(k_x a)}{[\epsilon_1\epsilon_2(k_{z1}^2 + k_{z2}^2) + k_{z1}b(\epsilon_1^2 k_{z2}^2 + \epsilon_2^2 k_{z1}^2)]\,e^{ik_x(l-r)T}} \\ \displaystyle +\int_{-\Gamma_3} f_a^R(\zeta)\mathrm{d}\zeta + \int_{-\Gamma_4} f_b^R(\zeta)\mathrm{d}\zeta \qquad \text{when } l < r\end{cases} \tag{37}$$

where k_x: zeros of $[\epsilon_2\kappa_1 - i\epsilon_1\kappa_2\tan(\kappa_2 b)]$, $k_{z1} = \sqrt{k_x^2 - k_1^2}$, and $k_{z2} = \sqrt{k_2^2 - k_x^2}$.

Note that

$$f_a^L(\zeta) = -\left[\frac{\epsilon_1\kappa_2 + i\epsilon_2\kappa_1\tan(\kappa_2 b)}{\epsilon_2\kappa_1 + i\epsilon_1\kappa_2\tan(\kappa_2 b)}\right]f_c(\zeta) \tag{38}$$

$$f_b^L(\zeta) = \left[\frac{\epsilon_1\kappa_2 - i\epsilon_2\kappa_1\tan(\kappa_2 b)}{\epsilon_2\kappa_1 - i\epsilon_1\kappa_2\tan(\kappa_2 b)}\right]f_c(\zeta) \tag{39}$$

$$f_a^R(\zeta) = -\left[\frac{\epsilon_1\kappa_2 + i\epsilon_2\kappa_1\tan(\kappa_2 b)}{\epsilon_2\kappa_1 + i\epsilon_1\kappa_2\tan(\kappa_2 b)}\right]f_c(\zeta) \tag{40}$$

$$f_b^R(\zeta) = \left[\frac{\epsilon_1\kappa_2 - i\epsilon_2\kappa_1\tan(\kappa_2 b)}{\epsilon_2\kappa_1 - i\epsilon_1\kappa_2\tan(\kappa_2 b)}\right]f_c(\zeta) \tag{41}$$

$$f_c(\zeta) = \frac{\zeta^2 a^4 F_m(\zeta a)F_n(-\zeta a)}{\kappa_2 e^{i\zeta(r-l)T}} . \tag{42}$$

1.4 EM Scattering from Rectangular Grooves in a Parallel-Plate Waveguide

Corrugated waveguide structures may find many practical applications for microwave filter, polarizer, and antenna [15]. In the next two subsections we will consider TE and TM scattering from a finite rectangular grooves periodically engraved in a perfectly-conducting parallel-plate waveguide. The present section analysis is applicable to scattering from a rectangular waveguide with multiple grooves, which is a typical microwave filter component.

1.4.1 TE Scattering [16]

Consider a TE wave $E_y^i(x, z)$ incident on rectangular grooves (width: $2a$, depth: d, period: T, and total number of grooves: N) in a parallel-plate waveguide. Regions (I) and (II) denote a parallel-plate waveguide interior and grooves with wavenumbers k_0 ($= \omega\sqrt{\mu_0\epsilon_0} = 2\pi/\lambda$) and k ($= \omega\sqrt{\mu\epsilon}$, $\mu = \mu_0\mu_r$, and $\epsilon = \epsilon_0\epsilon_r$), respectively. The total E-field in region (I) has the incident and scattered components

$$E_y^i(x, z) = e^{ik_{xs}x}\sin k_{zs}z \tag{1}$$

$$E_y^s(x, z) = \frac{1}{2\pi}\int_{-\infty}^{\infty}\left[\widetilde{E}_+^s(\zeta)e^{i\kappa_0 z} + \widetilde{E}_-^s(\zeta)e^{-i\kappa_0 z}\right]e^{-i\zeta x}\,\mathrm{d}\zeta \tag{2}$$

where $k_{zs} = s\pi/b$, s: integer, $k_{xs} = \sqrt{k_0^2 - k_{zs}^2}$, and $\kappa_0 = \sqrt{k_0^2 - \zeta^2}$. The E-field inside the lth groove is

$$E_y^t(x, z) = \sum_{m=1}^{\infty} c_m^l \sin a_m(x + a - lT)\sin\xi_m(z + d) \tag{3}$$

where $a_m = m\pi/(2a)$ and $\xi_m = \sqrt{k^2 - a_m^2}$.
    The continuity of tangential E-field at $z = b$ gives

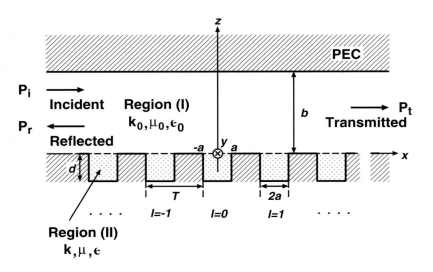

Fig. 1.6. Rectangular grooves in a parallel-plate waveguide

$$\widetilde{E}_-^s(\zeta) = -\mathrm{e}^{\mathrm{i}2\kappa_0 b}\,\widetilde{E}_+^s(\zeta)\ . \tag{4}$$

The continuity of tangential E-field at $z = 0$ is

$$E_y^s(x,0) = \begin{cases} E_y^t(x,0), & |x - lT| < a \\ 0, & |x - lT| > a\ . \end{cases} \tag{5}$$

Taking the Fourier transform of (5) yields

$$\widetilde{E}_+^s(\zeta)(1 - \mathrm{e}^{\mathrm{i}2\kappa_0 b}) = \sum_{l=-L_1}^{L_2}\sum_{m=1}^{\infty} c_m^l \mathrm{e}^{\mathrm{i}\zeta lT}\sin(\xi_m d)a_m a^2 F_m(\zeta a)\ . \tag{6}$$

We multiply the tangential H-field continuity along $(lT - a) < x < (lT + a)$ at $z = 0$ by $\sin a_n(x + a - rT)$ and integrate with respect to x from $(rT - a)$ to $(rT + a)$ to obtain

$$\boxed{\begin{aligned} &k_{zs}a_n \exp(\mathrm{i}k_{xs}rT)a^2 F_n(k_{xs}a) \\ &-\frac{a_n}{2\pi}\sum_{l=-L_1}^{L_2}\sum_{m=1}^{\infty} c_m^l a_m a^2 \sin(\xi_m d)\Lambda_5(k_0) = \frac{c_n^r \xi_n a}{\mu_r}\cos(\xi_n d) \end{aligned}} \tag{7}$$

where

$$\Lambda_5(k_0) = \int_{-\infty}^{\infty}\cot(\kappa_0 b)a^2 F_m(\zeta a)F_n(-\zeta a)\kappa_0 \exp[\mathrm{i}(l - r)\zeta T]\,\mathrm{d}\zeta\ . \tag{8}$$

Performing the residue calculus, we transform (8) into a rapidly-convergent series

$$\Lambda_5(k_0) = \frac{2\pi}{aa_m^2}\sqrt{k_0^2 - a_m^2}\cot\left(\sqrt{k_0^2 - a_m^2}\,b\right)\delta_{mn}\delta_{lr} + \bar{\Lambda}_5(k_0) \qquad (9)$$

$$\bar{\Lambda}_5(k_0) = -i\sum_{v=1}^{\infty}\frac{2\pi\kappa_0^2 A_1}{b(\zeta^2 - a_m^2)(\zeta^2 - a_n^2)\zeta a^2}\Bigg|_{\zeta=\sqrt{k_0^2 - (v\pi/b)^2}} \qquad (10)$$

$$\begin{aligned}
A_1 &= [(-1)^{m+n} + 1]\exp(i\zeta|l - r|T) \\
&\quad -(-1)^m\exp\left[i\zeta|(l - r)T + 2a|\right] \\
&\quad -(-1)^n\exp\left[i\zeta|(l - r)T - 2a|\right]\,.
\end{aligned} \qquad (11)$$

The scattered field at $x = \pm\infty$ is

$$E_y^s(\pm\infty, z) = \sum_v K_{\pm}^v \sin(k_{zv}z)e^{\pm ik_{xv}x} \qquad (12)$$

where v: integer, $1 \le v \le k_0 b/\pi$, $k_{zv} = v\pi/b$, $k_{xv} = \sqrt{k_0^2 - k_{zv}^2}$, and

$$K_{\pm}^v = \sum_{l=-L_1}^{L_2}\sum_{m=1}^{\infty} ic_m^l a_m \sin(\xi_m d)\frac{k_{zv}a^2 F_m(\mp k_{xv}a)\exp(\mp ik_{xv}lT)}{bk_{xv}}\,. \qquad (13)$$

The time-averaged incident, reflected, and transmitted powers are

$$P_i = \frac{k_{xs}b}{4\omega\mu_0} \qquad (14)$$

$$P_r = \frac{b}{4\omega\mu_0}\sum_v k_{xv}|K_-^v|^2 \qquad (15)$$

$$P_t = \frac{b}{4\omega\mu_0}\left\{k_{xs}[1 + 2Re(K_+^s) + |K_+^s|^2] + \sum_{v\neq s}k_{xv}|K_+^v|^2\right\}\,. \qquad (16)$$

1.4.2 TM Scattering [17]

An electromagnetic wave, which is transverse magnetic (TM) to the x-axis, is incident on an N number of grooves. The total H-field in region (I) is composed of the incident and scattered fields

$$H_y^i(x, z) = \cos(k_{zs}z)e^{ik_{xs}x} \qquad (17)$$

$$H_y^s(x, z) = \frac{1}{2\pi}\int_{-\infty}^{\infty}\cos\kappa_0(z - b)\tilde{H}_y^s(\zeta)e^{-i\zeta x}\,d\zeta \qquad (18)$$

where $0 \le s < k_0 b/\pi$, s: integer, $k_{zs} = s\pi/b$, $k_{xs} = \sqrt{k_0^2 - k_{zs}^2}$, and $\kappa_0 = \sqrt{k_0^2 - \zeta^2}$. In region (II) ($lT - a < x < lT + a$ and $-d < z < 0$: $l = -L_1, \ldots, L_2$) the H-field is

$$H_y^t(x, z) = \sum_{m=0}^{\infty} c_m^l \cos a_m(x - lT + a)\cos\xi_m(z + d) \qquad (19)$$

where $a_m = m\pi/(2a)$ and $\xi_m = \sqrt{k^2 - a_m^2}$.

The tangential E-field continuity at $z = 0$ yields

$$\tilde{H}_y^s(\zeta) = \sum_{l=-L_1}^{L_2} \sum_{m=0}^{\infty} c_m^l \left[\frac{i\epsilon_0 \xi_m \sin(\xi_m d)}{\epsilon} \right] \left[\frac{\zeta a^2 F_m(\zeta a) e^{i\zeta lT}}{\kappa_0 \sin \kappa_0 b} \right] . \tag{20}$$

The tangential H-field continuity at the aperture of grooves ($rT - a < x < rT + a, z = 0$: $r = -L_1, \ldots, L_2$) gives

$$\boxed{\begin{aligned} & -ik_{xs} a^2 F_n(k_{xs}a) \exp(ik_{xs}rT) \\ & = \sum_{l=-L_1}^{L_2} \sum_{m=0}^{\infty} c_m^l \left[\frac{\epsilon_0 \xi_m \sin(\xi_m d)}{2\pi\epsilon} \right] a^2 \Omega_5(k_0) + c_n^r \epsilon_n a \cos(\xi_n d) \end{aligned}} \tag{21}$$

where

$$\Omega_5(k_0) = \int_{-\infty}^{\infty} \cot(\kappa_0 b) a^2 F_m(\zeta a) F_n(-\zeta a) \kappa_0^{-1} \zeta^2 \exp[i(l-r)\zeta T] \, d\zeta . \tag{22}$$

Using the residue calculus, we transform $\Omega_5(k_0)$ into

$$\Omega_5(k_0) = \frac{2\pi\varepsilon_n \delta_{nm} \delta_{rl}}{a\sqrt{k_0^2 - a_m^2} \tan(\sqrt{k_0^2 - a_m^2} b)} - \bar{\Omega}_5(k_0) \tag{23}$$

where

$$\bar{\Omega}_5(k_0) = \sum_{w=0}^{\infty} \frac{i2\pi k_{xw} A_1}{\varepsilon_w b (k_{xw}^2 - a_m^2)(k_{xw}^2 - a_n^2) a^2} \tag{24}$$

$$A_1 = \left[1 + (-1)^{m+n}\right] \exp(ik_{xw}|l-r|T) \\ - (-1)^m \exp[ik_{xw}|(l-r)T + 2a|] \\ - (-1)^n \exp[ik_{xw}|(l-r)T - 2a|] . \tag{25}$$

We evaluate the total scattered fields at $x = \pm\infty$ as

$$H_y^s(\pm\infty, z) = \sum_v K_v^{\pm} \cos(k_{zv}z) e^{\pm ik_{xv}x} \tag{26}$$

where $0 \le v < k_0 b/\pi$, v: integer, and

$$K_v^{\pm} = \sum_{l=-L_1}^{L_2} \sum_{m=0}^{\infty} c_m^l$$
$$\cdot \frac{\mp\epsilon_0 \xi_m \sin(\xi_m d) a^2 F_m(\mp k_{xv} a) \exp(\mp ik_{xv}lT)}{\epsilon \varepsilon_v b} . \tag{27}$$

The transmission and reflection coefficients are

$$\tau = P_t/P_i = |1 + K_s^+|^2 + \frac{1}{\varepsilon_s k_{xs}} \sum_{v \ne s} \varepsilon_v k_{xv} |K_v^+|^2 \tag{28}$$

$$\varrho = P_r/P_i = \frac{1}{\varepsilon_s k_{xs}} \sum_v \varepsilon_v k_{xv} |K_v^-|^2 \tag{29}$$

where $0 \le v < (k_0 b/\pi)$, v: integer, and $P_i = k_{xs}\varepsilon_s b/(4\omega\epsilon_0)$.

1.5 EM Scattering from Double Grooves in Parallel Plates [18]

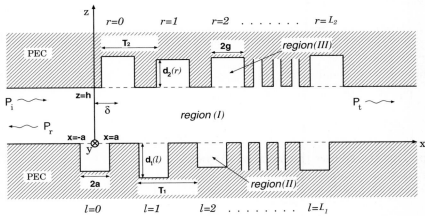

Fig. 1.7. Double rectangular grooves in a conducting parallel-plate waveguide

Corrugated waveguides are of fundamental interest in microwaves for their practical filter application. Electromagnetic wave scattering from rectangular grooves in a parallel-plate waveguide was studied in Sect. 1.4. In this section we will investigate scattering from double rectangular grooves in a metallic parallel-plate waveguide. A multiple groove scattering analysis given in this section is in continuation of the discussion given in Sect. 1.4.

1.5.1 TE Scattering

Consider a TE wave $E_y^i(x, z)$ incident on a finite number of double rectangular grooves in a conducting parallel-plate waveguide. Regions (I), (II), and (III) denote a parallel-plate waveguide interior, lower grooves, and upper grooves, respectively. The wavenumbers in regions (I), (II), and (III) are $k_0 \ (= \omega\sqrt{\mu_0\epsilon_0})$, $k_2 \ (= k_0\sqrt{\mu_2\epsilon_2})$, and $k_3 \ (= k_0\sqrt{\mu_3\epsilon_3})$, respectively. In region (I) the incident and scattered fields are

$$E_y^i(x, z) = \exp(ik_{xs}x)\sin(k_{zs}z) \tag{1}$$

$$E_y^s(x, z) = \frac{1}{2\pi}\int_{-\infty}^{\infty}\left[\widetilde{E}_+^s(\zeta)e^{i\kappa z} + \widetilde{E}_-^s(\zeta)e^{-i\kappa z}\right]e^{-i\zeta x}d\zeta . \tag{2}$$

where $k_{zs} = s\pi/h$, $1 \le s < k_0h/\pi$, s: integer, $k_{xs} = \sqrt{k_0^2 - k_{zs}^2}$, and $\kappa = \sqrt{k_0^2 - \zeta^2}$.
In the lth groove ($0 \le l \le L_1$) of region (II) the field is

$$E_y^{II}(x,z) = \sum_{m=1}^{\infty} b_m^l \sin a_m(x + a - lT_1) \sin \xi_m[z + d_1(l)] \tag{3}$$

where $a_m = \dfrac{m\pi}{2a}$ and $\xi_m = \sqrt{k_2^2 - a_m^2}$. In the rth groove $(0 \leq r \leq L_2)$ of region (III) the field is

$$E_y^{III}(x,z) = \sum_{n=1}^{\infty} c_n^r \sin g_n(x + g - rT_2 - \delta) \sin \eta_n[z - h - d_2(r)] \tag{4}$$

where $g_m = \dfrac{n\pi}{2g}$ and $\eta_n = \sqrt{k_3^2 - g_n^2}$.

The E_y field continuity at $z = 0$ requires

$$E_y^i(x,0) + E_y^s(x,0) = \begin{cases} E_y^{II}(x,0), & -a + lT_1 < x < a + lT_1 \\ 0, & \text{otherwise} \end{cases} \tag{5}$$

Applying the Fourier transform to (5) yields

$$\widetilde{E}_+^s(\zeta) + \widetilde{E}_-^s(\zeta) = \sum_{l=0}^{L_1} \sum_{m=1}^{\infty} b_m^l \sin[\xi_m d_1(l)] F_m^l(\zeta) \tag{6}$$

where

$$F_m^l(\zeta) = \frac{a_m e^{i\zeta lT_1}[(-1)^m e^{i\zeta a} - e^{-i\zeta a}]}{\zeta^2 - a_m^2}. \tag{7}$$

Similarly from the E_y field continuity at $z = h$, we obtain

$$\widetilde{E}_+^s(\zeta) e^{i\kappa h} + \widetilde{E}_-^s(\zeta) e^{-i\kappa h} = -\sum_{r=0}^{L_2} \sum_{n=1}^{\infty} c_n^r \sin[\eta_n d_2(r)] G_n^r(\zeta) \tag{8}$$

where

$$G_n^r(\zeta) = \frac{g_n e^{i\zeta(rT_2+\delta)}[(-1)^n e^{i\zeta g} - e^{-i\zeta g}]}{\zeta^2 - g_n^2}. \tag{9}$$

The H_x field continuity at $z = 0$ for $(-a + lT_1 < x < a + lT_1)$ is

$$k_{zs} \exp(ik_{xs}x) + \frac{i}{2\pi} \int_{-\infty}^{\infty} \kappa \left[\widetilde{E}_+^s(\zeta) - \widetilde{E}_-^s(\zeta)\right] e^{-i\zeta x} d\zeta$$

$$= \frac{1}{\mu_2} \sum_{m=1}^{\infty} \xi_m b_m^l \sin a_m(x + a - lT_1) \cos[\xi_m d_1(l)]. \tag{10}$$

Substituting $\widetilde{E}_+^s(\zeta)$ and $\widetilde{E}_-^s(\zeta)$ of (6) and (8) into (10), multiplying (10) by $\sin a_p(x + a - uT_1)$, and integrating over $(-a + uT_1 < x < a + uT_1)$, we get

$$k_{zs}F_p^u(k_{xs})$$

$$= \sum_{l=0}^{L_1} \sum_{m=1}^{\infty} \left\{ \sin[\xi_m d_1(l)]I_{1pm}^{ul} + \frac{a}{\mu_2} \cos[\xi_m d_1(l)]\delta_{ul}\delta_{pm} \right\} b_m^l$$

$$+ \sum_{r=0}^{L_2} \sum_{n=1}^{\infty} \left\{ \sin[\eta_n d_2(r)]I_{2pn}^{ur} \right\} c_n^r \qquad (11)$$

where

$$I_{1pm}^{ul} = \frac{1}{2\pi} \int_{-\infty}^{\infty} \frac{\kappa}{\tan(\kappa h)} F_m^l(\zeta)F_p^u(-\zeta)\,\mathrm{d}\zeta \qquad (12)$$

$$I_{2pn}^{ur} = \frac{1}{2\pi} \int_{-\infty}^{\infty} \frac{\kappa}{\sin(\kappa h)} G_n^r(\zeta)F_p^u(-\zeta)\,\mathrm{d}\zeta \ . \qquad (13)$$

Similarly from the H_x field continuity at $z = h$,

$$k_{zs}\cos(k_{zs}h)G_q^v(k_{xs}) = \sum_{l=0}^{L_1} \sum_{m=1}^{\infty} \left\{ \sin[\xi_m d_1(l)]I_{3qm}^{vl} \right\} b_m^l$$

$$+ \sum_{r=0}^{L_2} \sum_{n=1}^{\infty} \left\{ \sin[\eta_n d_2(r)]I_{4qn}^{vr} + \frac{g}{\mu_3} \cos[\eta_n d_2(r)]\delta_{vr}\delta_{qn} \right\} c_n^r \qquad (14)$$

where

$$I_{3qm}^{vl} = \frac{1}{2\pi} \int_{-\infty}^{\infty} \frac{\kappa}{\sin(\kappa h)} F_m^l(\zeta)G_q^v(-\zeta)\,\mathrm{d}\zeta \qquad (15)$$

$$I_{4qn}^{vr} = \frac{1}{2\pi} \int_{-\infty}^{\infty} \frac{\kappa}{\tan(\kappa h)} G_n^r(\zeta)G_q^v(-\zeta)\,\mathrm{d}\zeta \ . \qquad (16)$$

It is convenient to transform I_{1pm}^{ul}, I_{2pn}^{ur}, I_{3qm}^{vl}, and I_{4qn}^{vr} into rapidly-convergent series via the residue calculus. The results are summarized in Subsect. 1.5.3 Appendix.

The scattered field at $x = \pm\infty$ is given by

$$E_y^s(\pm\infty, z) = \sum_u K_u^{\pm} \sin(k_{zu}z) \exp(\pm ik_{xu}x) \qquad (17)$$

where $1 \leq u < k_0 h/\pi$, u: integer, and

$$K_u^{\pm} = \sum_{l=0}^{L_1} \sum_{m=1}^{\infty} \frac{ib_m^l k_{zu} \sin[\xi_m d_1(l)]F_m^l(\mp k_{xu})}{hk_{xu}}$$

$$+ \sum_{r=0}^{L_2} \sum_{n=1}^{\infty} \frac{ic_n^r k_{zu} \sin[\eta_n d_2(r)]G_n^r(\mp k_{xu})}{h(-1)^u k_{xu}} \ . \qquad (18)$$

Let the time-averaged incident, transmitted, and reflected powers be denoted by P_i, P_t, and P_r, then

$$\tau = P_t/P_i = |1 + K_s^+|^2 + \frac{1}{k_{xs}} \sum_{v \neq s} k_{xv} |K_v^+|^2 \qquad (19)$$

$$\varrho = P_r/P_i = \frac{1}{k_{xs}} \sum_{v} k_{xv} |K_v^-|^2 \qquad (20)$$

where $1 \leq v < k_0 h/\pi$, v: integer, and $P_i = \dfrac{k_{xs} h}{4 \omega \mu_0}$.

1.5.2 TM Scattering

In region (I) the incident and scattered fields are

$$H_y^i(x, z) = \exp(\mathrm{i} k_{xs} x) \cos(k_{zs} z) \qquad (21)$$

$$H_y^s(x, z) = \frac{1}{2\pi} \int_{-\infty}^{\infty} \left[\widetilde{H}_+^s(\zeta) \mathrm{e}^{\mathrm{i}\kappa z} + \widetilde{H}_-^s(\zeta) \mathrm{e}^{-\mathrm{i}\kappa z} \right] \mathrm{e}^{-\mathrm{i}\zeta x} \mathrm{d}\zeta \qquad (22)$$

where $k_{zs} = s\pi/h$, $0 \leq s < k_0 h/\pi$, s: integer, $k_{xs} = \sqrt{k_0^2 - k_{zs}^2}$, and $\kappa = \sqrt{k_0^2 - \zeta^2}$. In the lth groove ($0 \leq l \leq L_1$) of region (II) the scattered field is

$$H_y^{II}(x, z) = \sum_{m=0}^{\infty} b_m^l \cos a_m (x + a - l T_1) \cos \xi_m [z + d_1(l)] \qquad (23)$$

where $a_m = \dfrac{m\pi}{2a}$ and $\xi_m = \sqrt{k_2^2 - a_m^2}$. In the rth groove ($0 \leq r \leq L_2$) of region (III) the scattered field is

$$H_y^{III}(x, z) = \sum_{n=0}^{\infty} c_n^r \cos g_n (x + g - r T_2 - \delta) \cos \eta_n [z - h - d_2(r)] \qquad (24)$$

where $g_m = \dfrac{n\pi}{2g}$ and $\eta_n = \sqrt{k_3^2 - g_n^2}$.

The E_x field continuity at $z = 0$ requires

$$E_x^i(x, 0) + E_x^s(x, 0) = \begin{cases} E_x^{II}(x, 0), & -a + l T_1 < x < a + l T_1 \\ 0, & \text{otherwise} . \end{cases} \qquad (25)$$

Applying the Fourier transform to (25) gives

$$\widetilde{H}_+^s(\zeta) - \widetilde{H}_-^s(\zeta) = \frac{\mathrm{i}}{\epsilon_2 \kappa} \sum_{l=0}^{L_1} \sum_{m=0}^{\infty} b_m^l \xi_m \sin[\xi_m d_1(l)] F_m^l(\zeta) \qquad (26)$$

where

$$F_m^l(\zeta) = \frac{\mathrm{i}\zeta \mathrm{e}^{\mathrm{i}\zeta l T_1} [\mathrm{e}^{-\mathrm{i}\zeta a} - (-1)^m \mathrm{e}^{\mathrm{i}\zeta a}]}{\zeta^2 - a_m^2} . \qquad (27)$$

Similarly from the E_x field continuity at $z = h$,

$$\widetilde{H}_+^s(\zeta) \mathrm{e}^{\mathrm{i}\kappa h} - \widetilde{H}_-^s(\zeta) \mathrm{e}^{-\mathrm{i}\kappa h} = \frac{1}{\mathrm{i}\epsilon_3 \kappa} \sum_{r=0}^{L_2} \sum_{n=0}^{\infty} c_n^r \eta_n \sin[\eta_n d_2(r)] G_n^r(\zeta) \qquad (28)$$

where

$$G_n^r(\zeta) = \frac{\mathrm{i}\zeta e^{\mathrm{i}\zeta(rT_2+\delta)}[e^{-\mathrm{i}\zeta g} - (-1)^n e^{\mathrm{i}\zeta g}]}{\zeta^2 - g_n^2} . \tag{29}$$

The H_y field continuity at $z = 0$ for $(-a + lT_1 < x < a + lT_1)$ is

$$\exp(\mathrm{i}k_{xs}x) + \frac{1}{2\pi}\int_{-\infty}^{\infty}[\widetilde{H}_+^s(\zeta) + \widetilde{H}_-^s(\zeta)]e^{-\mathrm{i}\zeta x}\mathrm{d}\zeta$$

$$= \sum_{m=0}^{\infty} b_m^l \cos a_m(x + a - lT_1)\cos[\xi_m d_1(l)] . \tag{30}$$

Substituting $\widetilde{H}_+^s(\zeta)$ and $\widetilde{H}_-^s(\zeta)$ of (26) and (28) into (30), multiplying (30) by $\cos a_p(x + a - uT_1)$, and integrating over $(-a + uT_1 < x < a + uT_1)$, we get

$$\boxed{\begin{aligned}
&F_p^u(k_{xs})\\
&= \sum_{l=0}^{L_1}\sum_{m=0}^{\infty}\left\{\frac{\xi_m}{\epsilon_2}\sin[\xi_m d_1(l)]I_{1pm}^{ul} + a\varepsilon_m \cos[\xi_m d_1(l)]\delta_{ul}\delta_{pm}\right\}b_m^l\\
&+ \sum_{r=0}^{L_2}\sum_{n=0}^{\infty}\left\{\frac{\eta_n}{\epsilon_3}\sin[\eta_n d_2(r)]I_{2pn}^{ur}\right\}c_n^r
\end{aligned}} \tag{31}$$

where $\varepsilon_0 = 2$, $\varepsilon_m = 1$ $(m = 1, 2, 3, \ldots)$, δ_{ul} is the Kronecker delta, and

$$I_{1pm}^{ul} = \frac{1}{2\pi}\int_{-\infty}^{\infty}\frac{1}{\kappa\tan(\kappa h)}F_m^l(\zeta)F_p^u(-\zeta)\,\mathrm{d}\zeta \tag{32}$$

$$I_{2pn}^{ur} = \frac{1}{2\pi}\int_{-\infty}^{\infty}\frac{1}{\kappa\sin(\kappa h)}G_n^r(\zeta)F_p^u(-\zeta)\,\mathrm{d}\zeta . \tag{33}$$

Similarly from the H_y field continuity at $z = h$, we get

$$\boxed{\begin{aligned}
&\cos(k_{zs}h)G_q^v(k_{xs}) = \sum_{l=0}^{L_1}\sum_{m=0}^{\infty}\left\{\frac{\xi_m}{\epsilon_2}\sin[\xi_m d_1(l)]I_{3qm}^{vl}\right\}b_m^l\\
&+ \sum_{r=0}^{L_2}\sum_{n=0}^{\infty}\left\{\frac{\eta_n}{\epsilon_3}\sin[\eta_n d_2(r)]I_{4qn}^{vr} + g\varepsilon_n \cos[\eta_n d_2(r)]\delta_{vr}\delta_{qn}\right\}c_n^r
\end{aligned}} \tag{34}$$

where

$$I_{3qm}^{vl} = \frac{1}{2\pi}\int_{-\infty}^{\infty}\frac{1}{\kappa\sin(\kappa h)}F_m^l(\zeta)G_q^v(-\zeta)\,\mathrm{d}\zeta \tag{35}$$

$$I_{4qn}^{vr} = \frac{1}{2\pi}\int_{-\infty}^{\infty}\frac{1}{\kappa\tan(\kappa h)}G_n^r(\zeta)G_q^v(-\zeta)\,\mathrm{d}\zeta . \tag{36}$$

Rapidly-convergent series forms for I_{1pm}^{ul}, I_{2pn}^{ur}, I_{3qm}^{vl}, and I_{4qn}^{vr} are available in Subsect. 1.5.3.

The scattered field at $x = \pm\infty$ is

$$H_y^s(\pm\infty, z) = \sum_u K_u^\pm \cos(k_{zu}z)\exp(\pm ik_{xu}x) \tag{37}$$

where $0 \le u < k_0 h/\pi$, u: integer, and

$$K_u^\pm = \pm\sum_{l=0}^{L_1}\sum_{m=0}^{\infty} \frac{b_m^l \xi_m \sin[\xi_m d_1(l)]F_m^l(\mp k_{xu})}{i\epsilon_2\varepsilon_u h k_{xu}}$$

$$\pm\sum_{r=0}^{L_2}\sum_{n=0}^{\infty} \frac{c_n^r \eta_n \sin[\eta_n d_2(r)]G_n^r(\mp k_{xu})}{i\epsilon_3\varepsilon_u h(-1)^u k_{xu}} . \tag{38}$$

Let P_i, P_r, and P_t denote the time-averaged incident, reflected, and transmitted powers, respectively. Then the transmission (τ) and reflection (ϱ) coefficients are

$$\tau = P_t/P_i = |1 + K_s^+|^2 + \frac{1}{\varepsilon_s k_{xs}}\sum_{v\ne s}\varepsilon_v k_{xv}|K_v^+|^2 \tag{39}$$

$$\varrho = P_r/P_i = \frac{1}{\varepsilon_s k_{xs}}\sum_v \varepsilon_v k_{xv}|K_v^-|^2 \tag{40}$$

where $0 \le v < k_0 h/\pi$, v: integer, and $P_i = k_{xs}\varepsilon_s h/(4\omega\epsilon_0)$.

1.5.3 Appendix

- TE wave
  Utilizing the technique of contour integration, we evaluate I_{1pm}^{ul}, I_{2pn}^{ur}, I_{3qm}^{vl}, and I_{4qn}^{vr} analytically in the complex ζ-plane.

$$I_{1pm}^{ul} = \Lambda_{1m}\delta_{ul}\delta_{pm} + J_{1pm}^{ul} \tag{41}$$

$$I_{2pn}^{ur} = X_1(\zeta) + \begin{cases} 2\pi i P_1(\zeta)|_{\zeta=a_p} + 2\pi i Q_1(\zeta)|_{\zeta=g_n}, & a_p \ne g_n \\ 2\pi i \dfrac{f_1'(\zeta)s_1(\zeta) - f_1(\zeta)s_1'(\zeta)}{s_1^2(\zeta)}\bigg|_{\zeta=a_p}, & a_p = g_n \end{cases} \tag{42}$$

$$I_{3qm}^{vl} = X_2(\zeta) + \begin{cases} 2\pi i P_2(\zeta)|_{\zeta=a_m} + 2\pi i Q_2(\zeta)|_{\zeta=g_q}, & a_m \ne g_q \\ 2\pi i \dfrac{f_2'(\zeta)s_2(\zeta) - f_2(\zeta)s_2'(\zeta)}{s_2^2(\zeta)}\bigg|_{\zeta=a_m}, & a_m = g_q \end{cases} \tag{43}$$

$$I_{4qn}^{vr} = \Lambda_{2n}\delta_{qn}\delta_{vr} + J_{2qn}^{vr} \tag{44}$$

where

$$\Lambda_{1m} = a\sqrt{k_0^2 - a_m^2}\cot\left(\sqrt{k_0^2 - a_m^2}\,h\right) \tag{45}$$

$$J_{1pm}^{ul} = -\sum_{\alpha=1}^{\infty} i a_m a_p \kappa^2 \frac{w_1(\zeta)}{h\zeta(\zeta^2 - a_m^2)(\zeta^2 - a_p^2)}\bigg|_{\zeta=\sqrt{k_0^2 - (\alpha\pi/h)^2}} \tag{46}$$

$$w_1(\zeta) = \Big[(-1)^{m+p} + 1\Big]e^{i\zeta|l-u|T_1} - (-1)^m e^{i\zeta|(l-u)T_1+2a|}$$
$$-(-1)^p e^{i\zeta|(l-u)T_1-2a|} \tag{47}$$

$$P_1(\zeta) = \frac{f_1(\zeta)}{2\pi \sin(\kappa h)(\zeta + a_p)(\zeta^2 - g_n^2)} \tag{48}$$

$$Q_1(\zeta) = \frac{f_1(\zeta)}{2\pi \sin(\kappa h)(\zeta + g_n)(\zeta^2 - a_p^2)} \tag{49}$$

$$f_1(\zeta) = \kappa a_p g_n \Big[(-1)^{n+p} e^{i\zeta|rT_2-uT_1+g-a+\delta|}$$
$$-(-1)^p e^{i\zeta|rT_2-uT_1-g-a+\delta|}$$
$$-(-1)^n e^{i\zeta|rT_2-uT_1+g+a+\delta|} + e^{i\zeta|rT_2-uT_1-g+a+\delta|}\Big] \tag{50}$$

$$s_1(\zeta) = 2\pi \sin(\kappa h)(\zeta + a_p)^2 \tag{51}$$

$$X_1(\zeta) = -\sum_{\beta=1}^{\infty} \frac{i\kappa f_1(\zeta)}{h\zeta(-1)^{\beta}(\zeta^2 - a_p^2)(\zeta^2 - g_n^2)}\Bigg|_{\zeta=\sqrt{k_0^2-(\beta\pi/h)^2}} \tag{52}$$

$$P_2(\zeta) = \frac{f_2(\zeta)}{2\pi \sin(\kappa h)(\zeta + a_m)(\zeta^2 - g_q^2)} \tag{53}$$

$$Q_2(\zeta) = \frac{f_2(\zeta)}{2\pi \sin(\kappa h)(\zeta + g_q)(\zeta^2 - a_m^2)} \tag{54}$$

$$f_2(\zeta) = \kappa a_m g_q \Big[(-1)^{m+q} e^{i\zeta|lT_1-vT_2+a-g-\delta|}$$
$$-(-1)^m e^{i\zeta|lT_1-vT_2+a+g-\delta|}$$
$$-(-1)^q e^{i\zeta|lT_1-vT_2-a-g-\delta|} + e^{i\zeta|lT_1-vT_2-a+g-\delta|}\Big] \tag{55}$$

$$s_2(\zeta) = 2\pi \sin(\kappa h)(\zeta + a_m)^2 \tag{56}$$

$$X_2(\zeta) = -\sum_{\beta=1}^{\infty} \frac{i\kappa f_2(\zeta)}{h\zeta(-1)^{\beta}(\zeta^2 - a_m^2)(\zeta^2 - g_q^2)}\Bigg|_{\zeta=\sqrt{k_0^2-(\beta\pi/h)^2}} \tag{57}$$

$$\Lambda_{2n} = g\sqrt{k_0^2 - g_n^2} \cot\left(\sqrt{k_0^2 - g_n^2}\, h\right) \tag{58}$$

$$J_{2qn}^{vr} = -\sum_{\alpha=1}^{\infty} ig_n g_q \kappa^2 \frac{w_2(\zeta)}{h\zeta(\zeta^2 - g_n^2)(\zeta^2 - g_q^2)}\Bigg|_{\zeta=\sqrt{k_0^2-(\alpha\pi/h)^2}} \tag{59}$$

$$w_2(\zeta) = \Big[(-1)^{q+n} + 1\Big]e^{i\zeta|r-v|T_2} - (-1)^n e^{i\zeta|(r-v)T_2+2g|}$$
$$-(-1)^q e^{i\zeta|(r-v)T_2-2g|} \ . \tag{60}$$

- TM wave

Similarly we evaluate I_{1pm}^{ul}, I_{2pn}^{ur}, I_{3qm}^{vl}, and I_{4qn}^{vr} analytically in the complex ζ-plane.

$$I_{1pm}^{ul} = \Lambda_{1m}\delta_{ul}\delta_{pm} + J_{1pm}^{ul} \tag{61}$$

$$I_{2pn}^{ur} = X_1(\zeta) + \begin{cases} 2\pi i P_1(\zeta)|_{\zeta=a_p} + 2\pi i Q_1(\zeta)|_{\zeta=g_n}, & a_p \neq g_n \\[2mm] 2\pi i \dfrac{f_{01}'(\zeta)s_1(\zeta) - f_{01}(\zeta)s_1'(\zeta)}{\varepsilon_p s_1^2(\zeta)}\Bigg|_{\zeta=a_p}, & a_p = g_n \end{cases} \tag{62}$$

$$I_{3qm}^{vl} = X_2(\zeta) + \begin{cases} 2\pi i P_2(\zeta)|_{\zeta=a_m} + 2\pi i Q_2(\zeta)|_{\zeta=g_q}, & a_m \neq g_q \\[2mm] 2\pi i \dfrac{f_{02}'(\zeta)s_2(\zeta) - f_{02}(\zeta)s_2'(\zeta)}{\varepsilon_m s_2^2(\zeta)}\Bigg|_{\zeta=a_m}, & a_m = g_q \end{cases} \tag{63}$$

$$I_{4qn}^{vr} = \Lambda_{2n}\delta_{vr}\delta_{qn} + J_{2qn}^{vr} \tag{64}$$

where

$$\Lambda_{1m} = \frac{a\varepsilon_m}{\sqrt{k_0^2 - a_m^2}\,\tan\left(\sqrt{k_0^2 - a_m^2}\,h\right)} \tag{65}$$

$$J_{1pm}^{ul} = -\sum_{\alpha=0}^{\infty} \frac{i\zeta w_1(\zeta)}{\varepsilon_\alpha h(\zeta^2 - a_m^2)(\zeta^2 - a_p^2)}\Bigg|_{\zeta=\sqrt{k_0^2-(\alpha\pi/h)^2}} \tag{66}$$

$$w_1(\zeta) = \left[(-1)^{m+p} + 1\right]e^{i\zeta|l-u|T_1} - (-1)^m e^{i\zeta|(l-u)T_1+2a|} \\ -(-1)^p e^{i\zeta|(l-u)T_1-2a|} \tag{67}$$

$$P_1(\zeta) = \frac{\zeta^2 f_1(\zeta)}{2\pi\kappa\sin(\kappa h)(\zeta + a_p)(\zeta^2 - g_n^2)} \tag{68}$$

$$Q_1(\zeta) = \frac{\zeta^2 f_1(\zeta)}{2\pi\kappa\sin(\kappa h)(\zeta + g_n)(\zeta^2 - a_p^2)} \tag{69}$$

$$f_1(\zeta) = \left[(-1)^{n+p}e^{i\zeta|rT_2-uT_1+g-a+\delta|} - (-1)^p e^{i\zeta|rT_2-uT_1-g-a+\delta|} \\ -(-1)^n e^{i\zeta|rT_2-uT_1+g+a+\delta|} + e^{i\zeta|rT_2-uT_1-g+a+\delta|}\right] \tag{70}$$

$$f_{01}(\zeta) = \begin{cases} \zeta^2 f_1(\zeta), & p \neq 0 \\ f_1(\zeta), & p = 0 \end{cases} \tag{71}$$

$$s_1(\zeta) = \begin{cases} 2\pi\kappa\sin(\kappa h)(\zeta + a_p)^2, & p \neq 0 \\ 2\pi\kappa\sin(\kappa h), & p = 0 \end{cases} \tag{72}$$

$$X_1(\zeta) = -\sum_{\beta=0}^{\infty} \frac{i\zeta f_1(\zeta)}{\varepsilon_\beta h(-1)^\beta(\zeta^2 - a_p^2)(\zeta^2 - g_n^2)}\Bigg|_{\zeta=\sqrt{k_0^2-(\beta\pi/h)^2}} \tag{73}$$

$$P_2(\zeta) = \frac{\zeta^2 f_2(\zeta)}{2\pi\kappa \sin(\kappa h)(\zeta + a_m)(\zeta^2 - g_q^2)} \tag{74}$$

$$Q_2(\zeta) = \frac{\zeta^2 f_2(\zeta)}{2\pi\kappa \sin(\kappa h)(\zeta + g_q)(\zeta^2 - a_m^2)} \tag{75}$$

$$f_2(\zeta) = \left[(-1)^{m+q} e^{i\zeta|lT_1 - vT_2 + a - g - \delta|} - (-1)^m e^{i\zeta|lT_1 - vT_2 + a + g - \delta|}\right.$$
$$\left. -(-1)^q e^{i\zeta|lT_1 - vT_2 - a - g - \delta|} + e^{i\zeta|lT_1 - vT_2 - a + g - \delta|}\right] \tag{76}$$

$$f_{02}(\zeta) = \begin{cases} \zeta^2 f_2(\zeta), & m \neq 0 \\ f_2(\zeta), & m = 0 \end{cases} \tag{77}$$

$$s_2(\zeta) = \begin{cases} 2\pi\kappa \sin(\kappa h)(\zeta + a_m)^2, & m \neq 0 \\ 2\pi\kappa \sin(\kappa h), & m = 0 \end{cases} \tag{78}$$

$$X_2(\zeta) = -\sum_{\beta=0}^{\infty} \left. \frac{i\zeta f_2(\zeta)}{\varepsilon_\beta h(-1)^\beta(\zeta^2 - a_m^2)(\zeta^2 - g_q^2)}\right|_{\zeta = \sqrt{k_0^2 - (\beta\pi/h)^2}} \tag{79}$$

$$\Lambda_{2n} = \frac{g\varepsilon_n}{\sqrt{k_0^2 - g_n^2} \tan\left(\sqrt{k_0^2 - g_n^2}\,h\right)} \tag{80}$$

$$J_{2qn}^{vr} = -\sum_{\alpha=0}^{\infty} \left. \frac{i\zeta w_2(\zeta)}{\varepsilon_\alpha h(\zeta^2 - g_n^2)(\zeta^2 - g_q^2)}\right|_{\zeta = \sqrt{k_0^2 - (\alpha\pi/h)^2}} \tag{81}$$

$$w_2(\zeta) = \left[(-1)^{q+n} + 1\right] e^{i\zeta|r - v|T_2} - (-1)^n e^{i\zeta|(r-v)T_2 + 2g|}$$
$$-(-1)^q e^{i\zeta|(r-v)T_2 - 2g|} \;. \tag{82}$$

1.6 Water Wave Scattering from Rectangular Grooves in a Plane

A modeling of water flow over obstacles is of importance due to its practical applications in many fluid-mechanic related areas. In this section we will consider a problem of small-water wave scattering from a horizontal water bed consisting of an N number of rectangular grooves [19,20]. An incident wave velocity potential is given by $\Phi^i(x, y) = A_i \cosh k_0(y+h)e^{ik_0 x}$. It is convenient to use an approximate linear modeling approach [21] for small-water wave scattering. For small waves, a dispersion relation $\omega^2 = gk_0 \tanh(k_0 h)$ satisfies the surface boundary condition

$$-\omega^2 \Phi^i(x, y) + g\frac{\partial \Phi^i(x, y)}{\partial y} = 0 \qquad \text{at } y = 0 \tag{1}$$

where g and k_0 are the gravitational acceleration and the wavenumber, respectively. In region (I) $(-h < y < 0)$ the scattered wave is

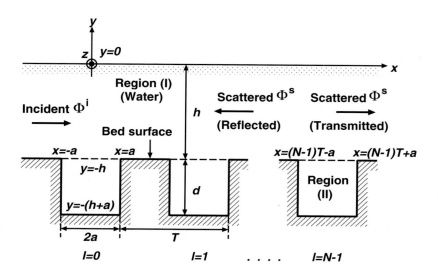

Fig. 1.8. Rectangular grooves on a horizontal water bed

$$\Phi^s(x,y) = \frac{1}{2\pi} \int_{-\infty}^{\infty} [A \cosh \zeta(y+h) + B \sinh \zeta(y+h)] e^{-i\zeta x}\, d\zeta . \qquad (2)$$

Inside the lth groove of region (II) ($|x - lT| < a$ and $-(h+d) < y < -h$: $l = 0, 1, 2, \ldots , N-1$) the water wave is

$$\Phi^t(x,y) = \sum_{m=0}^{\infty} C_m^l \cos a_m(x + a - lT) \cosh a_m(y + h + d) \qquad (3)$$

where $a_m = m\pi/(2a)$.

The surface boundary condition at $y = 0$

$$-\omega^2 \Phi^s(x,y) + g\frac{\partial \Phi^s(x,y)}{\partial y}\bigg|_{y=0} = 0 \qquad (4)$$

gives

$$A = \left[\frac{g\zeta - \omega^2 \tanh(\zeta h)}{\omega^2 - g\zeta \tanh(\zeta h)}\right] B . \qquad (5)$$

The boundary condition at a bed surface ($y = -h$)

$$\frac{\partial \Phi^s(x,y)}{\partial y} = \begin{cases} 0, & |x - lT| > a \\ \dfrac{\partial \Phi^t(x,y)}{\partial y}, & |x - lT| < a \end{cases} \qquad (6)$$

is rewritten as

$$\frac{1}{2\pi} \int_{-\infty}^{\infty} B\zeta e^{-i\zeta x} d\zeta$$

$$= \begin{cases} 0, & |x - lT| > a \\ \sum_{m=0}^{\infty} C_m^l a_m \cos a_m (x + a - lT) \sinh(a_m d), & |x - lT| < a . \end{cases} \tag{7}$$

Applying the Fourier transform to (7) yields

$$B = -\sum_{l=0}^{N-1} \sum_{m=0}^{\infty} i C_m^l a_m \sinh(a_m d) F_m(\zeta a) a^2 e^{i\zeta lT} . \tag{8}$$

Multiplying the boundary condition, $\Phi^i(x, -h) + \Phi^s(x, -h) = \Phi^t(x, -h)$ for $|x - lT| < a$, by $\cos a_n(x + a - rT)$, and integrating from $(rT - a)$ to $(rT + a)$, we obtain

$$\boxed{\begin{aligned} -ik_0 A_i a^2 F_n(k_0 a) e^{ik_0 rT} + \frac{1}{2\pi} \sum_{l=0}^{N-1} \sum_{m=0}^{\infty} C_m^l a_m \sinh(a_m d) I \\ = C_n^r a \varepsilon_n \cosh(a_n d) \delta_{mn} \delta_{lr} \end{aligned}} \tag{9}$$

where

$$I = \int_{-\infty}^{\infty} \left[\frac{g\zeta - \omega^2 \tanh(\zeta h)}{\omega^2 - g\zeta \tanh(\zeta h)} \right] \\ \cdot \zeta a^4 F_m(\zeta a) F_n(-\zeta a) \exp[i\zeta(l-r)T] d\zeta . \tag{10}$$

Using the residue calculus, we transform I into

$$I = \frac{-4\pi i k_0^2 E(k_0)}{\sinh(2k_0 h) + 2k_0 h} + \frac{2\pi}{g} \sum_{v=1}^{\infty} \frac{\zeta_v E(i\zeta_v)[g\zeta_v - \omega^2 \tan(\zeta_v h)]}{\zeta_v h \sec^2(\zeta_v h) + \tan(\zeta_v h)}$$

$$+ \begin{cases} \dfrac{4\pi a}{\omega^2}(g - \omega^2 h)\delta_{mn}\delta_{lr}, & m = 0 \\ \\ \dfrac{2a\pi}{a_m}\left[\dfrac{ga_m - \omega^2 \tanh(a_m h)}{\omega^2 - ga_m \tanh(a_m h)}\right]\delta_{mn}\delta_{lr}, & \text{otherwise} \end{cases} \tag{11}$$

with

$$E(\zeta) = \frac{A_1}{(\zeta^2 - a_m^2)(\zeta^2 - a_n^2)} \tag{12}$$

$$\begin{aligned} A_1 = &[(-1)^{m+n} + 1] \exp[i\zeta|(l-r)T|] \\ &-(-1)^m \exp[i\zeta|(l-r)T + 2a|] \\ &-(-1)^n \exp[i\zeta|(l-r)T - 2a|] \end{aligned} \tag{13}$$

and the positive real numbers ζ_v ($v = 1, 2, \ldots$) are determined by the characteristic equation $\omega^2 = -g\zeta_v \tan(\zeta_v h)$.

It is possible to obtain the scattered field in terms of rapidly-convergent series using the residue calculus.

For $x > (N-1)T + a$,

$$\Phi^s(x,y) = \sum_{l=0}^{N-1} \sum_{m=0}^{\infty} C_m^l a_m \sinh(a_m d)$$

$$\cdot \left\{ \frac{2[\omega^2 \sinh(k_0 y) + gk_0 \cosh(k_0 y)]}{g[2k_0 h + \sinh(2k_0 h)]} \right.$$

$$\cdot \cosh(k_0 h) a^2 F_m(-k_0 a) \exp(-ik_0 lT + ik_0 x)$$

$$+ \sum_{v=1}^{\infty} \frac{2[\omega^2 \sin(\zeta_v y) + g\zeta_v \cos(\zeta_v y)]}{g[2\zeta_v h + \sin(2\zeta_v h)]}$$

$$\left. \cdot \cos(\zeta_v h) a^2 F_m(-i\zeta_v a) \exp(\zeta_v lT - \zeta_v x) \right\} . \tag{14}$$

For $x < -a$,

$$\Phi^s(x,y) = \sum_{l=0}^{N-1} \sum_{m=0}^{\infty} C_m^l a_m \sinh(a_m d)$$

$$\cdot \left\{ - \frac{2[\omega^2 \sinh(k_0 y) + gk_0 \cosh(k_0 y)]}{g[2k_0 h + \sinh(2k_0 h)]} \right.$$

$$\cdot \cosh(k_0 h) a^2 F_m(k_0 a) \exp(ik_0 lT - ik_0 x)$$

$$- \sum_{v=1}^{\infty} \frac{2[\omega^2 \sin(\zeta_v y) + g\zeta_v \cos(\zeta_v y)]}{g[2\zeta_v h + \sin(2\zeta_v h)]}$$

$$\left. \cdot \cos(\zeta_v h) a^2 F_m(i\zeta_v a) \exp(-\zeta_v lT + \zeta_v x) \right\} . \tag{15}$$

References for Chapter 1

1. J. M. Jin and J. L. Volakis, "TM scattering by an inhomogeneously filled aperture in a thick conducting plane," *IEE Proceedings*, pt. H, vol. 137, no. 3, pp. 153-159, June 1990.
2. T. B. A. Senior, K. Sarabandi, and J. R. Natzke, "Scattering by a narrow gap," *IEEE Trans. Antennas Propagat.*, vol. 38, no. 7, pp. 1102-1110, July 1990.
3. S. K. Jeng, "Scattering from a cavity-backed slit on a ground plane- TE case," *IEEE Trans. Antennas Propagat.*, pp. 1523-1529, Oct. 1990.
4. S. K. Jeng and S. T. Tzeng, "Scattering from a cavity-backed slit on a ground plane- TM case," *IEEE Trans. Antennas Propagat.*, vol. 39, no. 3, pp. 661-663, May 1991.
5. K. Yoshidomi, "Scattering of an electromagnetic beam wave by rectangular grooves on a perfect conductor," *Trans. Inst. Electron. Commun. Eng. Jpn.*, vol. E. 67, no. 8, pp. 447-448, Aug. 1984.
6. T. J. Park, H. J. Eom, and K. Yoshitomi, "An analytic solution for transverse magnetic scattering from a rectangular channel in a conducting plane," *J. Appl. Phys.*, vol. 73, no. 7, pp. 3571-3573, 1. April 1993.

7. T. J. Park, H. J. Eom, and K. Yoshitomi, "An analysis of transverse electric scattering from a rectangular channel in a conducting plane," *Radio Sci.*, vol. 28, no. 5, pp. 663-673, Sept.-Oct. 1993.

8. A. P. Prudnikov, Yu. A. Brychkov, and O. I. Marichev, *Integrals and Series,* vol. I *Elementary Functions,* pp. 105-325, Gordon and Breach, New York, 1986.

9. R. Petit, *Electromagnetic Theory of Gratings*, vol. X of Topics in Current Physics, New York, Springer-Verlag, 1980.

10. J. W. Heath and E. V. Jull, "Perfectly blazed reflection grating with rectangular grooves," *J. Opt. Soc. Am.*, vol. 66, pp. 772-775, 1976.

11. Y. L. Kok, "Boundary-value solution to electromagnetic scattering by a rectangular groove in a ground plane," *J. Opt. Soc. Am. A.*, vol. 9, pp. 302-311, 1992.

12. T. J. Park, H. J. Eom, and K. Yoshitomi, "Analysis of TM scattering from finite rectangular grooves in a conducting plane," *J. Opt. Soc. Am. A*, vol. 10, no. 5, pp. 905-911, May 1993.

13. K. Uchida, "Numerical analysis of surface-wave scattering by finite periodic notches in a ground plane," *IEEE Trans. Microwave Theory Tech.*, vol. 35, no. 5, pp. 481-486, May 1987.

14. K. H. Park, H. J. Eom, and T. J. Park, "Surface wave scattering from a notch in a dielectric-covered ground plane: TE-mode analysis," *IEEE Trans. Antennas Propagat.*, vol. 42, no. 2, pp. 286-288, Feb. 1994.

15. I. L. Verbitskii, "Dispersion relation for comb-type slow-wave structures," *IEEE Trans. Microwave Theory Tech.*, vol. 28, no. 1, pp. 48-50, Jan. 1980.

16. J. H. Lee, H. J. Eom, J. W. Lee, and K. Yoshitomi, "Transverse electric mode scattering from a rectangular grooves in parallel-plate," *Radio Sci.*, vol. 29, no. 5, pp. 1215-1218, Sept.-Oct. 1994.

17. K. H. Park, H. J. Eom, and K. Uchida, "TM-scattering from notches in a parallel-plate waveguide," *IEICE Trans. Commun.*, vol. E79-B, no.2, pp. 202-204, Feb. 1996.

18. S. B. Park, *Electromagnetic Scattering from a Parallel-Plate with Double Rectangular Corrugations*, Master Thesis, Department of Electrical Engineering, Korea Advanced Institute of Science and Technology, Taejon, Korea, Dec. 1998.

19. J. J. Lee and R. M. Ayer, "Wave propagation over a rectangular trench," *J. Fluid Mech.*, vol. 110, pp. 335-347, 1981.

20. J. W. Miles, "On surface-wave diffraction by a trench," *J. Fluid Mech.*, vol. 115, pp. 315-325, 1982.

21. G. D. Crapper, *Introduction to Water Waves*, pp. 154-157, Ellis Horwood Limited, 1984.

2. Flanged Parallel-Plate Waveguide Array

2.1 EM Radiation from a Flanged Parallel-Plate Waveguide

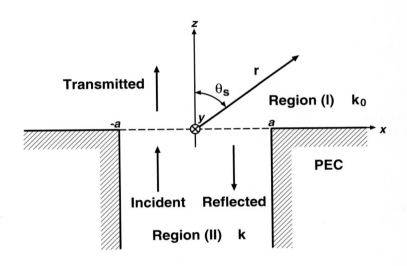

Fig. 2.1. A flanged parallel-plate waveguide

Electromagnetic wave scattering from a conducting double wedge and a flanged parallel-plate waveguide was considered in [1-7]. A flanged parallel-plate waveguide is one of the basic radiating structures used for various aperture antennas. A study of radiation from a flanged parallel-plate waveguide is applicable to radiation from a flanged conducting rectangular waveguide that is a practical radiating element. In the next two subsections we will formulate TE and TM radiation from a flanged parallel-plate waveguide.

2.1.1 TE Radiation [8,9]

Consider a TE wave $E_y^i(x, z)$ radiating from a flanged parallel-plate waveguide with width $2a$. The incident and reflected E-fields inside a parallel-plate waveguide (region (II) with wavenumber k) are

$$E_y^i(x, z) = \sin a_p(x + a) \exp(i\xi_p z) \tag{1}$$

$$E_y^r(x, z) = \sum_{m}^{\infty} c_m \sin a_m(x + a) \exp(-i\xi_m z) \tag{2}$$

where $a_m = m\pi/(2a)$ and $\xi_m = \sqrt{k^2 - a_m^2}$. Note that $m = 1, 3, 5, \ldots$ for odd p and $m = 2, 4, 6, \ldots$ for even p. In region (I) $(z > 0$ with wavenumber $k_0)$ the transmitted E-field is

$$E_y^t(x, z) = \frac{1}{2\pi} \int_{-\infty}^{\infty} \widetilde{E}_y^t(\zeta) \exp(-i\zeta x + i\kappa_0 z)\, d\zeta \tag{3}$$

where $\kappa_0 = \sqrt{k_0^2 - \zeta^2}$.

Applying the Fourier transform to the tangential E-field continuity at $z = 0$ yields

$$\widetilde{E}_y^t(\zeta) = a_p a^2 F_p(\zeta a) + \sum_{m}^{\infty} c_m a_m a^2 F_m(\zeta a) . \tag{4}$$

Multiplying the tangential H-field continuity at $(-a < x < a$ and $z = 0)$ by $\sin a_n(x + a)$ and integrating over $(-a < x < a)$, we get

$$\boxed{\frac{a^2 a_n}{2\pi} \sum_{m}^{\infty} (\delta_{mp} + c_m) a_m \Lambda_1(k_0) = \xi_p a \delta_{np} - \xi_n c_n a} \tag{5}$$

where $\Lambda_1(k_0)$ is given in Subsect. 1.1.3.

The far-zone transmitted field at distance r from the origin is

$$E_y^{\,t}(\theta_s) = \sqrt{\frac{2k_0}{\pi r}} \exp(ik_0 r + i3\pi/4) \cos\theta_s$$

$$\cdot \sum_{m}^{\infty} (\delta_{mp} + c_m) \left[\frac{a_m \cos(k_0 a \sin\theta_s)}{(k_0 \sin\theta_s)^2 - a_m^2} \right] . \tag{6}$$

2.1.2 TM Radiation [10]

Assume that a TM wave $H_y^i(x, z)$ radiates from the aperture of a parallel-plate waveguide (region (II) with wavenumber k). The incident and reflected H-fields within a parallel-plate waveguide are

$$H_y^i(x, z) = \cos a_p(x + a) \exp(i\xi_p z) \tag{7}$$

$$H_y^r(x, z) = \sum_{m=0}^{\infty} c_m \cos a_m(x + a) \exp(-i\xi_m z) . \tag{8}$$

In region (I) (wavenumber k_0) the transmitted H-field is

$$H_y^t(x, z) = \frac{1}{2\pi} \int_{-\infty}^{\infty} \widetilde{H}_y^t(\zeta) \exp(-i\zeta x + i\kappa_0 z) \, d\zeta \qquad (9)$$

where $\kappa_0 = \sqrt{k_0^2 - \zeta^2}$.

The tangential E-field continuity at $z = 0$ yields

$$\kappa_0 \widetilde{H}_y^t(\zeta) = -i\zeta \xi_p a^2 F_p(\zeta a) + \sum_{m=0}^{\infty} i\zeta \xi_m c_m a^2 F_m(\zeta a) \,. \qquad (10)$$

The tangential H-field continuity along the aperture ($-a < x < a$ and $z = 0$) gives

$$\boxed{\sum_{m=0}^{\infty} (\delta_{mp} - c_m)\xi_m a \Omega_1(k_0) = 2\pi(\delta_{np} + c_n)\varepsilon_n} \qquad (11)$$

where $\Omega_1(k_0)$ is given in Subsect. 1.1.3. When the operating frequency approaches infinity ($k_0 a \to \infty$), $\Omega_1(k_0) \to 2\pi \varepsilon_n / (a\sqrt{k_0^2 - a_m^2})\delta_{mn}$, thereby leading (11) to a solution in the Kirchhoff approximation.

The far-zone transmitted field at r is

$$H_y^t(\theta_s) = \sqrt{\frac{k_0}{2\pi r}} \exp(ik_0 r + i\pi/4) \sin\theta_s$$
$$\cdot \sum_{m=0}^{\infty} (\delta_{mp} - c_m)\xi_m a^2 F_m(-k_0 a \sin\theta_s) \,. \qquad (12)$$

2.2 EM Radiation from a Parallel-Plate Waveguide into a Dielectric Slab

Electromagnetic wave radiation from a parallel-plate waveguide and a rectangular waveguide into a dielectric slab was studied in [11-16]. A study of radiation from a parallel-plate waveguide into a dielectric slab is useful for the design of radome and microwave permittivity sensors. In the next two subsections we will investigate TE and TM wave radiation from a flanged parallel-plate waveguide into a displaced dielectric slab.

2.2.1 TE Radiation

Consider a problem of radiation from a flanged parallel-plate waveguide into a dielectric slab. A wave transverse electric (TE) to the z-axis, $E_y^i(x, z)$, is incident on a dielectric slab from inside a parallel-plate waveguide. Regions (I), (II), (III), and (IV), respectively, denote the air (wavenumber: $k_1 = \omega\sqrt{\mu\epsilon_1} = 2\pi/\lambda$ and λ: wavelength), a dielectric slab (wavenumber: $k_2 =$

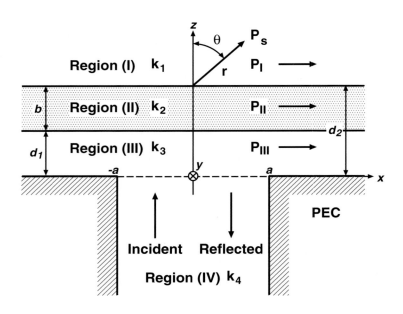

Fig. 2.2. A flanged parallel-plate waveguide radiating into a dielectric slab

$\omega\sqrt{\mu\epsilon_2}$), a background medium (wavenumber: $k_3 = \omega\sqrt{\mu\epsilon_3}$), and an aperture (wavenumber: $k_4 = \omega\sqrt{\mu\epsilon_4}$). In region (I) the total E-field is

$$E_y^I(x, z) = \frac{1}{2\pi} \int_{-\infty}^{\infty} \tilde{E}_I(\zeta) \exp(-i\zeta x + i\kappa_1 z) \, d\zeta \tag{1}$$

where $\kappa_1 = \sqrt{k_1^2 - \zeta^2}$. In regions (II) and (III) each E-field is represented as

$$E_y^{II}(x, z) = \frac{1}{2\pi} \int_{-\infty}^{\infty} \left[\tilde{E}_{II}^+(\zeta)e^{i\kappa_2 z} + \tilde{E}_{II}^-(\zeta)e^{-i\kappa_2 z} \right] e^{-i\zeta x} \, d\zeta \tag{2}$$

$$E_y^{III}(x, z) = \frac{1}{2\pi} \int_{-\infty}^{\infty} \left[\tilde{E}_{III}^+(\zeta)e^{i\kappa_3 z} + \tilde{E}_{III}^-(\zeta)e^{-i\kappa_3 z} \right] e^{-i\zeta x} \, d\zeta \tag{3}$$

where $\kappa_2 = \sqrt{k_2^2 - \zeta^2}$ and $\kappa_3 = \sqrt{k_3^2 - \zeta^2}$. In region (IV) $(-a < x < a)$ the total incident and reflected fields are

$$E_y^i(x, z) = \sin a_p(x + a)e^{i\xi_p z} \tag{4}$$

$$E_y^r(x, z) = \sum_{m=1}^{\infty} c_m \sin a_m(x + a)e^{-i\xi_m z} \tag{5}$$

where $\xi_m = \sqrt{k_4^2 - a_m^2}$ and $a_m = m\pi/(2a)$.

The tangential field continuities at $z = d_2 = d_1 + b$ and $z = d_1$ give, respectively,

$$\widetilde{E}_{II}^{-}(\zeta) = e^{i2\kappa_2 d_2} \left(\frac{\kappa_2 - \kappa_1}{\kappa_2 + \kappa_1} \right) \widetilde{E}_{II}^{+}(\zeta) \tag{6}$$

$$\widetilde{E}_{III}^{-}(\zeta) = e^{i2\kappa_3 d_1}$$
$$\cdot \left[\frac{(\kappa_3 - \kappa_2)(\kappa_2 + \kappa_1)e^{2i\kappa_2 d_1} + (\kappa_3 + \kappa_2)(\kappa_2 - \kappa_1)e^{2i\kappa_2 d_2}}{(\kappa_3 + \kappa_2)(\kappa_2 + \kappa_1)e^{2i\kappa_2 d_1} + (\kappa_3 - \kappa_2)(\kappa_2 - \kappa_1)e^{2i\kappa_2 d_2}} \right]$$
$$\cdot \widetilde{E}_{III}^{+}(\zeta)$$
$$\equiv \left(\frac{\alpha_1}{\alpha_2} \right) \widetilde{E}_{III}^{+}(\zeta)$$
$$\equiv \alpha \widetilde{E}_{III}^{+}(\zeta) . \tag{7}$$

The tangential E-field continuity at $z = 0$ gives

$$\widetilde{E}_{III}^{+}(\zeta) \left(1 + \frac{\alpha_1}{\alpha_2} \right) = aK_p(\zeta a) + \sum_{m=1}^{\infty} c_m aK_m(\zeta a) \tag{8}$$

where

$$K_m(\zeta a) = \frac{m\pi}{2} \left[\frac{e^{i\zeta a}(-1)^m - e^{-i\zeta a}}{(\zeta a)^2 - (m\pi/2)^2} \right] . \tag{9}$$

Substituting (8) into the tangential H-field continuity at the aperture ($-a < x < a$ and $z = 0$) gives

$$\boxed{\frac{1}{2\pi} \left(I_{np} + \sum_{m=1}^{\infty} c_m I_{nm} \right) = \xi_p a \delta_{np} - \xi_n c_n a} \tag{10}$$

where

$$I_{nm} = \int_{-\infty}^{\infty} \kappa_3 \left(\frac{\alpha_2 - \alpha_1}{\alpha_2 + \alpha_1} \right) a^2 K_m(\zeta a) K_n(-\zeta a) \, d\zeta . \tag{11}$$

When $m + n$ is odd, $I_{nm} = 0$. When $m + n$ is even, I_{nm} is

$$I_{nm} = \int_{-\infty}^{\infty} 2\kappa_3 \left(\frac{\alpha_2 - \alpha_1}{\alpha_2 + \alpha_1} \right) a_m a_n \frac{[1 - (-1)^m e^{i2\zeta a}]}{(\zeta^2 - a_m^2)(\zeta^2 - a_n^2)} \, d\zeta . \tag{12}$$

It is expedient to transform (12) into a numerically-efficient integral based on the residue calculus. Let's consider a complex ζ-plane in Fig. 2.3. For analytic convenience, ϵ_2 is assumed to have a small positive imaginary part. Integrating along the deformed contour Γ_1, Γ_2, Γ_3, and Γ_4 in the upper-half plane, we obtain

$$I_{nm} = h_m \delta_{nm} + l_{nm} + r_{nm} \tag{13}$$

where δ_{nm} is the Kronecker delta and

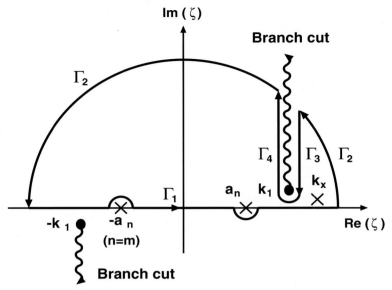

Fig. 2.3. Contour path in the ζ-plane

$$h_m = 2\pi a\sqrt{k_3^2 - a_m^2}\left(\frac{\alpha_2 - \alpha_1}{\alpha_2 + \alpha_1}\right)\bigg|_{\zeta=a_m} \tag{14}$$

$$l_{nm} = \sum_{k_x} \frac{2\pi i a_m a_n(\alpha_2 - \alpha_1)[1 - (-1)^m \mathrm{e}^{\mathrm{i}2k_x a}]}{(\alpha_2 + \alpha_1)'(\zeta^2 - a_m^2)(\zeta^2 - a_n^2)}\bigg|_{\zeta=k_x} \tag{15}$$

$$r_{nm} = -\int_0^\infty \mathrm{d}v f(v) a_m a_n$$

$$\cdot \frac{2\mathrm{i}[1 - (-1)^m \mathrm{e}^{\mathrm{i}2ak_1}\mathrm{e}^{-2ak_1 v}]\sqrt{k_3^2 - k_1^2(1 + \mathrm{i}v)^2}}{k_1^3[(1 + \mathrm{i}v)^2 - (a_m/k_1)^2][(1 + \mathrm{i}v)^2 - (a_n/k_1)^2]} \tag{16}$$

$$f(v) = -\left(\frac{\alpha_2 - \alpha_1}{\alpha_2 + \alpha_1}\right)\bigg|_{\kappa_1 \to -\kappa_1, \zeta=k_1+\mathrm{i}k_1 v} + \left(\frac{\alpha_2 - \alpha_1}{\alpha_2 + \alpha_1}\right)\bigg|_{\zeta=k_1+\mathrm{i}k_1 v} \tag{17}$$

and $(\cdot)'$ denotes differentiation with respect to ζ. Note that h_m and l_{nm} are the residue contributions at $\zeta = \pm a_m$ and k_x, respectively, where k_x is a zero of $(\alpha_2 + \alpha_1)$, and r_{nm} is a branch-cut integration along Γ_3 and Γ_4 associated with a branch-point at $\zeta = k_1$.

The far-zone field $E_y^s(r, \theta)$ and the total radiated power (P_s) normalized by the incident one (P_i) are

$$E_y^s(r,\theta) = \sqrt{\frac{k_1 a^2}{2\pi r}} \exp(\mathrm{i}k_1 r - \mathrm{i}\pi/4)\kappa_2 \cos\theta$$

$$\cdot \Lambda \left[K_p(\zeta a) + \sum_{m=1}^{\infty} c_m K_m(\zeta a) \right] \Bigg|_{\zeta = -k_1 \sin\theta} \tag{18}$$

$$\sigma = P_s/P_i = \frac{k_1}{k_{zs}a} \int_{-\frac{\pi}{2}}^{\frac{\pi}{2}} \left| E_y^s(r,\theta) \right|^2 r\, d\theta \tag{19}$$

where

$$\Lambda = \left\{ \left[-\mathrm{i}\frac{\cos\theta}{\sqrt{(\epsilon_2/\epsilon_1) - \sin^2\theta}} \cos(k_1 d_1 \cos\theta) \right. \right.$$

$$\left. -\frac{\sqrt{(\epsilon_2/\epsilon_1) - \sin^2\theta}}{\cos\theta} \sin(k_1 d_1 \cos\theta) \right] \sin\left(k_1 b\sqrt{(\epsilon_2/\epsilon_1) - \sin^2\theta}\right)$$

$$\left. + \exp(-\mathrm{i}k_1 d_1 \cos\theta)\cos\left(k_1 b\sqrt{(\epsilon_2/\epsilon_1) - \sin^2\theta}\right) \right\}^{-1}. \tag{20}$$

The reflection and transmission coefficients associated with the surface waves in regions (I), (II), and (III) are

$$\varrho = P_r/P_i$$

$$= \frac{1}{\xi_p}\sum_m \xi_m \left| c_m \right|^2, \qquad\qquad 1 \le m < \frac{2k_4 a}{\pi} \tag{21}$$

$$\tau_1 = P_I/P_i$$

$$= \frac{k_x}{\xi_p a}\frac{\mathrm{e}^{-2k_{z1}d_2}}{2k_{z1}}|K_I|^2 \tag{22}$$

$$\tau_2 = P_{II}/P_i$$

$$= \frac{k_x}{2\xi_p a}\left\{ |K_{II}^c|^2 b + |K_{II}^c|^2\frac{[\sin(2k_{z2}d_2) - \sin(2k_{z2}d_1)]}{2k_{z2}} + |K_{II}^s|^2 b \right.$$

$$\left. - |K_{II}^s|^2\frac{[\sin(2k_{z2}d_2) - \sin(2k_{z2}d_1)]}{2k_{z2}} \right\}$$

$$- \frac{k_x}{4\xi_p a k_{z2}}(\cos k_{z2}d_2 - \cos k_{z2}d_1)\,\mathrm{Re}\left[K_{II}^c (K_{II}^s)^* + K_{II}^s (K_{II}^c)^* \right] \tag{23}$$

$$\tau_3 = P_{III}/P_i$$

$$= \frac{k_x}{2\xi_p a}\left(|K_{III}^c|^2 d_1 - |K_{III}^c|^2 \frac{1}{2k_{z3}}\sinh 2k_{z3}d_1 + |K_{III}^s|^2 d_1 \right.$$

$$\left. + |K_{III}^s|^2 \frac{1}{2k_{z3}}\sinh 2k_{z3}d_1 \right)$$

$$- \frac{k_x}{2\xi_p a k_{z3}}[1 + \cosh(2k_{z3}d_1)]$$

$$\cdot Re\left[iK_{III}^c(K_{III}^s)^* - iK_{III}^s(K_{III}^c)^*\right] \qquad (24)$$

where $k_{z1} = \sqrt{k_1^2 - k_x^2}$, $k_{z2} = \sqrt{k_2^2 - k_x^2}$, $k_{z3} = \sqrt{k_3^2 - k_x^2}$, and the power conservation requires $\sigma + \tau_1 + \tau_2 + \tau_3 + \varrho = 1$. We further note that

$$K_I = -2iA\exp[i(k_{z2} - ik_{z1})d_2]k_{z2} \qquad (25)$$

$$K_{II}^c = -2iAe^{ik_{z2}d_2}[k_{z2}\cos(k_{z2}d_2) + k_{z1}\sin(k_{z2}d_2)] \qquad (26)$$

$$K_{II}^s = -2iAe^{ik_{z2}d_2}[k_{z2}\sin(k_{z2}d_2) - k_{z1}\cos(k_{z2}d_2)] \qquad (27)$$

$$K_{III}^c = 2Be^{-k_{z3}d_1}k_{z3}$$

$$\cdot \left[ik_{z1}(e^{i2k_{z2}d_1} - e^{i2k_{z2}d_2}) + k_{z2}(e^{i2k_{z2}d_1} + e^{i2k_{z2}d_2})\right] \qquad (28)$$

$$K_{III}^s = 2Be^{-k_{z3}d_1}k_{z2}$$

$$\cdot \left[k_{z2}(e^{i2k_{z2}d_1} - e^{i2k_{z2}d_2}) + ik_{z1}(e^{i2k_{z2}d_1} + e^{i2k_{z2}d_2})\right] \qquad (29)$$

$$A = \left\{\left[aK_p(\zeta a) + \sum_{m=1}^{\infty} c_m aK_m(\zeta a)\right]\left(\alpha_2 e^{i\kappa_3 d_1} + \alpha_1 e^{-i\kappa_3 d_1}\right)e^{k_{z3}d_1}\right\}$$

$$\left\{\left[e^{i\kappa_2 d_1}(\kappa_2 + \kappa_1) + \exp(i2\kappa_2 d_2 - i\kappa_2 d_1)(\kappa_2 - \kappa_1)\right]\right.$$

$$\left.\cdot\left[(\alpha_2 + \alpha_1)e^{-i\kappa_3 d_1}\right]'\right\}^{-1}\Bigg|_{\zeta=-k_x} \qquad (30)$$

$$B = \frac{\left[aK_p(\zeta a) + \sum\limits_{m=1}^{\infty} c_m aK_m(\zeta a)\right]}{(\alpha_2 + \alpha_1)'}\Bigg|_{\zeta=-k_x} \qquad (31)$$

where $(\cdot)'$ denotes differentiation with respect to ζ.

2.2.2 TM Radiation [17]

Consider a flanged parallel-plate waveguide radiating a TM wave into a dielectric slab as discussed in Subsect. 2.2.1. Regions (I), (II), (III), and (IV), respectively, denote a half-space (wavenumber: $k_1 = \omega\sqrt{\mu\epsilon_0\epsilon_1} = 2\pi/\lambda_1$), a dielectric slab (wavenumber: $k_2 = \omega\sqrt{\mu\epsilon_0\epsilon_2} = 2\pi/\lambda_2$), a background medium (wavenumber: $k_3 = \omega\sqrt{\mu\epsilon_0\epsilon_3} = 2\pi/\lambda_3$), and an aperture (wavenumber: $k_4 = \omega\sqrt{\mu\epsilon_0\epsilon_4} = 2\pi/\lambda_4$). A wave transverse magnetic (TM) to the z-axis, $H_y^i(x, z)$, impinges on a dielectric slab from inside a parallel-plate waveguide. In region (I) the H-field is

$$H_y^I(x, z) = \frac{1}{2\pi} \int_{-\infty}^{\infty} \tilde{H}_I(\zeta) \exp(-i\zeta x + i\kappa_1 z) \, d\zeta \tag{32}$$

where $\kappa_1 = \sqrt{k_1^2 - \zeta^2}$. In region (II) the total H-field is

$$H_y^{II}(x, z) = \frac{1}{2\pi} \int_{-\infty}^{\infty} \left[\tilde{H}_{II}^+(\zeta) e^{i\kappa_2 z} + \tilde{H}_{II}^-(\zeta) e^{-i\kappa_2 z} \right] e^{-i\zeta x} \, d\zeta \tag{33}$$

where $\kappa_2 = \sqrt{k_2^2 - \zeta^2}$. In region (III) the total H-field is

$$H_y^{III}(x, z) = \frac{1}{2\pi} \int_{-\infty}^{\infty} \left[\tilde{H}_{III}^+(\zeta) e^{i\kappa_3 z} + \tilde{H}_{III}^-(\zeta) e^{-i\kappa_3 z} \right] e^{-i\zeta x} \, d\zeta \tag{34}$$

where $\kappa_3 = \sqrt{k_3^2 - \zeta^2}$. In region (IV) ($-a < x < a$) the incident and reflected fields are

$$H_y^i(x, z) = \cos a_p(x + a) e^{i\xi_p z} \tag{35}$$

$$H_y^r(x, z) = \sum_{m=0}^{\infty} c_m \cos a_m(x + a) e^{-i\xi_m z} \tag{36}$$

where $\xi_m = \sqrt{k_4^2 - a_m^2}$ and $a_m = m\pi/(2a)$.

The tangential E-field and H-field continuities at $z = d_2 = d_1 + b$ yield

$$\tilde{H}_{II}^-(\zeta) = e^{i2\kappa_2 d_2} \left(\frac{\epsilon_1 \kappa_2 - \epsilon_2 \kappa_1}{\epsilon_1 \kappa_2 + \epsilon_2 \kappa_1} \right) \tilde{H}_{II}^+(\zeta) . \tag{37}$$

Similarly the tangential E-field and H-field continuities at $z = d_1$ yield

$$\tilde{H}_{III}^-(\zeta) = \Big[(\epsilon_2\kappa_3 - \epsilon_3\kappa_2)(\epsilon_1\kappa_2 + \epsilon_2\kappa_1) e^{2i\kappa_2 d_1}$$
$$+ (\epsilon_2\kappa_3 + \epsilon_3\kappa_2)(\epsilon_1\kappa_2 - \epsilon_2\kappa_1) e^{2i\kappa_2 d_2} \Big]$$
$$\cdot \Big[(\epsilon_2\kappa_3 + \epsilon_3\kappa_2)(\epsilon_1\kappa_2 + \epsilon_2\kappa_1) e^{2i\kappa_2 d_1}$$
$$+ (\epsilon_2\kappa_3 - \epsilon_3\kappa_2)(\epsilon_1\kappa_2 - \epsilon_2\kappa_1) e^{2i\kappa_2 d_2} \Big]^{-1} e^{i2\kappa_3 d_1} \tilde{H}_{III}^+(\zeta)$$
$$\equiv \left(\frac{\alpha_1}{\alpha_2} \right) \tilde{H}_{III}^+(\zeta)$$
$$\equiv \alpha \tilde{H}_{III}^+(\zeta) . \tag{38}$$

The tangential E-field continuity at $z = 0$ yields

$$\tilde{H}_{III}^{+}(\zeta)\left(1 - \frac{\alpha_1}{\alpha_2}\right) = \frac{1}{\kappa_3}\left[\xi_p K_p(\zeta) - \sum_{m=0}^{\infty} c_m \xi_m K_m(\zeta)\right]\frac{\epsilon_3}{\epsilon_4} \qquad (39)$$

where

$$K_m(\zeta) = -\mathrm{i}\zeta a^2 \frac{\mathrm{e}^{\mathrm{i}\zeta a}(-1)^m - \mathrm{e}^{-\mathrm{i}\zeta a}}{(\zeta a)^2 - (m\pi/2)^2} \ . \qquad (40)$$

From the tangential H-field continuity along the aperture ($-a < x < a$ and $z = 0$), we obtain

$$\boxed{\frac{\epsilon_3}{2\pi\epsilon_4}\left(\xi_p J_{np} - \sum_{m=0}^{\infty} c_m \xi_m J_{nm}\right) = \varepsilon_n a \delta_{np} + \varepsilon_n c_n a} \qquad (41)$$

where

$$J_{nm} = \int_{-\infty}^{\infty}\left[\frac{\alpha_2 + \alpha_1}{\kappa_3(\alpha_2 - \alpha_1)}\right]K_m(\zeta)K_n(-\zeta)\,\mathrm{d}\zeta \ . \qquad (42)$$

When $m + n$ is odd, $J_{nm} = 0$. When $m + n$ is even, J_{nm} is rewritten as

$$J_{nm} = \int_{-\infty}^{\infty} 2\left(\frac{\alpha_2 + \alpha_1}{\alpha_2 - \alpha_1}\right)\frac{\zeta^2[1 - (-1)^m \mathrm{e}^{\mathrm{i}2\zeta a}]}{(\zeta^2 - a_m^2)(\zeta^2 - a_n^2)\kappa_3}\,\mathrm{d}\zeta \ . \qquad (43)$$

Performing a contour integration in view of Fig. 2.3, we obtain

$$J_{nm} = \varepsilon_m h_m \delta_{nm} - r_{nm} + l_{nm} \ . \qquad (44)$$

We note that h_m and l_{nm} are the residue contributions at $\zeta = \pm a_m$ and k_x, respectively, where k_x is a zero of $(\alpha_2 - \alpha_1)$, and r_{nm} is a branch-cut integration associated with a branch-point at $\zeta = k_1$. They are given as

$$h_m = 2\pi a/\sqrt{k_3^2 - a_m^2}\left(\frac{\alpha_2 + \alpha_1}{\alpha_2 - \alpha_1}\right)\Bigg|_{\zeta = a_m} \qquad (45)$$

$$l_{nm} = \sum_{k_x}\frac{4\pi\mathrm{i}\zeta^2(\alpha_2 + \alpha_1)[1 - (-1)^m \mathrm{e}^{\mathrm{i}2k_x a}]}{(\alpha_2 - \alpha_1)'(\zeta^2 - a_m^2)(\zeta^2 - a_n^2)\kappa_3}\Bigg|_{\zeta = k_x} \qquad (46)$$

$$r_{nm} = \int_0^{\infty} \mathrm{d}v\Big\{2\mathrm{i}[1 - (-1)^m \mathrm{e}^{\mathrm{i}2ak_1}\mathrm{e}^{-2ak_1 v}](1 + \mathrm{i}v)^2 f(v)\Big\}$$
$$\cdot\Big\{k_1[(1 + \mathrm{i}v)^2 - (a_m/k_1)^2][(1 + \mathrm{i}v)^2 - (a_n/k_1)^2]$$
$$\cdot\sqrt{k_3^2 - k_1^2(1 + \mathrm{i}v)^2}\Big\}^{-1} \qquad (47)$$

$$f(v) = -\left(\frac{\alpha_2 + \alpha_1}{\alpha_2 - \alpha_1}\right)\Bigg|_{\kappa_1 \to -\kappa_1, \zeta = k_1 + \mathrm{i}k_1 v} + \left(\frac{\alpha_2 + \alpha_1}{\alpha_2 - \alpha_1}\right)\Bigg|_{\zeta = k_1 + \mathrm{i}k_1 v} \ . \qquad (48)$$

The far-zone field $H_y^s(r, \theta)$ and the normalized radiated power are

$$H_y^s(r,\theta) = \sqrt{\frac{k_1}{2\pi r}}\, \exp(ik_1 r - i\pi/4)\epsilon_1\epsilon_2 \cos\theta\, \Lambda\, L_s(\zeta)\, \Big|_{\zeta=-k_1 \sin\theta} \tag{49}$$

$$\sigma = P_s/P_i = \frac{k_1}{\xi_p \mathcal{E}_p a} \int_{-\frac{\pi}{2}}^{\frac{\pi}{2}} \left|H_y^s(r,\theta)\right|^2 r\, d\theta \tag{50}$$

where

$$L_s(\zeta) = \frac{1}{\kappa_3}\left[\xi_p K_p(\zeta) - \sum_{m=0}^{\infty} c_m \xi_m K_m(\zeta)\right]\frac{\epsilon_3}{\epsilon_4} \tag{51}$$

$$\Lambda = \left\{\left[-i\epsilon_3\epsilon_1 \frac{\beta}{\cos\theta}\cos(k_1 d_1 \cos\theta) - \epsilon_2{}^2 \frac{\cos\theta}{\beta}\sin(k_1 d_1 \cos\theta)\right]\right.$$
$$\left. \cdot \sin(k_1 b\beta) + \exp(-ik_1 d_1 \cos\theta)\epsilon_3\epsilon_2 \cos(k_1 b\beta)\right\}^{-1} \tag{52}$$

and $\beta = \sqrt{(\epsilon_2/\epsilon_1) - \sin^2\theta}$.

The reflected (P_r) and transmitted powers $(P_I, P_{II}, \text{ and } P_{III})$ associated with the surface waves in regions (I), (II), and (III) are

$$\varrho = P_r/P_i$$
$$= \frac{1}{\xi_p \mathcal{E}_p}\sum_m \xi_m \mathcal{E}_m \left|c_m\right|^2, \qquad 0 \leq m < \frac{2k_4 a}{\pi} \tag{53}$$

$$\tau_1 = P_I/P_i$$
$$= \frac{k_x \epsilon_4}{\epsilon_1 \xi_p \mathcal{E}_p a}\left(\frac{1}{2k_{z1}}e^{-2k_{z1}d_2}\right)|K_I|^2 \tag{54}$$

$$\tau_2 = P_{II}/P_i$$
$$= \frac{k_x \epsilon_4}{2\epsilon_2 \xi_p \mathcal{E}_p a}\left\{|K_{II}^c|^2 b + |K_{II}^c|^2\frac{[\sin(2k_{z2}d_2) - \sin(2k_{z2}d_1)]}{2k_{z2}}\right.$$
$$\left. + |K_{II}^s|^2 b - |K_{II}^s|^2\frac{[\sin(2k_{z2}d_2) - \sin(2k_{z2}d_1)]}{2k_{z2}}\right\}$$
$$- \frac{k_x \epsilon_4}{4\xi_p \epsilon_2 \mathcal{E}_p k_{z2} a}(\cos k_{z2}d_2 - \cos k_{z2}d_1)$$
$$\cdot Re\left[K_{II}^c(K_{II}^s)^* + K_{II}^s(K_{II}^c)^*\right] \tag{55}$$

$$\tau_3 = P_{III}/P_i$$

$$= \frac{k_x \epsilon_4}{2\epsilon_3 \xi_p \varepsilon_p a} \left(|K_{III}^c|^2 d_1 - |K_{III}^c|^2 \frac{1}{2k_{z3}} \sinh 2k_{z3} d_1 \right.$$

$$\left. + |K_{III}^s|^2 d_1 + |K_{III}^s|^2 \frac{1}{2k_{z3}} \sinh 2k_{z3} d_1 \right)$$

$$- \frac{k_x \epsilon_4}{4\xi_p \epsilon_3 \varepsilon_p k_{z3} a} [2 + 2\cosh(2k_{z3} d_1)]$$

$$\cdot Re\left[iK_{III}^c (K_{III}^s)^* - iK_{III}^s (K_{III}^c)^* \right] \tag{56}$$

where $k_{z1} = \sqrt{k_x^2 - k_1^2}$, $k_{z2} = \sqrt{k_2^2 - k_x^2}$, $k_{z3} = \sqrt{k_x^2 - k_3^2}$, and the power conservation requires $\sigma + \tau_1 + \tau_2 + \tau_3 + \varrho = 1$.
Note that

$$K_I = -2iA \exp[i(k_{z2} - ik_{z1})d_2]\epsilon_1 k_{z2} \tag{57}$$

$$K_{II}^c = -2iAe^{ik_{z2}d_2} [\epsilon_1 k_{z2} \cos(k_{z2}d_2) + \epsilon_2 k_{z1} \sin(k_{z2}d_2)] e^{k_{z3}d_1} \tag{58}$$

$$K_{II}^s = -2iAe^{ik_{z2}d_2} [\epsilon_1 k_{z2} \sin(k_{z2}d_2) - \epsilon_2 k_{z1} \cos(k_{z2}d_2)] e^{k_{z3}d_1} \tag{59}$$

$$K_{III}^c = 2Be^{-k_{z3}d_1} \epsilon_2 k_{z3} \left[i\epsilon_2 k_{z1} (e^{i2k_{z2}d_1} - e^{i2k_{z2}d_2}) \right.$$

$$\left. + \epsilon_1 k_{z2} (e^{i2k_{z2}d_1} + e^{i2k_{z2}d_2}) \right] \tag{60}$$

$$K_{III}^s = 2Be^{-k_{z3}d_1} \epsilon_3 k_{z2} \left[\epsilon_1 k_{z2} (e^{i2k_{z2}d_1} - e^{i2k_{z2}d_2}) \right.$$

$$\left. + i\epsilon_2 k_{z1} (e^{i2k_{z2}d_1} + e^{i2k_{z2}d_2}) \right] \tag{61}$$

$$A = \left\{ (\epsilon_3/\epsilon_4) \left[\xi_p K_p(\zeta) - \sum_{m=0}^{\infty} c_m \xi_m K_m(\zeta) \right] \right.$$

$$\cdot \left(\alpha_2 e^{i\kappa_3 d_1} + \alpha_1 e^{-i\kappa_3 d_1} \right) \right\}$$

$$\cdot \left\{ \kappa_3 [e^{i\kappa_2 d_1} (\epsilon_1 \kappa_2 + \epsilon_2 \kappa_1) + e^{i2\kappa_2 d_2 - i\kappa_2 d_1} (\epsilon_1 \kappa_2 - \epsilon_2 \kappa_1)] \right.$$

$$\left. \cdot \left[(\alpha_2 - \alpha_1) e^{-i\kappa_3 d_1} \right]' \right\}^{-1}_{\zeta = -k_x} \tag{62}$$

$$B = \frac{\left[\xi_p K_p(\zeta) - \sum\limits_{m=0}^{\infty} c_m \xi_m K_m(\zeta) \right] (\epsilon_3/\epsilon_4)}{(\alpha_2 - \alpha_1)' \kappa_3} \Bigg|_{\zeta = -k_x} \tag{63}$$

where the symbol $(\cdot)'$ denotes differentiation with respect to ζ.

2.3 TE Scattering from a Parallel-Plate Waveguide Array [18]

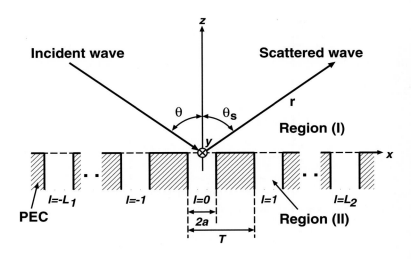

Fig. 2.4. A parallel-plate waveguide array

Electromagnetic wave radiation from a parallel-plate waveguide array has been extensively studied in [19-21] due to its practical applications in array antenna design. In this section we will consider TE wave scattering from a flanged parallel-plate waveguide array when a uniform plane wave $E_y^i(x,z)$ is incident on a waveguide array (width $2a$, period T, and array number N). This section is a continuation of Sect. 2.1 where radiation from a single parallel-plate waveguide was considered. Regions (I) and (II) denote the air and a waveguide interior with wavenumbers $k_0 = \omega\sqrt{\mu_0\epsilon_0} = 2\pi/\lambda$ and $k = \omega\sqrt{\mu\epsilon}$ ($\mu = \mu_0\mu_r$ and $\epsilon = \epsilon_0\epsilon_r$), respectively. In region (I) the E-field has the incident, reflected, and scattered waves

$$E_y^i(x,z) = \exp(\mathrm{i}k_x x - \mathrm{i}k_z z) \tag{1}$$

$$E_y^r(x,z) = -\exp(\mathrm{i}k_x x + \mathrm{i}k_z z) \tag{2}$$

$$E_y^s(x,z) = \frac{1}{2\pi}\int_{-\infty}^{\infty} \widetilde{E}_y^s(\zeta)\exp(-\mathrm{i}\zeta x + \mathrm{i}\kappa_0 z)\,\mathrm{d}\zeta \tag{3}$$

where $k_x = k_0\sin\theta$, $k_z = k_0\cos\theta$, and $\kappa_0 = \sqrt{k_0^2 - \zeta^2}$. The transmitted E-field inside the lth waveguide of region (II) is

$$E_y^t(x,z) = \sum_{m=1}^{\infty} c_m^l \sin a_m(x+a-lT)e^{-i\xi_m z} \tag{4}$$

where $a_m = m\pi/(2a)$ and $\xi_m = \sqrt{k^2 - a_m^2}$.

Applying the Fourier transform to the tangential E-field continuity, we obtain

$$\widetilde{E}_y^s(\zeta) = \sum_{l=-L_1}^{L_2} \sum_{m=1}^{\infty} c_m^l a_m a^2 F_m(\zeta a)e^{i\zeta lT} \ . \tag{5}$$

Multiplying the tangential H-field continuity at $(lT - a) < x < (lT + a)$ and $z = 0$ by $\sin a_n(x+a-rT)$, and integrating with respect to x from $(rT - a)$ to $(rT + a)$, we obtain

$$
\begin{aligned}
&2ik_z a_n \exp(ik_x rT)a^2 F_n(k_x a) \\
&= \frac{ia_n}{2\pi} \sum_{l=-L_1}^{L_2} \sum_{m=1}^{\infty} c_m^l a_m a^2 \Lambda_2(k_0) + i\frac{c_n^r \xi_n a}{\mu_r} \ .
\end{aligned}
\tag{6}
$$

where the expression for $\Lambda_2(k_0)$ is given in Subsect. 1.2.3.

The far-zone scattered field at distance r from the origin is

$$E_y^s(\theta_s, \theta) = \sqrt{\frac{k_0}{2\pi r}} \exp(ik_0 r - i\pi/4) \cos\theta_s \widetilde{E}_y^s(-k_0 \sin\theta_s) \ . \tag{7}$$

The transmission coefficient, which is a ratio of the transmitted power to the incident one over the apertures, is

$$\tau = \frac{1}{2\omega\mu_0} \sum_{l=-L_1}^{L_2} \sum_{m=1}^{<\lambda/4a} |c_m^l|^2 \xi_m^* a/\mu_r^* \ . \tag{8}$$

2.4 EM Radiation from Obliquely-Flanged Parallel Plates

Electromagnetic radiation from a flanged parallel-plate waveguide is an important subject due to its flush-mounted antenna applications. Radiation from a right-angled and infinitely-flanged parallel-plate waveguide has been extensively studied in Sects. 2.1 through 2.3. Radiation from an obliquely-flanged parallel-plate waveguide is of some theoretical interest, but its radiation study is relatively very little [22,23]. In this section we intend to study radiation from an obliquely-flanged parallel-plate waveguide. Consider a TE (transverse electric to the z-axis) wave radiating from an obliquely-flanged parallel-plate waveguide into a conducting plane. When a conducting plane is removed, the scattering geometry becomes a half-space radiation problem. The wavenumber inside the waveguide is k $(= 2\pi/\lambda = \omega\sqrt{\mu\epsilon}$ and λ: wavelength). In region (I) the total incident and reflected E-fields are

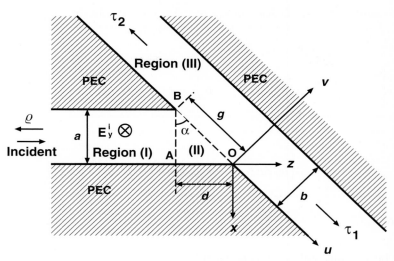

Fig. 2.5. An obliquely flanged parallel-plate waveguide

$$E_y^i(x, z) = \sin(a_p x) e^{i k_p z} \tag{1}$$

$$E_y^r(x, z) = \sum_{m=1}^{\infty} B_m \sin(a_m x) e^{-i k_m z} \tag{2}$$

where $a_p = p\pi/a$ and $k_p = \sqrt{k^2 - a_p^2}$. In region (III) the transmitted E-field is

$$E_y^{III}(u, v) = \frac{1}{2\pi} \int_{-\infty}^{\infty} \left[\tilde{E}_+(\zeta) e^{i\kappa v} + \tilde{E}_-(\zeta) e^{-i\kappa v} \right] e^{-i\zeta u} \, d\zeta \tag{3}$$

where $\kappa = \sqrt{k^2 - \zeta^2}$.

We apply the second Green's formula

$$\oint_c \left(E_y \frac{\partial G}{\partial \nu} - G \frac{\partial E_y}{\partial \nu} \right) dl = 0 \tag{4}$$

to the contour OAB with the auxiliary functions satisfying the Helmholtz equation in region (I) as

$$G_q^{\pm} = \sin(a_q x) e^{\pm i k_q z}, \quad q = 1, 2, \ldots. \tag{5}$$

The fields on the line OB are assumed on the interval $(-g, 0)$ as

$$E_y|_{OB} = \sum_{n=-\infty}^{\infty} C_n \exp\left(i \frac{2\pi n}{g} u \right) \tag{6}$$

$$\left. \frac{\partial E_y}{\partial \nu} \right|_{OB} = \sum_{n=-\infty}^{\infty} D_n \exp\left(i \frac{2\pi n}{g} u \right). \tag{7}$$

As a result of the application of Green's formula, we obtain

$$
\sum_{m=1}^{\infty} B_m \mathrm{i} a k_q \delta_{mq}
$$
$$
= \sum_{n=-\infty}^{\infty} \left[C_n \mathrm{i}\left(a_q \sin \alpha \chi_n^+ + k_q g_q \cos \alpha \right) - D_n g_q \right] \Upsilon_q(\chi_n^+) \tag{8}
$$

$$
-\mathrm{i} a k_q \delta_{pq}
$$
$$
= \sum_{n=-\infty}^{\infty} \left[C_n \mathrm{i}\left(a_q \sin \alpha \chi_n^- - k_q g_q \cos \alpha \right) - D_n g_q \right] \Upsilon_q(\chi_n^-) \tag{9}
$$

where $\Upsilon_q(\zeta) = \dfrac{1 - (-1)^q \mathrm{e}^{-\mathrm{i} g \zeta}}{g_q^2 - \zeta^2}$, $g_q = \dfrac{q\pi}{g}$, $\chi_n^{\pm} = \dfrac{2\pi n}{g} \pm k_q \sin \alpha$, and $q = 1, 2, 3 \dots$.

The E_y continuity at $v = b$ yields

$$
\widetilde{E}_-(\zeta) = -\widetilde{E}_+(\zeta) \mathrm{e}^{\mathrm{i} 2\kappa b} . \tag{10}
$$

The E_y continuity at $v = 0$ between regions (II) and (III) yields

$$
\widetilde{E}_+(\zeta) + \widetilde{E}_-(\zeta) = -\mathrm{i} \sum_{n=-\infty}^{\infty} C_n \Xi_n^-(\zeta) \tag{11}
$$

where $\Xi_n^{\pm}(\zeta) = \dfrac{1 - \mathrm{e}^{\pm \mathrm{i} g \zeta}}{\zeta + 2\pi n/g}$. The H_u continuity at $v = 0$ is given as

$$
\sum_{n=-\infty}^{\infty} D_n \exp\left(\mathrm{i} \frac{2\pi n}{g} u \right) = \frac{1}{2\pi} \int_{-\infty}^{\infty} \mathrm{i}\kappa \left[\widetilde{E}_+(\zeta) - \widetilde{E}_-(\zeta) \right] \mathrm{e}^{-\mathrm{i}\zeta u} \mathrm{d}\zeta . \tag{12}
$$

We multiply (12) by $\exp\left(-\mathrm{i} \dfrac{2\pi m}{g} u \right)$ and perform integration with respect to u from $-g$ to 0 to get

$$
D_m = - \sum_{n=-\infty}^{\infty} \frac{C_n}{g} J_{mn} \tag{13}
$$

where

$$
J_{mn} = \frac{1}{2\pi} \int_{-\infty}^{\infty} \kappa \cot(\kappa b) \Xi_m^+(\zeta) \Xi_n^-(\zeta) \, \mathrm{d}\zeta . \tag{14}
$$

Note that

$$
J_{mn} = -\mathrm{i} \sum_{t=0}^{\infty} \frac{2b_t^2 (1 - \mathrm{e}^{\mathrm{i} g \zeta_t})}{b \zeta_t} \frac{\zeta_t^2 + (2\pi/g)^2 mn}{\left[\zeta_t^2 - (2\pi m/g)^2 \right] \left[\zeta_t^2 - (2\pi n/g)^2 \right]}
$$
$$
+ g \delta_{mn} \sqrt{k^2 - (2\pi n/g)^2} \cot \left[b \sqrt{k^2 - (2\pi n/g)^2} \right] \tag{15}
$$

where $b_t = t\pi/b$ and $\zeta_t = \sqrt{k^2 - (t\pi/b)^2}$. When b goes to ∞, it is expedient to transform (14) into a numerically-efficient radiation integral based on the residue calculus. The result is given by

$$J_{mn} = -\frac{2i}{\pi}k^2 \int_0^\infty \sqrt{v(v-2i)}$$

$$\cdot \frac{[1 - \exp[igk(1+iv)]]\,[k^2(1+iv)^2 + (2\pi/g)^2 mn]}{[k^2(1+iv)^2 - (2\pi m/g)^2]\,[k^2(1+iv)^2 - (2\pi n/g)^2]}\,dv$$

$$+\delta_{mn}\,g\,\sqrt{k^2 - (2\pi m/g)^2}\;. \tag{16}$$

The scattered field at $u = \pm\infty$ in region (III) is

$$E_y^{III}(\pm\infty, v) = \sum_t K_t^\pm \sin b_t(v-b)\mathrm{e}^{\pm i\zeta_t u} \tag{17}$$

where

$$K_t^\pm = \sum_{n=-\infty}^{\infty} C_n \frac{b_t(1 - \mathrm{e}^{\pm i\zeta_t g})}{b\zeta_t(-1)^t(\mp\zeta_t + 2\pi n/g)} \tag{18}$$

$0 < t$, and t: integer. The transmission (τ_1 and τ_2) and reflection (ϱ) coefficients are

$$\tau_1 = \frac{b}{ak_p}\sum_t \zeta_t |K_t^+|^2 \tag{19}$$

$$\tau_2 = \frac{b}{ak_p}\sum_t \zeta_t |K_t^-|^2 \tag{20}$$

$$\varrho = \sum_m \frac{k_m}{k_p}|B_m|^2 \tag{21}$$

where $0 < t < kb/\pi$, t: integer, $0 < m < ka/\pi$, and m: integer. When b goes to ∞, the far-zone scattered field in region (III) is

$$E_y^{III}(r, \theta) = \sqrt{\frac{1}{2\pi kr}}\,\exp(ikr - i\pi/4)k\cos\theta\,\widetilde{E}_+(-k\sin\theta)\;. \tag{22}$$

2.5 EM Radiation from Parallel Plates with a Window [24]

A use of a window in an open-ended waveguide can reduce an undesirable reflection when an open-ended waveguide is used as a radiating element in phased array antennas. The effect of a window in an open-ended waveguide on radiation characteristics has been studied theoretically in [25,26]. In this section we will investigate a TM wave radiating from a flanged parallel-plate waveguide with a window. In region (I) ($z > 0$) an incident field $H_y^i(x, z)$ impinges on a slit (width: $2a$ and depth: d) in a thick perfectly conducting

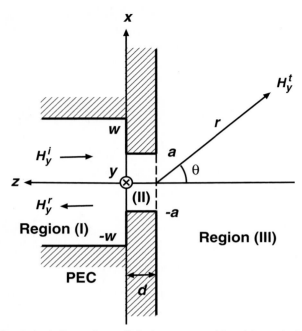

Fig. 2.6. A flanged parallel-plate waveguide with a window

plane. Region (II) ($-d < z < 0$ and $-a < x < a$) denotes a slit and region (III) denotes a half-space ($z < -d$). The wavenumber in regions (I), (II), and (III) is $k = \omega\sqrt{\mu\epsilon} = 2\pi/\lambda$. The fields in regions (I) (H_y^i and H_y^r), (II) (H_y^d), and (III) (H_y^t) are represented as

$$H_y^i(x, z) = \cos w_p(x + w) \exp(-i\eta_p z) \tag{1}$$

$$H_y^r(x, z) = \sum_{m=0}^{\infty} d_m \cos w_m(x + w) \exp(i\eta_m z) \tag{2}$$

$$H_y^d(x, z) = \sum_{m=0}^{\infty} (b_m \cos \xi_m z + c_m \sin \xi_m z) \cos a_m(x + a) \tag{3}$$

$$H_y^t(x, z) = \frac{1}{2\pi} \int_{-\infty}^{\infty} \tilde{H}_y^t(\zeta) \exp\left[-i\zeta x - i\kappa(z + d)\right] d\zeta \tag{4}$$

where $w_m = m\pi/(2w)$, $\eta_m = \sqrt{k^2 - w_m^2}$, $a_m = m\pi/(2a)$, $\xi_m = \sqrt{k^2 - a_m^2}$, and $\kappa = \sqrt{k^2 - \zeta^2}$.

The tangential E-field continuity at $-w < x < w$ and $z = 0$ yields

$$
\begin{aligned}
&i\eta_n(-\delta_{np} + d_n)\omega\varepsilon_n \\
&= \sum_{m=0}^{\infty} c_m \xi_m w_n \left[\frac{(-1)^m \sin w_n(a+w) + \sin w_n(a-w)}{-a_m^2 + w_n^2} \right].
\end{aligned}
\tag{5}
$$

The tangential H-field continuity at $-a < x < a$ and $z = 0$ gives

$$
\begin{aligned}
&\sum_{m=0}^{\infty} (d_m + \delta_{mp}) \left[\frac{(-1)^n \sin w_m(a+w) + \sin w_m(a-w)}{w_m^2 - a_n^2} \right] w_m \\
&= b_n a \varepsilon_n .
\end{aligned}
\tag{6}
$$

The tangential field continuities at $z = -d$ give

$$
\begin{aligned}
&(b_n \cos \xi_n d - c_n \sin \xi_n d) a \varepsilon_n \\
&= \frac{i}{2\pi} \sum_{m=0}^{\infty} \xi_m (b_m \sin \xi_m d + c_m \cos \xi_m d) a^2 \Omega_1(k)
\end{aligned}
\tag{7}
$$

where $\Omega_1(k)$ is

$$
\Omega_1(k) = \int_{-\infty}^{\infty} a^2 F_m(\zeta a) F_n(-\zeta a) \zeta^2 \kappa^{-1} \, d\zeta .
\tag{8}
$$

An explicit evaluation of $\Omega_1(k)$ is available in Subsect. 1.1.3.

The far-zone transmitted field at distance r is

$$
\begin{aligned}
H_y^t(r,\theta) &= e^{i(kr-\pi/4)} \sqrt{\frac{k}{2\pi r}} \sin \theta \\
&\cdot \sum_{m=0}^{\infty} [b_m \sin(\xi_m d) + c_m \cos(\xi_m d)] \xi_m a^2 F_m(-ka \sin \theta)
\end{aligned}
\tag{9}
$$

where $\theta = \sin^{-1}(x/r)$ and $r = \sqrt{x^2 + (z+d)^2}$. The transmission coefficient τ, a ratio of the time-averaged power transmitted through a slit to that incident on a slit, is given by

$$
\begin{aligned}
\tau = -\mathrm{Im}\Bigg\{ &\frac{a}{\varepsilon_p \eta_p w} \sum_{m=0}^{\infty} \varepsilon_m \xi_m \big[|b_m|^2 \sin \xi_m d (\cos \xi_m d)^* \\
&+ b_m^* c_m |\cos \xi_m d|^2 - b_m c_m^* |\sin \xi_m d|^2 \\
&- |c_m|^2 \cos \xi_m d (\sin \xi_m d)^* \big] \Bigg\} .
\end{aligned}
\tag{10}
$$

References for Chapter 2

1. H. M. Nussenzveig, "Solution of diffraction problem, 1. The wide double wedge, 2. The narrow double wedge," *Phil. Trans. Royal Soc. London*, ser. A, vol. 252, pp. 1-51, Oct. 1959.

2. K. Hongo, "Diffraction by a flanged parallel-plate waveguide," *Radio Sci.*, vol. 7, no. 10, pp. 955-963, Oct. 1972.
3. M. S. Leong, P. S. Kooi, and Chandra, "Radiation from a flanged parallel-plate waveguide: Solution by moment method with inclusion of edge condition," *IEE Proceedings*, vol. 135, pt. H, no. 4, pp. 249-255, Aug. 1988.
4. S. W. Lee and L. Grun, "Radiation from flanged waveguide: Comparison of solutions," *IEEE Trans. Antennas Propagat.*, vol. 30, no. 1, Jan. 1982.
5. A. Michaeli, "A new asymptotic high-frequency analysis of electromagnetic scattering by a pair of parallel wedge: Closed form results," *Radio Sci.*, vol. 20, no. 6, pp. 1537-1548, Nov.-Dec. 1985.
6. M. Schneider and R. J. Luebbers, "A general uniform double wedge diffraction coefficient," *IEEE Trans. Antennas Propagat.*, no. 1, pp. 8-14, Jan. 1991.
7. K. Hongo and Y. Ogawa, "Receiving characteristics of a flanged parallel plate waveguide," *IEEE Trans. Antennas Propagat.*, pp. 424-425, Mar. 1977.
8. T. J. Park and H. J. Eom, "Analytic solution for TE-mode radiation from a flanged parallel-plate waveguide," *IEE Proceedings-H*, vol. 140, no. 5, pp.387-389, Oct. 1993.
9. T. J. Park and H. J. Eom, "Scattering and reception by a flanged parallel-plate waveguide: TE-mode analysis," *IEEE Trans. Microwave Theory Tech.*, vol. 41, no. 8, pp. 1458-1460, Aug. 1993.
10. C. H. Kim, H. J. Eom, and T. J. Park, "A series solution for TM-mode radiation from a flanged parallel-plate waveguide," *IEEE Trans. Antennas Propagat.*, vol. 41, no. 10, pp. 1469-1471, Oct. 1993.
11. C. P. Wu, "Integral Equation solutions for the radiation from a waveguide through a dielectric slab," *IEEE Trans. Antennas Propagat.*, vol. 17, no. 6, pp. 733-739, Nov. 1969.
12. W. F. Croswell, R. G. Rudduck, and D. M. Hatcher, "The admittance of a rectangular waveguide radiating into a dielectric layer," *IEEE Trans. Antennas Propagat.*, vol. 15, no. 5, pp. 627-633, Sept. 1967.
13. J. Galejs, *Antennas in Inhomogeneous Media*, First Ed., 1969, Pergamon Press, pp. 104-119.
14. W. D. Burnside, R. C. Rudduck, L. L. Tsai, and J. E. Jones, "Reflection coefficient of a TEM mode symmetric parallel-plate waveguide illuminating a dielectric layer," *Radio Sci.*, vol. 4, no. 6, pp. 545-556, June 1969.
15. Y. Sugio, T. Makimoto, and T. Tsugawa, "Two dimensional-analysis for gain enhancement of dielectric loaded antenna with a ground plane," *IEICE Trans. Commun.* (in Japanese), vol. J73-B-II, no. 8, pp. 405-412, Aug. 1990.
16. G. B. Gentili, G. Manara, G. Pelosi, and R. Tiberio, "Radiation of open-ended waveguides into stratified media," *Microwave Opt. Technol. Lett.*, vol. 4, no. 10, pp. 401-403, Sept. 1991.
17. J. W. Lee, H. J. Eom, and J. H. Lee, "TM-wave radiation from flanged parallel-plate into a dielectric slab," *IEE Proceedings-H Antennas Propagat.*, vol. 143, no. 3, pp. 207-210, 1996.
18. T. J. Park and H. J. Eom, "TE-scattering and reception by a parallel-plate waveguide array," *IEEE Trans. Antennas Propagat.*, vol. 42, no. 8, pp. 862-865, June 1994.
19. S. W. Lee, "Radiation from an infinite aperiodic array of parallel-plate waveguides," *IEEE Trans. Antennas Propagat.*, vol. 15, pp. 598-606, Sept. 1967.
20. C. P. Wu, "Analysis of finite parallel-plate waveguide arrays," *IEEE Trans. Antennas Propagat.*, vol. 18, pp. 328-334, May 1970.
21. G. F. Vanblaricum and R. Mittra, "A modified residue-calculus technique for solving a class of boundary value problems-part II: waveguide phased arrays,

modulated surfaces, and diffraction gratings," *IEEE Trans. Microwave Theory Tech.*, vol. 17, no. 6, pp. 310-319, June 1969.

22. R. C. Rudduck and L. L. Tsai, "Aperture reflection coefficient of TEM and TE_{01} mode parallel-plate waveguides," *IEEE Trans. Antennas Propagat.*, vol. 16, no. 1, pp. 83-89, Jan. 1968.

23. C. Thompson, "Radiation from a tapered open end of a waveguide with an infinite flange," *Radio Eng. Electron. Phys.*, vol. 22, no. 1, pp. 23-30, Jan. 1977.

24. H. J. Eom and T. J. Park "Radiation from a parallel-plate fed slit in a thick conducting screen," *IEICE Trans. Commun.*, vol. E78-c, no. 8, pp. 1131-1133, Aug. 1995.

25. R.W. Scharstein, "Two numerical solutions for the parallel plate-fed slot antenna," *IEEE Trans. Antennas Propagat.*, vol. 37, no. 11, pp. 1415-1425, Nov. 1989.

26. B.N. Das, A. Chakraborty, and S. Gupta, "Analysis of waveguide-fed thick radiating rectangular windows in a ground plane," *IEE Proceedings-H,* vol. 138, no. 2, pp. 142-146, Apr. 1991

3. Slits in a Plane

3.1 Electrostatic Potential Distribution Through a Slit in a Plane [1]

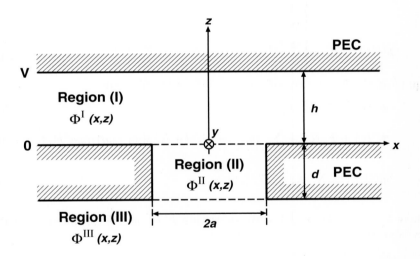

Fig. 3.1. A thick slit near a conducting plane at potential V

Electromagnetic interference (EMI) problems often require an estimation of field strength penetrating into a slit in a conducting plane. Many analytic and numerical approaches have been used to predict the effect of field leakage on electric circuit and system performance. For instance, a low-frequency field penetration into a slit is approximately modeled in terms of the electrostatic potential that is governed by Laplace's equation [2]. In this section we will analyze an electrostatic potential distribution through a slit in a thick perfectly-conducting plane. Consider a thick and perfect conducting slit at zero potential placed near a conducting plane at potential V. The electrostatic potentials in regions (I), (II), and (III) are represented as

$$\Phi^I(x,z) = \frac{1}{2\pi} \int_{-\infty}^{\infty} \left[\tilde{\Phi}^+(\zeta)e^{-\zeta z} + \tilde{\Phi}^-(\zeta)e^{\zeta z}\right] e^{-i\zeta x}\, d\zeta \tag{1}$$

$$\Phi^{II}(x,z) = \sum_{m=1}^{\infty} \left[b_m \cosh a_m(z+d) + c_m \sinh a_m(z+d)\right]$$
$$\cdot \sin a_m(x+a) \tag{2}$$

$$\Phi^{III}(x,z) = \frac{1}{2\pi} \int_{-\infty}^{\infty} \tilde{\Phi}^{III}(\zeta) \exp[|\zeta|(z+d) - i\zeta x]\, d\zeta \tag{3}$$

where $a_m = m\pi/(2a)$ and $m = 1, 3, 5, \ldots$.

The boundary conditions on the field continuities require

$$\Phi^I(x,h) = V \tag{4}$$

$$\Phi^I(x,0) = \begin{cases} \Phi^{II}(x,0), & |x| < a \\ 0, & |x| > a \end{cases} \tag{5}$$

$$\Phi^{III}(x,-d) = \begin{cases} \Phi^{II}(x,-d), & |x| < a \\ 0, & |x| > a\,. \end{cases} \tag{6}$$

Applying the Fourier transform to (4) through (6) yields

$$\tilde{\Phi}^+(\zeta)e^{-\zeta h} + \tilde{\Phi}^-(\zeta)e^{\zeta h} = 2\pi V\delta(\zeta) \tag{7}$$

$$\tilde{\Phi}^+(\zeta) + \tilde{\Phi}^-(\zeta) = \sum_{m=1}^{\infty} (b_m \cosh a_m d + c_m \sinh a_m d)\, a_m a^2 F_m(\zeta a) \tag{8}$$

$$\tilde{\Phi}^{III}(\zeta) = \sum_{m=1}^{\infty} b_m a_m a^2 F_m(\zeta a) \tag{9}$$

where $\delta(\zeta)$ is the Dirac delta. We substitute $\tilde{\Phi}^+(\zeta)$ and $\tilde{\Phi}^-(\zeta)$ into the boundary condition

$$\left.\frac{\partial\Phi^I(x,z)}{\partial z}\right|_{z=0} = \left.\frac{\partial\Phi^{II}(x,z)}{\partial z}\right|_{z=0}, \qquad |x| < a \tag{10}$$

multiply (10) by $\sin a_n(x+a)$, and integrate with respect to x from $-a$ to a to get

$$\boxed{\frac{V[1-(-1)^n]}{ha_n} = \sum_{m=1}^{\infty} (b_m \cosh a_m d + c_m \sinh a_m d)I \\ + a_n a(b_n \sinh a_n d + c_n \cosh a_n d)\,.} \tag{11}$$

Similarly from the boundary condition

$$\left.\frac{\partial\Phi^{III}(x,z)}{\partial z}\right|_{z=-d} = \left.\frac{\partial\Phi^{II}(x,z)}{\partial z}\right|_{z=-d}, \qquad |x| < a \tag{12}$$

we obtain

$$\boxed{\sum_{m=1}^{\infty} b_m J = a_n a c_n} \tag{13}$$

where I and J are

$$I = \frac{1}{2\pi} \int_{-\infty}^{\infty} \frac{\zeta}{\tanh \zeta h} F_m(\zeta a) F_n(-\zeta a) a_m a_n a^4 \, d\zeta \tag{14}$$

$$J = \frac{1}{2\pi} \int_{-\infty}^{\infty} |\zeta| F_m(\zeta a) F_n(-\zeta a) a_m a_n a^4 \, d\zeta \ . \tag{15}$$

Performing the residue calculus, we get the rapidly-convergent series

$$I = \frac{a_n a}{\tanh a_n h} \delta_{mn} - \bar{I} \tag{16}$$

$$J = a_n a \delta_{mn} - \bar{J} \tag{17}$$

where

$$\bar{I} = \sum_{l=1}^{\infty} \left[\frac{2v a_m a_n}{h} \frac{1 + e^{-2av}}{(v^2 + a_m^2)(v^2 + a_n^2)} \right]_{v=l\pi/h}, \qquad m+n = \text{even} \tag{18}$$

$$\bar{J} = \begin{cases} \dfrac{4 a_m a_n}{2\pi(a_n^2 - a_m^2)} [\ln(a_n/a_m) + \mathrm{ci}(m\pi) - \mathrm{ci}(n\pi)], & m+n = \text{even} \\[2mm] \dfrac{a_m}{\pi a_n} [2 - n\pi \mathrm{si}(n\pi)], & m = n \ . \end{cases} \tag{19}$$

Note that $\mathrm{ci}(\cdot)$ and $\mathrm{si}(\cdot)$ are the cosine and sine integrals defined as

$$\mathrm{si}(x) = -\int_x^{\infty} \frac{\sin t}{t} dt \tag{20}$$

$$\mathrm{ci}(x) = -\int_x^{\infty} \frac{\cos t}{t} dt \ . \tag{21}$$

For $m+n = \text{odd}$, $I = J = 0$.

3.2 Electrostatic Potential Distribution due to a Potential Across a Slit [3]

Section 3.1 discusses an electrostatic potential distribution through a slit when a charged perfectly-conducting plane is placed nearby a slit. In this section we will consider an electrostatic potential distribution through a slit when a potential V is applied across a slit with thickness d and width $2a$. In regions (I) ($z > 0$) and (III) ($z < -d$) the scattered potentials are written as

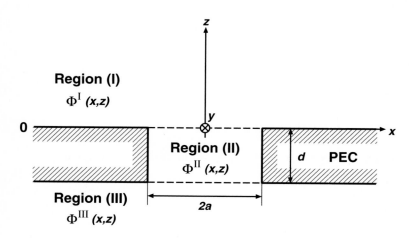

Fig. 3.2. A thick slit with a potential difference V

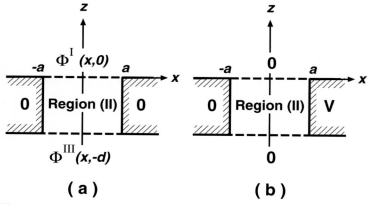

Fig. 3.3. An equivalent problem based on the superposition principle

$$\Phi^I(x,z) = \frac{1}{2\pi} \int_{-\infty}^{\infty} \widetilde{\Phi}^I(\zeta) \exp(-|\zeta|z - i\zeta x)d\zeta \tag{1}$$

$$\Phi^{III}(x,z) = \frac{1}{2\pi} \int_{-\infty}^{\infty} \widetilde{\Phi}^I(\zeta) \exp[|\zeta|(z+d) - i\zeta x]d\zeta . \tag{2}$$

In (1) and (2) we note that $\Phi^I(x,0) = \Phi^{III}(x,-d)$. Based on the superposition principle, the original problem in Fig. 3.2 may be decomposed into two different cases (a) and (b) as shown in Fig. 3.3. Hence the total potential in region (II) is a sum of two solutions to problems (a) and (b) as

$$\Phi^{II}(x,z) = \sum_{m=1}^{\infty} b_m \cosh\left[a_m\left(z + \frac{d}{2}\right)\right] \sin a_m(x+a)$$

$$-\frac{4V}{\pi} \sum_{k=1}^{\infty} \frac{\sinh\dfrac{k\pi(x+a)}{d}}{k\sinh\dfrac{k\pi 2a}{d}} \sin\frac{k\pi z}{d} \tag{3}$$

where $a_m = \dfrac{m\pi}{2a}$, $m = 1, 2, 3, \ldots$, and $k = 1, 3, 5, \ldots$.

Due to a symmetry of the problem geometry, $\Phi^I(x,0) = \Phi^{III}(x,-d)$ and $\left.\dfrac{\partial \Phi^I(x,z)}{\partial z}\right|_{z=0} = -\left.\dfrac{\partial \Phi^{III}(x,z)}{\partial z}\right|_{z=-d}$. This indicates that only the boundary conditions at $z = 0$ need to be enforced as

$$\Phi^I(x,0) = \begin{cases} 0, & x < -a \\ \Phi^{II}(x,0), & |x| < a \\ V, & x > a \end{cases} \tag{4}$$

$$\left.\frac{\partial \Phi^I(x,z)}{\partial z}\right|_{z=0} = \left.\frac{\partial \Phi^{II}(x,z)}{\partial z}\right|_{z=0}, \qquad |x| < a. \tag{5}$$

Applying the Fourier transform to (4) yields

$$\tilde{\Phi}^I(\zeta) = \sum_{m=1}^{\infty} b_m \cosh\left(\frac{a_m d}{2}\right) \Xi_m(\zeta) + V e^{i\zeta a}\left[\pi\delta(\zeta) - \frac{1}{i\zeta}\right] \tag{6}$$

where $\delta(\zeta)$ is the Dirac delta and

$$\Xi_m(\zeta) = \frac{a_m[(-1)^m e^{i\zeta a} - e^{-i\zeta a}]}{\zeta^2 - a_m^2}. \tag{7}$$

Rewriting (5) gives

$$-\frac{1}{2\pi} \int_{-\infty}^{\infty} \tilde{\Phi}^I(\zeta)|\zeta| e^{-i\zeta x}\, d\zeta$$

$$= \sum_{m=1}^{\infty} a_m b_m \sinh\left(\frac{a_m d}{2}\right) \sin a_m(x+a)$$

$$-\frac{4V}{d} \sum_{k=1}^{\infty} \frac{\sinh\dfrac{k\pi(x+a)}{d}}{\sinh\dfrac{k\pi 2a}{d}}. \tag{8}$$

We multiply (8) by $\sin a_n(x+a)$ and integrate with respect to x from $-a$ to a to get

$$-\frac{1}{2\pi}\int_{-\infty}^{\infty}\tilde{\varPhi}^{I}(\zeta)|\zeta|\varXi_{n}(-\zeta)\mathrm{d}\zeta$$

$$=aa_{n}b_{n}\sinh\left(\frac{a_{n}d}{2}\right)+\frac{4V}{d}\sum_{k=1}^{\infty}\frac{a_{n}\cos n\pi}{(k\pi/d)^{2}+a_{n}^{2}}\cdot \tag{9}$$

Substituting (6) into (9) gives

$$\frac{V}{2\pi}\int_{-\infty}^{\infty}\mathrm{e}^{\mathrm{i}\zeta a}\frac{|\zeta|}{\mathrm{i}\zeta}\varXi_{n}(-\zeta)\mathrm{d}\zeta$$

$$-\sum_{m=1}^{\infty}b_{m}\cosh\left(\frac{a_{m}d}{2}\right)\frac{1}{2\pi}\int_{-\infty}^{\infty}|\zeta|\varXi_{m}(\zeta)\varXi_{n}(-\zeta)\mathrm{d}\zeta$$

$$=aa_{n}b_{n}\sinh\left(\frac{a_{n}d}{2}\right)+\frac{4V}{d}\sum_{k=1}^{\infty}\frac{a_{n}\cos n\pi}{(k\pi/d)^{2}+a_{n}^{2}}\cdot \tag{10}$$

Using the residue calculus, we evaluate

$$\frac{V}{2\pi}\int_{-\infty}^{\infty}\mathrm{e}^{\mathrm{i}\zeta a}\frac{|\zeta|}{\mathrm{i}\zeta}\varXi_{n}(-\zeta)\mathrm{d}\zeta=-(-1)^{n}\left[\frac{V}{2}+\frac{V}{\pi}\mathrm{si}(n\pi)\right] \tag{11}$$

$$\frac{1}{2\pi}\int_{-\infty}^{\infty}|\zeta|\varXi_{m}(\zeta)\varXi_{n}(-\zeta)\mathrm{d}\zeta=aa_{n}\delta_{nm}-\bar{J} \tag{12}$$

where δ_{nm} is the Kronecker delta and $\bar{J}$ is given by (19) in Sect. 3.1. Substituting (11) and (12) into (10), we obtain a matrix equation

$$\boxed{\mathbf{B}=\boldsymbol{\varPsi}^{-1}\boldsymbol{\varGamma}} \tag{13}$$

where B is a column vector of b_{n} and the matrix elements are

$$\psi_{nm}=\delta_{nm}-\frac{\exp[(a_{m}-a_{n})d/2]+\exp[-(a_{m}+a_{n})d/2]}{2aa_{n}}\bar{J} \tag{14}$$

$$\gamma_{n}=(-1)^{n+1}\frac{\exp(-a_{n}d/2)}{aa_{n}}$$

$$\cdot\left[\frac{V}{2}+\frac{V}{\pi}\mathrm{si}(n\pi)+\frac{4V}{d}\sum_{k=1}^{\infty}\frac{a_{n}}{(k\pi/d)^{2}+a_{n}^{2}}\right]\cdot \tag{15}$$

It is of practical interest to represent the scattered potential and the surface charge distribution in fast convergent series. Substituting (6) into (1) and performing a contour integration, we get a series form

$$\varPhi^{I}(x,z)$$

$$=\sum_{m=1}^{\infty}\frac{a_{m}b_{m}}{\pi}\cosh\frac{a_{m}d}{2}$$

$$\cdot\mathrm{Im}\left\{(-1)^{m}K(a_{m},a-x+\mathrm{i}z)-K(a_{m},-(a+x)+\mathrm{i}z)\right\}$$

$$+\frac{V}{2}-\frac{V}{\pi}\arctan\left(\frac{a-x}{z}\right),\qquad\qquad x<-a \tag{16}$$

$$= \sum_{m=1}^{\infty} \frac{a_m b_m}{\pi} \cosh \frac{a_m d}{2}$$

$$\cdot Im\left\{(-1)^m K(a_m, a-x+iz) - K(a_m, a+x+iz)\right\}$$

$$+ b_m \cosh \frac{a_m d}{2} \exp(-a_m z) \sin a_m(a+x)$$

$$+ \frac{V}{2} - \frac{V}{\pi} \arctan\left(\frac{a-x}{z}\right), \qquad\qquad |x| \le a \qquad\qquad (17)$$

$$= \sum_{m=1}^{\infty} \frac{a_m b_m}{\pi} \cosh \frac{a_m d}{2}$$

$$\cdot Im\left\{(-1)^m K(a_m, x-a+iz) - K(a_m, a+x+iz)\right\}$$

$$+ \frac{V}{2} + \frac{V}{\pi} \arctan\left(\frac{x-a}{z}\right), \qquad\qquad x > a \qquad\qquad (18)$$

where $K(\beta, \mu) = \frac{1}{\beta}\left[ci(\beta\mu)\sin(\beta\mu) - si(\beta\mu)\cos(\beta\mu)\right]$. The surface charge density on a slit is

$$\frac{|\rho_s(x,z)|}{\epsilon} =$$

$$\begin{cases} \sum_{m=1}^{\infty} \frac{a_m b_m}{\pi} \cosh\left(a_m \frac{d}{2}\right) \left[(-1)^m P(a_m, a-x) - P(a_m, -a-x)\right] \\ \quad + \frac{V}{\pi} \frac{1}{a-x}, \qquad\qquad x < -a, \ z = 0 \\ \sum_{m=1}^{\infty} a_m b_m \cosh\left[a_m\left(z + \frac{d}{2}\right)\right] \\ \quad - \frac{4V}{d} \sum_{k=1,3}^{\infty} \frac{\sin(k\pi z/d)}{\sinh(k\pi 2a/d)}, \qquad -d < z < 0, \ x = -a \end{cases} \qquad (19)$$

where $P(\beta, \mu) = ci(\beta\mu)\cos(\beta\mu) + si(\beta\mu)\sin(\beta\mu)$ and ϵ is the medium permittivity.

3.3 EM Scattering from a Slit in a Conducting Plane

Electromagnetic wave scattering from a slit in a conducting plane is a canonical problem [4-9] often encountered in electromagnetic interference and compatibility-related area. In the next two subsections we will consider TE and TM scattering from a two-dimensional slit in a perfectly-conducting plane. Since the slit axis (y) is chosen to be perpendicular to the plane of incidence, the scattering analysis becomes a two-dimensional problem. An analytical formulation given in this section is similar to that in Sect. 1.1.

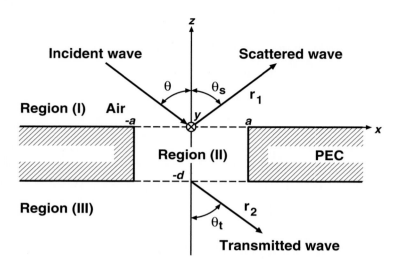

Fig. 3.4. A slit in a thick conducting plane

3.3.1 TE Scattering [10]

In region (I) $(z > 0)$ a field $E_y^i(x, z)$ is incident on a slit (width: $2a$ and depth: d) in a thick perfectly conducting plane. Regions (II) $(-d < z < 0$ and $-a < x < a$) and (III) $(z < -d)$ are a slit and a lossless half-space, respectively. The wavenumbers in regions (I), (II), and (III) are k_0, k_1, and k_2, respectively, where $k_0 = \omega\sqrt{\mu_0\epsilon_0}$, $k_1 = \omega\sqrt{\mu_1\epsilon_1}$, and $k_2 = \omega\sqrt{\mu_2\epsilon_2}$. The E-field in region (I) has the incident, reflected, and scattered waves

$$E_y^i(x, z) = \exp(ik_x x - ik_z z) \tag{1}$$

$$E_y^r(x, z) = -\exp(ik_x x + ik_z z) \tag{2}$$

$$E_y^s(x, z) = \frac{1}{2\pi} \int_{-\infty}^{\infty} \tilde{E}_y^s(\zeta) \exp(-i\zeta x + i\kappa_0 z)\, d\zeta \tag{3}$$

where $k_x = k_0 \sin\theta$, $k_z = k_0 \cos\theta$, and $\kappa_0 = \sqrt{k_0^2 - \zeta^2}$. In region (II) the total E-field is

$$E_y^d(x, z) = \sum_{m=1}^{\infty} (b_m \cos\xi_m z + c_m \sin\xi_m z) \sin a_m(x + a) \tag{4}$$

where $a_m = m\pi/(2a)$ and $\xi_m = \sqrt{k_1^2 - \zeta^2}$. In region (III) the transmitted field is

$$E_y^t(x, z) = \frac{1}{2\pi} \int_{-\infty}^{\infty} \tilde{E}_y^t(\zeta) \exp[-i\zeta x - i\kappa_2(z + d)]\, d\zeta \tag{5}$$

where $\kappa_2 = \sqrt{k_2^2 - \zeta^2}$.

Applying the Fourier transform to the tangential E-field continuity along the x-axis ($z = 0$) yields

$$\tilde{E}_y^s(\zeta) = \sum_{m=1}^{\infty} b_m a_m a^2 F_m(\zeta a) \ . \tag{6}$$

The tangential H-field continuity along $(-a < x < a)$ at $z = 0$ yields

$$\boxed{\frac{2\mathrm{i}k_z a_n}{\mu_0} a^2 F_n(k_x a) = \frac{\mathrm{i}a_n}{2\pi\mu_0} \sum_{m=1}^{\infty} b_m a_m a^2 \Lambda_1(k_0) - \frac{c_n \xi_n a}{\mu_1}} \tag{7}$$

where the explicit expression for $\Lambda_1(k_0)$ is available in Subsect. 1.1.3. Similarly the tangential E-field continuity at $z = -d$ gives

$$\tilde{E}_y^t(\zeta) = \sum_{m=1}^{\infty} (b_m \cos \xi_m d - c_m \sin \xi_m d) a_m a^2 F_m(\zeta a) \ . \tag{8}$$

The tangential H-field continuity along $(-a < x < a)$ and $z = -d$ gives

$$\boxed{\begin{aligned} \mathrm{i}\frac{\mu_2}{\mu_1}&(b_n \sin \xi_n d + c_n \cos \xi_n d)\xi_n a \\ &= \frac{a_n}{2\pi} \sum_{m=1}^{\infty} a_m a^2 (b_m \cos \xi_m d - c_m \sin \xi_m d)\Lambda_1(k_2) \end{aligned}} \tag{9}$$

where

$$\Lambda_1(k_2) = \int_{-\infty}^{\infty} a^2 F_m(\zeta a) F_n(-\zeta a) \kappa_2 \, \mathrm{d}\zeta \ . \tag{10}$$

The far-zone scattered and transmitted fields at distances r_1 and r_2 are

$$E_y^s(\theta_s, \theta) = \sqrt{\frac{k_0}{2\pi r_1}} \exp(\mathrm{i}k_0 r_1 - \mathrm{i}\pi/4) \cos \theta_s \tilde{E}_y^s(-k_0 \sin \theta_s) \tag{11}$$

$$E_y^t(\theta_t, \theta) = \sqrt{\frac{k_2}{2\pi r_2}} \exp(\mathrm{i}k_2 r_2 - \mathrm{i}\pi/4) \cos \theta_t \tilde{E}_y^t(-k_2 \sin \theta_t) \ . \tag{12}$$

The transmission (reflection) coefficient is a ratio of the transmitted (reflected) power to that incident on a slit. The transmission coefficient τ and the reflection coefficient ϱ are

$$\tau = Im\left\{ \frac{\mu_0}{2k_0\mu_1} \sum_{m=1}^{\infty} \xi_m^* \left[|b_m|^2 \cos \xi_m d(\sin \xi_m d)^* + b_m c_m^* |\cos \xi_m d|^2 \right. \right.$$

$$\left. \left. - b_m^* c_m |\sin \xi_m d|^2 - |c_m|^2 \sin \xi_m d(\cos \xi_m d)^* \right] \right\} \tag{13}$$

$$\varrho = \frac{\mu_0}{2k_0} Im\left\{ \sum_{m=1}^{\infty} \frac{1}{\mu_1} b_m c_m^* \xi_m^* \right\} \ . \tag{14}$$

3.3.2 TM Scattering [11]

Consider an incident $H_y^i(x, z)$ impinging on a slit in a conducting plane. The wavenumbers in regions (I), (II), and (III) are $k_0 = \omega\sqrt{\mu_0\epsilon_0} = 2\pi/\lambda$, $k_1 = \omega\sqrt{\mu_0\mu_{r1}\epsilon_0\epsilon_{r1}}$, and $k_2 = \omega\sqrt{\mu_0\mu_{r2}\epsilon_0\epsilon_{r2}}$, respectively. In region (I) the field consists of the incident, reflected, and scattered components as

$$H_y^i(x, z) = \exp(ik_x x - ik_z z) \tag{15}$$
$$H_y^r(x, z) = \exp(ik_x x + ik_z z) \tag{16}$$
$$H_y^s(x, z) = \frac{1}{2\pi} \int_{-\infty}^{\infty} \tilde{H}_y^s(\zeta) \exp(-i\zeta x + i\kappa_0 z)\, d\zeta \ . \tag{17}$$

In region (II) the H-field is

$$H_y^d(x, z) = \sum_{m=0}^{\infty} (b_m \cos\xi_m z + c_m \sin\xi_m z) \cos a_m(x + a) \ . \tag{18}$$

In region (III) the total transmitted H-field is

$$H_y^t(x, z) = \frac{1}{2\pi} \int_{-\infty}^{\infty} \tilde{H}_y^t(\zeta) \exp[-i\zeta x - i\kappa_2(z + d)]\, d\zeta \ . \tag{19}$$

Applying the Fourier transform to the tangential E-field continuity along the x-axis yields

$$\tilde{H}_y^s(\zeta) = -\sum_{m=0}^{\infty} \xi_m c_m \frac{\zeta a^2 F_m(\zeta a)}{\kappa_0 \epsilon_{r1}} \ . \tag{20}$$

The tangential H-field continuity along $(-a < x < a)$ at $z = 0$ gives

$$\boxed{-2ik_x a^2 F_n(k_x a) = \frac{i}{2\pi\epsilon_{r1}} \sum_{m=0}^{\infty} \xi_m a^2 c_m \Omega_1(k_0) + b_n a\varepsilon_n \ .} \tag{21}$$

The explicit expression for $\Omega_1(k_0)$ is given in Subsect. 1.1.3. The tangential E-field continuity at $z = -d$ similarly gives

$$\tilde{H}_y^t(\zeta) = \frac{\epsilon_{r2}}{\epsilon_{r1}\kappa_2} \sum_{m=0}^{\infty} \xi_m (b_m \sin\xi_m d + c_m \cos\xi_m d) \zeta a^2 F_m(\zeta a) \ . \tag{22}$$

The tangential H-field continuity along $(-a < x < a)$ at $z = -d$ gives

$$\boxed{\begin{aligned} &(b_n \cos\xi_n d - c_n \sin\xi_n d)a\varepsilon_n \\ &= \frac{i\epsilon_{r2}}{2\pi\epsilon_{r1}} \sum_{m=0}^{\infty} \xi_m a^2 (b_m \sin\xi_m d + c_m \cos\xi_m d) \Omega_1(k_2) \end{aligned}} \tag{23}$$

where

$$\Omega_1(k_2) = \int_{-\infty}^{\infty} a^2 F_m(\zeta a) F_n(-\zeta a) \zeta^2 \kappa_2^{-1}\, d\zeta \ . \tag{24}$$

The far-zone scattered and transmitted fields at distances r_1 and r_2 are

$$H_y^s(\theta_s, \theta) = \sqrt{\frac{k_0}{2\pi r_1}} \exp(ik_0 r_1 - i\pi/4)\cos\theta_s \tilde{H}_y^s(-k_0 \sin\theta_s) \qquad (25)$$

$$H_y^t(\theta_t, \theta) = -\sqrt{\frac{k_2}{2\pi r_2}} \exp(ik_2 r_2 - i\pi/4)\cos\theta_t \tilde{H}_y^t(-k_2 \sin\theta_t) \ . \qquad (26)$$

The transmission coefficient τ is

$$\begin{aligned}
\tau = -Im\Bigg\{ &\frac{1}{2k_0\epsilon_{r1}} \sum_{m=0}^{\infty} \varepsilon_m \xi_m \big[|b_m|^2 \sin\xi_m d(\cos\xi_m d)^* \\
&+ b_m^* c_m |\cos\xi_m d|^2 - b_m c_m^* |\sin\xi_m d|^2 \\
&- |c_m|^2 \cos\xi_m d(\sin\xi_m d)^* \big] \Bigg\} \ .
\end{aligned} \qquad (27)$$

3.4 Magnetostatic Potential Distribution Through Slits in a Plane

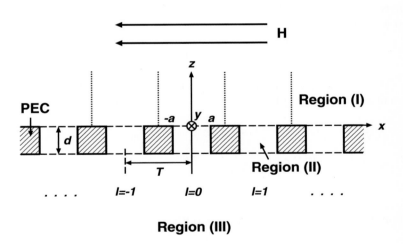

Fig. 3.5. Multiple slits in a thick conducting plane

A study of magnetostatic potential distribution through apertures in a conducting plane is of interest in EMI/EMC-related problems. In the present section we will study a problem of magnetostatic potential distribution through multiple slits in a thick conducting plane. Consider an incident magnetostatic potential $\Phi^i(x, z)$ impinging on a finite number of slits (thickness: d, width: $2a$, period: T, total number of slits: N) in a thick conducting plane. Regions (II) $(-d < z < 0$ and $|x - lT| < a)$ and (III) $(z < -d)$ denote a slit interior and a half-space, respectively. In region (I) $(z > 0)$ the total magnetostatic potential consists of the incident and scattered potentials

$$\Phi^{i,l}(x, z) = x - A^l, \qquad |x - lT| < T/2, \quad l = 0, \pm 1, \pm 2, \ldots \tag{1}$$

$$\Phi^s(x, z) = \frac{1}{2\pi} \int_{-\infty}^{\infty} \widetilde{\Phi}^s(\zeta) e^{-i\zeta x} e^{-|\zeta|z} \, d\zeta \tag{2}$$

The magnetostatic potential for a multiply-connected domain may be a multi-valued function of position; hence, it is necessary to introduce a cut in a domain to make the magnetostatic potential a single-valued function [12]. This implies that a constant A^l is yet to be determined later when the boundary conditions are matched. In region (II) of the lth slit the magnetostatic potential is

$$\Phi^{d,l}(x, z) = c_0^l + \sum_{m=1}^{\infty} [b_m^l \sinh a_m(z + d) + c_m^l \cosh a_m(z + d)]$$
$$\cdot \cos a_m(x + a - lT) \tag{3}$$

where $a_m = m\pi/(2a)$ and $m = 1, 2, 3, \ldots$. In region (III) the total transmitted potential is

$$\Phi^t(x, z) = \frac{1}{2\pi} \int_{-\infty}^{\infty} \widetilde{\Phi}^t(\zeta) \exp[-i\zeta x + |\zeta|(z + d)] \, d\zeta . \tag{4}$$

The boundary condition on the continuity of normal derivative of the potential at $z = 0$ requires

$$\frac{\partial}{\partial z} \left[\Phi^{i,l}(x, z) + \Phi^s(x, z)\right]_{z=0}$$
$$= \begin{cases} \dfrac{\partial}{\partial z} \left[\Phi^{d,l}(x, z)\right]_{z=0}, & |x - lT| < a \\ \\ 0, & |x - lT| > a . \end{cases} \tag{5}$$

Applying the Fourier transform to (5) gives

$$|\zeta| \widetilde{\Phi}^s(\zeta)$$
$$= i\zeta \sum_{l=-L_1}^{L_2} \sum_{m=1}^{\infty} (b_m^l \cosh a_m d + c_m^l \sinh a_m d) e^{i\zeta lT} a^2 a_m F_m(\zeta a) . \tag{6}$$

The boundary condition on the continuity of the potential across the slits requires

$$\Phi^{i,l}(x, 0) + \Phi^s(x, 0) = \Phi^{d,l}(x, 0), \qquad |x - lT| < a . \tag{7}$$

Multiplying (7) by $\cos a_n(x + a - rT)$ $(n = 1, 2, 3, \dots)$ and integrating with respect to x from $(rT - a)$ to $(rT + a)$, we get

$$\frac{(-1)^n - 1}{a_n} = \sum_{l=-L_1}^{L_2} \sum_{m=1}^{\infty} (b_m^l \cosh a_m d + c_m^l \sinh a_m d) I$$
$$+ a_n a\, (b_n^r \sinh a_n d + c_n^r \cosh a_n d) \tag{8}$$

where

$$I = \frac{1}{2\pi} \int_{-\infty}^{\infty} e^{i\zeta(l-r)T} |\zeta| a^4 a_m a_n F_m(\zeta a) F_n(-\zeta a)\, d\zeta . \tag{9}$$

Using the contour integration, we evaluate

$$I = a_n a\, \delta_{mn} \delta_{lr} + \bar{I} . \tag{10}$$

When $l = r$,

$$\bar{I} = -\begin{cases} \dfrac{4a_m a_n}{2\pi(a_n^2 - a_m^2)} [\ln(a_n/a_m) + \mathrm{ci}(m\pi) - \mathrm{ci}(n\pi)], & m + n = \text{even} \\[3mm] \dfrac{a_m}{\pi a_n} [2 - (n\pi)\mathrm{si}(n\pi)], & m = n \end{cases} \tag{11}$$

where $\mathrm{ci}(\cdot)$ and $\mathrm{si}(\cdot)$ are the cosine and sine integrals, respectively.
When $l \ne r$,

$$\bar{I} = \frac{a_m a_n}{2\pi} \{[(-1)^{m+n} + 1] I_{mn}(qT) \\ - (-1)^m I_{mn}(qT + 2a) - (-1)^n I_{mn}(qT - 2a)\} \tag{12}$$

where $q = l - r$ and

$$I_{mn}(c) = \int_{-\infty}^{\infty} \frac{|\zeta| e^{i\zeta|c|}}{(\zeta^2 - a_m^2)(\zeta^2 - a_n^2)}\, d\zeta . \tag{13}$$

Utilizing the residue calculus, we obtain

$$I_{mn}(c)$$
$$= -\begin{cases} \dfrac{2}{(a_n^2 - a_m^2)} [-\mathrm{ci}(a_m c)\cos(a_m c) + \mathrm{ci}(a_n c)\cos(a_n c) \\ \qquad - \mathrm{si}(a_m c)\sin(a_m c) + \mathrm{si}(a_n c)\sin(a_n c)], & m \ne n \\[3mm] \dfrac{1}{a_n^2} \{1 - a_n c[\mathrm{ci}(a_n c)\sin(a_n c) - \mathrm{si}(a_n c)\cos(a_n c)]\}, & m = n . \end{cases} \tag{14}$$

It is necessary to obtain another set of simultaneous equations for c_0^l by integrating directly (7) with respect to x from $(lT - a)$ to $(lT + a)$. The result is

$$\boxed{\begin{aligned}2ac_0^l &= 2alT - 2aA^l \\ &+ \sum_{l=-L_1}^{L_2} \sum_{m=1}^{\infty} \left(b_m^l \cosh a_m d + c_m^l \sinh a_m d\right) J\end{aligned}} \tag{15}$$

where

$$J = \frac{1}{2\pi} \int_{-\infty}^{\infty} \frac{e^{i\zeta a} - e^{-i\zeta a}}{|\zeta|} a^2 a_m F_m(\zeta a) \, d\zeta \, . \tag{16}$$

Similarly from the boundary conditions at $z = -d$

$$\frac{\partial}{\partial z}\left[\Phi^t(x,z)\right]_{z=-d} = \begin{cases} \dfrac{\partial}{\partial z}\left[\Phi^{d,l}(x,z)\right]_{z=-d}, & |x - lT| < a \\[2mm] 0, & |x - lT| > a \end{cases} \tag{17}$$

$$\Phi^t(x,-d) = \Phi^{d,l}(x,-d), \qquad |x - lT| < a \tag{18}$$

we get

$$\boxed{\begin{aligned}\sum_{l=-L_1}^{L_2} \sum_{m=1}^{\infty} b_m^l I &= a_n a c_n^r, \qquad n = 1,2,3\ldots \\ 2ac_0^l &= -\sum_{l=-L_1}^{L_2} \sum_{m=1}^{\infty} b_m^l J \, .\end{aligned}} \tag{19}$$

The magnetic polarizability of the lth slit at $z = -d$ is

$$\begin{aligned}\psi^l &\equiv L \int_{lT-a}^{lT+a} x \frac{\partial}{\partial z}\left[\Phi^{d,l}(x,z)\right]_{z=-d} dx \\ &= \frac{4aL}{\pi} \sum_{m=1}^{\infty} \frac{b_m^l}{m}\end{aligned} \tag{20}$$

where L is a slit length in the y-direction.

3.5 EM Scattering from Slits in a Conducting Plane [13]

Electromagnetic scattering from a strip-grating was studied in [14-18] for microwave-optical filter and polarizer applications. A study of electromagnetic scattering from multiple slits is also useful to solve the field-leakage problems in electromagnetic interference. In the next two subsections we will study TE and TM scattering from multiple slits in a thick and perfectly-conducting plane. A scattering analysis given in this section is an extension of the single-slit case considered in Sect. 3.3.

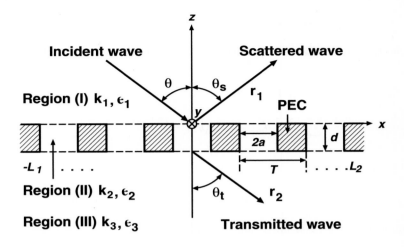

Fig. 3.6. Multiple slits in a thick conducting plane

3.5.1 TE Scattering

Consider an incident wave $E_y^i(x, z)$ impinging on a perfect conducting plane with multiple slits (width: $2a$, depth: d, period: T, and total number: $N = L_1 + L_2 + 1$). Regions (I), (II), and (III) are the air, slits, and a lower half-space where their respective wavenumbers are $k_1(= \omega\sqrt{\mu_1\epsilon_1} = 2\pi/\lambda)$, $k_2(= \omega\sqrt{\mu_2\epsilon_2})$, and $k_3(= \omega\sqrt{\mu_3\epsilon_3})$. The E-fields in regions (I) (incident E_y^i, reflected E_y^r, and scattered E_y^s), (II) (E_y^d), and (III) (transmitted E_y^t) are assumed to be

$$E_y^i(x, z) = \exp(ik_x x - ik_z z) \tag{1}$$

$$E_y^r(x, z) = -\exp(ik_x x + ik_z z) \tag{2}$$

$$E_y^s(x, z) = \frac{1}{2\pi}\int_{-\infty}^{\infty} \tilde{E}_y^s(\zeta)\exp(-i\zeta x + i\kappa_1 z)\,d\zeta \tag{3}$$

$$E_y^d(x, z) = \sum_{m=1}^{\infty} \left(b_m^l \cos\xi_m z + c_m^l \sin\xi_m z\right)\sin a_m(x + a - lT) \tag{4}$$

$$E_y^t(x, z) = \frac{1}{2\pi}\int_{-\infty}^{\infty} \tilde{E}_y^t(\zeta)\exp[-i\zeta x - i\kappa_3(z + d)]\,d\zeta \tag{5}$$

where $k_x = k_1 \sin\theta$, $k_z = k_1\cos\theta$, $\kappa_1 = \sqrt{k_1^2 - \zeta^2}$, $a_m = m\pi/(2a)$, $\xi_m = \sqrt{k_2^2 - a_m^2}$, and $\kappa_3 = \sqrt{k_3^2 - \zeta^2}$.

Applying the Fourier transform to the tangential E-field continuity condition at $z = 0$ gives

$$\widetilde{E}_y^s(\zeta) = \sum_{l=-L_1}^{L_2} \sum_{m=1}^{\infty} b_m^l e^{i\zeta lT} a_m a^2 F_m(\zeta a) \ . \tag{6}$$

The tangential H-field continuity over $(lT - a) < x < (lT + a)$ $(l = 0, \pm 1, \pm 2, \dots)$ at $z = 0$ yields

$$
\begin{aligned}
&2ik_z a_n e^{ik_z rT} a^2 F_n(k_x a) \\
&= \frac{ia_n}{2\pi} \sum_{l=-L_1}^{L_2} \sum_{m=1}^{\infty} b_m^l a_m a^2 \Lambda_2(k_1) - \frac{\mu_1 c_n^r \xi_n a}{\mu_2} \ .
\end{aligned}
\tag{7}
$$

Note that the expression for $\Lambda_2(k_1)$ is available in Subsect. 1.2.3 as

$$\Lambda_2(k_1) = \int_{-\infty}^{\infty} a^2 F_m(\zeta a) F_n(-\zeta a)\kappa_1 \exp\left[i(l-r)\zeta T\right] d\zeta \ . \tag{8}$$

Similarly the tangential E- and H-field continuities at $z = -d$ yield

$$
\begin{aligned}
&i\frac{\mu_3}{\mu_2}(b_n^r \sin \xi_n d + c_n^r \cos \xi_n d)\xi_n a \\
&= \frac{a_n}{2\pi} \sum_{l=-L_1}^{L_2} \sum_{m=1}^{\infty} a_m (b_m^l \cos \xi_m d - c_m^l \sin \xi_m d) a^2 \Lambda_2(k_3) \ .
\end{aligned}
\tag{9}
$$

The far-zone scattered fields at distances r_1 and r_2 are

$$E_y^s(\theta_s, \theta) = \sqrt{\frac{k_1}{2\pi r_1}} \exp(ik_1 r_1 - i\pi/4) \cos\theta_s \widetilde{E}_y^s(-k_1 \sin\theta_s) \tag{10}$$

$$
\begin{aligned}
E_y^t(\theta_t, \theta) &= \sqrt{\frac{k_3}{2\pi r_2}} \exp(ik_3 r_2 - i\pi/4) \cos\theta_t \\
&\quad \cdot \sum_{l=-L_1}^{L_2} \sum_{m=1}^{\infty} [b_m^l \cos(\xi_m d) - c_m^l \sin(\xi_m d)] \\
&\quad \cdot a_m \exp(-ik_3 \sin\theta_t lT) a^2 F_m(-k_3 a \sin\theta_t) \ .
\end{aligned}
\tag{11}
$$

The transmission coefficient (τ) is shown to be

$$\tau = \frac{\mu_1}{2k_1} \frac{1}{N} Im \left\{ \sum_{l=-L_1}^{L_2} \sum_{m=1}^{\infty} \frac{1}{\mu_2} b_m^l c_m^{l*} \xi_m^* \right\} \tag{12}$$

where $N = L_1 + L_2 + 1$.

3.5.2 TM Scattering

Asume that a TM wave $H_y^i(x, z)$ is incident on multiple thick slits in a perfectly-conducting plane. The H-fields in regions (I), (II), and (III) are

$$H_y^i(x, z) = \exp(\mathrm{i}k_x x - \mathrm{i}k_z z) \tag{13}$$

$$H_y^r(x, z) = \exp(\mathrm{i}k_x x + \mathrm{i}k_z z) \tag{14}$$

$$H_y^s(x, z) = \frac{1}{2\pi} \int_{-\infty}^{\infty} \widetilde{H}_y^s(\zeta) \exp(-\mathrm{i}\zeta x + \mathrm{i}\kappa_1 z)\, \mathrm{d}\zeta \tag{15}$$

$$H_y^d(x, z) = \sum_{m=1}^{\infty} \left(b_m^l \cos \xi_m z + c_m^l \sin \xi_m z\right) \cos a_m(x + a - lT) \tag{16}$$

$$H_y^t(x, z) = \frac{1}{2\pi} \int_{-\infty}^{\infty} \widetilde{H}_y^t(\zeta) \exp[-\mathrm{i}\zeta x - \mathrm{i}\kappa_3(z + d)]\, \mathrm{d}\zeta \tag{17}$$

where $k_x = k_1 \sin \theta$ and $k_z = k_1 \cos \theta$.

The tangential E- and H-field continuities at $z = 0$ yield

$$
\begin{aligned}
&-\mathrm{i}2k_x \mathrm{e}^{\mathrm{i}k_x rT} a^2 F_n(k_x a) \\
&= \frac{\mathrm{i}\epsilon_1}{2\pi\epsilon_2} \sum_{l=-L_1}^{L_2} \sum_{m=0}^{\infty} \xi_m^l c_m^l a^2 \Omega_2(k_1) + b_n^r a\varepsilon_n
\end{aligned} \tag{18}
$$

where the explicit expression for $\Omega_2(k_1)$ is given in Subsect. 1.2.2 as

$$\Omega_2(k_1) = \int_{-\infty}^{\infty} a^2 F_m(\zeta a) F_n(-\zeta a)\zeta^2 \exp[\mathrm{i}(l - r)\zeta T]\kappa_1^{-1}\, \mathrm{d}\zeta . \tag{19}$$

The tangential E- and H-field continuities at $z = -d$ also yield

$$
\begin{aligned}
&(b_n^r \cos \xi_n d - c_n^r \sin \xi_n d)a\varepsilon_n \\
&= \frac{\mathrm{i}\epsilon_3}{2\pi\epsilon_2} \sum_{l=-L_1}^{L_2} \sum_{m=0}^{\infty} \xi_m (b_m^l \sin \xi_m d + c_m^l \cos \xi_m d)a^2 \Omega_2(k_3)
\end{aligned} \tag{20}
$$

where

$$\Omega_2(k_3) = \int_{-\infty}^{\infty} a^2 F_m(\zeta a) F_n(-\zeta a)\zeta^2 \exp[\mathrm{i}(l - r)\zeta T]\kappa_3^{-1}\, \mathrm{d}\zeta . \tag{21}$$

The far-zone fields are

$$H_y^s(\theta_s, \theta) = \sqrt{\frac{k_1}{2\pi r_1}} \exp(\mathrm{i}k_1 r_1 - \mathrm{i}\pi/4) \sin \theta_s$$

$$\cdot \frac{\epsilon_1}{\epsilon_2} \sum_{l=-L_1}^{L_2} \sum_{m=0}^{\infty} \xi_m c_m^l \exp(-\mathrm{i}k_1 \sin \theta_s lT)a^2 F_m(k_1 a \sin \theta_s) \tag{22}$$

$$H_y^t(\theta_t, \theta) = \sqrt{\frac{k_3}{2\pi r_2}} \exp(\mathrm{i}k_3 r_2 - \mathrm{i}\pi/4) \sin \theta_t \frac{\epsilon_3}{\epsilon_2}$$

$$\cdot \sum_{l=-L_1}^{L_2} \sum_{m=0}^{\infty} \xi_m [b_m^l \sin(\xi_m d) + c_m^l \cos(\xi_m d)]$$

$$\cdot \exp(-\mathrm{i}k_3 \sin \theta_t lT)a^2 F_m(k_3 a \sin \theta_t) . \tag{23}$$

The transmission coefficient (τ) is

$$\tau = -Im\left\{\frac{\epsilon_1}{2k_1\epsilon_2}\frac{1}{N}\sum_{l=-L1}^{L2}\sum_{m=0}^{\infty}\varepsilon_n\xi_m b_m^{l*}c_m^l\right\} . \qquad (24)$$

3.6 EM Scattering from Slits in a Parallel-Plate Waveguide

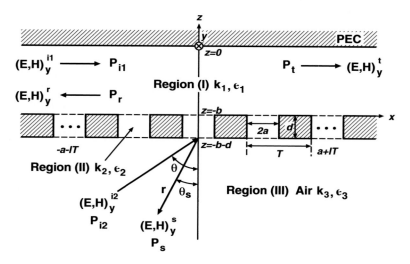

Fig. 3.7. Multiple slits in a parallel-plate waveguide

Electromagnetic wave radiation and scattering from a slitted parallel-plate waveguide was studied in [19-25] for aperture array antenna applications. A slitted waveguide is an important array antenna element that enables us to realize the narrow beamwidth and wideband radiation characteristics. In this section we will consider TE and TM scattering from a finite number of slits in a parallel-plate waveguide. An analytical formulation in this section is similar to the discussion in Sect. 3.5.

3.6.1 TE Scattering [26]

Consider a parallel-plate waveguide with N slits of width $2a$ and depth d. Regions (I), (II), and (III) respectively denote a parallel-plate waveguide interior (wavenumber $k_1 = \omega\sqrt{\mu_1\epsilon_1}$), an $N(= L_1 + L_2 + 1)$ number of slits

$(k_2 = \omega\sqrt{\mu_2\epsilon_2})$, and a lower half-space $(k_3 = \omega\sqrt{\mu_3\epsilon_3} = 2\pi/\lambda)$. Two incident TE waves, $E_y^{i1}(x,z)$ and $E_y^{i2}(x,z)$, are assumed to excite regions (I) and (III), respectively. The total E-field in region (I) has the incident and scattered fields

$$E_y^{i1}(x,z) = A_1 e^{ik_{xs}x}\sin k_{zs}(z+b) \tag{1}$$

$$E_y^I(x,z) = \frac{i}{\pi}\int_{-\infty}^{\infty} \widetilde{E}_y^I(\zeta)\sin(\kappa_1 z)e^{-i\zeta x}\,d\zeta \tag{2}$$

where $0 < s < k_1 b/\pi$ (s: integer), $k_{zs} = s\pi/b$, $k_{xs} = \sqrt{k_1^2 - k_{zs}^2}$, and $\kappa_1 = \sqrt{k_1^2 - \zeta^2}$. In region (II) ($lT - a < x < lT + a$ and $-d - b < z < -b$: $l = -L_1,\dots,L_2$) the E-field is

$$E_y^{II}(x,z) = \sum_{m=1}^{\infty} \sin a_m(x + a - lT)$$
$$\cdot [b_m^l \cos\xi_m(z+b) + c_m^l \sin\xi_m(z+b)] \tag{3}$$

where $a_m = m\pi/(2a)$ and $\xi_m = \sqrt{k_2^2 - a_m^2}$. In region (III) the total E-field is composed of

$$E_y^{i2}(x,z) = A_2 \exp[ik_x x + ik_z(z + b + d)] \tag{4}$$

$$E_y^r(x,z) = -A_2 \exp[ik_x x - ik_z(z + b + d)] \tag{5}$$

$$E_y^{III}(x,z) = \frac{1}{2\pi}\int_{-\infty}^{\infty}\widetilde{E}_y^{III}(\zeta)\exp[-i\zeta x - i\kappa_3(z + b + d)]\,d\zeta \tag{6}$$

where $k_x = k_3\sin\theta$, $k_z = k_3\cos\theta$, and $\kappa_3 = \sqrt{k_3^2 - \zeta^2}$.

Applying the Fourier transform to the tangential E-field continuity at $z = -b$ gives

$$-2i\sin(\kappa_1 b)\widetilde{E}_y^I(\zeta) = \sum_{l=-L_1}^{L_2}\sum_{m=1}^{\infty} b_m^l a_m e^{i\zeta lT} a^2 F_m(a\zeta). \tag{7}$$

Multiplying the tangential H-field continuity at the slit apertures ($rT - a < x < rT + a$ and $z = -b$: $r = -L_1,\dots,L_2$) by $\sin a_n(x + a - rT)$ and integrating, we get

$$\boxed{\begin{aligned}&\frac{k_{zs}A_1}{\mu_1}a_n\exp(ik_{xs}rT)a^2 F_n(k_{xs}a)\\[1em]&-\frac{1}{2\pi\mu_1}\sum_{l=-L_1}^{L_2}\sum_{m=1}^{\infty} b_m^l a_m a_n a^2 \Lambda_5(k_1) = \frac{1}{\mu_2}c_n^r\xi_n a\end{aligned}} \tag{8}$$

where

$$\Lambda_5(k_1) = \int_{-\infty}^{\infty}\kappa_1\cot(\kappa_1 b)a^2 F_m(\zeta a)F_n(-\zeta a)\exp[i(l - r)\zeta T]\,d\zeta \tag{9}$$

and a fast-convergent series for $\Lambda_5(k_1)$ is available in Subsect. 1.4.1. Applying the Fourier transform to the tangential E-field continuity at $z = -b - d$ gives

$$\tilde{E}_y^{III}(\zeta)$$
$$= \sum_{l=-L_1}^{L_2} \sum_{m=1}^{\infty} [b_m^l \cos(\xi_m d) - c_m^l \sin(\xi_m d)] a_m a^2 \mathrm{e}^{\mathrm{i}\zeta lT} F_m(\zeta a) . \tag{10}$$

The tangential H-field continuity at the slit apertures $(rT - a < x < rT + a$ and $z = -b - d$: $r = -L_1, \ldots, L_2)$ gives

$$\boxed{\begin{aligned} &-\frac{2k_z A_2}{\omega\mu_3} a_n \exp(\mathrm{i}k_x rT) a^2 F_n(k_x a) \\ &+ \frac{1}{2\pi\mu_3} \sum_{l=-L_1}^{L_2} \sum_{m=1}^{\infty} a_m a_n \left[b_m^l \cos(\xi_m d) - c_m^l \sin(\xi_m d) \right] a^2 \Lambda_2(k_3) \\ &= \frac{\mathrm{i}}{\mu_2} a \xi_n [b_n^r \sin(\xi_n d) + c_n^r \cos(\xi_n d)] \end{aligned}} \tag{11}$$

where $\Lambda_2(k_3)$ is given in Subsect. 1.2.3 as

$$\Lambda_2(k_3) = \int_{-\infty}^{\infty} a^2 F_m(\zeta a) F_n(-\zeta a) \kappa_3 \exp\left[\mathrm{i}(l - r)\zeta T\right] \mathrm{d}\zeta . \tag{12}$$

The total scattered field at $x = \pm\infty$ is

$$E_y^I(\pm\infty, z) = \sum_v K_v^{\pm} \sin k_{zv}(z + b) \exp(\pm\mathrm{i}k_{xv}x) \tag{13}$$

where $0 < v < k_1 b/\pi$, v: integer, $k_{zv} = v\pi/b$, $k_{xv} = \sqrt{k_1^2 - k_{zv}^2}$, and

$$K_v^{\pm} = \mathrm{i} \sum_{l=-L_1}^{L_2} \sum_{m=1}^{\infty} b_m^l a_m k_{zv} \exp(\mp\mathrm{i}k_{xv}lT) a^2 F_m(\mp k_{xv}a)/(k_{xv}b) . \tag{14}$$

When $A_1 = 1$ and $A_2 = (P_{i2} =) 0$, the time-averaged incident, reflected, transmitted, and radiated (scattered into region (III)) powers are, respectively

$$P_{i1} = \frac{k_{xs}b}{4\omega\mu_1} \tag{15}$$

$$P_r = \frac{b}{4\omega\mu_1} \sum_v k_{xv}|K_v^-|^2 \tag{16}$$

$$P_t = \frac{b}{4\omega\mu_1} \left(k_{xs}|1 + K_s^+|^2 + \sum_{v \neq s} k_{xv}|K_v^+|^2 \right) \tag{17}$$

$$P_s = \sum_{l=-L1}^{L_2} \sum_{m=1}^{\infty} Re \left\{ \frac{-i\xi_m^* a}{2\omega\mu_2} \right.$$

$$\left. \cdot [b_m^l \cos(\xi_m d) - c_m^l \sin(\xi_m d)][b_m^l \sin(\xi_m d) + c_m^l \cos(\xi_m d)]^* \right\} \quad (18)$$

where $0 < v < (k_1 b/\pi)$, v: integer, and $P_r + P_t + P_s = P_{i1}$.
When $A_1 = (P_{i1} =) 0$ and $A_2 = 1$, then

$$P_{i2} = \frac{aNk_3}{\omega\mu_3} \quad (19)$$

$$P_r = \frac{b}{4\omega\mu_1} \sum_v k_{xv} |K_v^-|^2 \quad (20)$$

$$P_t = \frac{b}{4\omega\mu_1} \sum_v k_{xv} |K_v^+|^2 \ . \quad (21)$$

The far-zone scattered field at distance r is

$$E_y^{III}(r,\theta_s) = \sqrt{\frac{k_3}{2\pi r}} \exp\left(ik_3 r - i\frac{\pi}{4}\right) \cos\theta_s \widetilde{E}_y^{III}(k_3 \sin\theta_s) \ . \quad (22)$$

3.6.2 TM Scattering [27]

Assume that two incident TM waves propagate in regions (I) and (III). In region (I) the total field consists of the incident and scattered components

$$H_y^{i1}(x,z) = A_1 e^{ik_{xs}x} \cos k_{zs}(z+b) \quad (23)$$

$$H_y^I(x,z) = \frac{1}{\pi} \int_{-\infty}^{\infty} \widetilde{H}_y^I(\zeta) \cos(\kappa_1 z) e^{-i\zeta x} \, d\zeta \ . \quad (24)$$

In region (II) the field is

$$H_y^{II}(x,z) = \sum_{m=0}^{\infty} \cos a_m(x+a-lT)$$

$$\cdot [b_m^l \cos \xi_m(z+b) + c_m^l \sin \xi_m(z+b)] \ . \quad (25)$$

In region (III) the incident, reflected, and scattered fields are

$$H_y^{i2}(x,z) = A_2 \exp[ik_x x + ik_z(z+b+d)] \quad (26)$$

$$H_y^r(x,z) = A_2 \exp[ik_x x - ik_z(z+b+d)] \quad (27)$$

$$H_y^{III}(x,z) = \frac{1}{2\pi} \int_{-\infty}^{\infty} \widetilde{H}_y^{III}(\zeta) \exp[-i\zeta x - i\kappa_3(z+b+d)] \, d\zeta \ . \quad (28)$$

The tangential E- and H-field continuities at $z = -b$ yield

$$-iA_1 k_{xs} \exp(ik_{xs}rT)a^2 F_n(k_{xs}a)$$

$$= - \sum_{l=-L_1}^{L_2} \sum_{m=0}^{\infty} c_m^l \frac{\epsilon_1 \xi_m}{2\pi\epsilon_2} a^2 \Omega_5(k_1) + b_n^r \varepsilon_n a \quad (29)$$

where

$$\Omega_5(k_1)$$
$$= \int_{-\infty}^{\infty} \kappa_1^{-1} \zeta^2 \cot(\kappa_1 b) a^2 F_m(\zeta a) F_n(-\zeta a) \exp[i(l-r)\zeta T] \, d\zeta \ . \qquad (30)$$

A fast convergent series representation for $\Omega_5(k_1)$ is given in Subsect. 1.4.2. The tangential E- and H-field continuities at $z = -b - d$ yield

$$\begin{array}{|l|}
\hline
-i2A_2 k_x \exp(ik_x rT) a^2 F_n(k_x a) \\[2mm]
\quad - \displaystyle\sum_{l=-L_1}^{L_2} \sum_{m=0}^{\infty} \frac{\epsilon_3 \xi_m [b_m^l \sin(\xi_m d) + c_m^l \cos(\xi_m d)]}{2i\pi \epsilon_2} a^2 \Omega_2(k_3) \\[2mm]
= \varepsilon_n a [b_n^r \cos(\xi_n d) - c_n^r \sin(\xi_n d)] \\
\hline
\end{array} \qquad (31)$$

where $\Omega_2(k_3)$ is given in Sect. 1.2 as

$$\Omega_2(k_3) = \int_{-\infty}^{\infty} a^2 F_m(\zeta a) F_n(-\zeta a) \zeta^2 \exp[i(l-r)\zeta T] \kappa_3^{-1} \, d\zeta \ . \qquad (32)$$

3.7 EM Scattering from Slits in a Rectangular Cavity

A study of electromagnetic wave penetration into a cavity with multiple slits has been of importance in electromagnetic wave interference and compatibility problems [28]. In this section we will study an electromagnetic field penetration into a two-dimensional rectangular cavity with multiple slits. Scattering analyses for the TM and TE wave incidences are, respectively, presented in the next two subsections.

3.7.1 TM Scattering [29]

A TM (transverse magnetic to the wave propagation direction) uniform plane wave is obliquely incident on a cavity. In region (I) the incident and reflected H-fields are

$$H_z^i(x,y) = -\exp(ik_x x + ik_y y) \qquad (1)$$
$$H_z^r(x,y) = -\exp(ik_x x - ik_y y) \qquad (2)$$

where $k_x = k_0 \sin\theta$, $k_y = k_0 \cos\theta$, and the free-space wavenumber is $k_0 = \omega\sqrt{\mu_0 \epsilon_0} = 2\pi/\lambda$. The scattered H-field in region (I) is given by

$$H_z^s(x,y) = \frac{1}{2\pi} \int_{-\infty}^{\infty} \tilde{H}_z^s(\zeta) \exp(-i\zeta x + i\kappa_0 y) \, d\zeta \qquad (3)$$

where $\kappa_0 = \sqrt{k_0^2 - \zeta^2}$. In region (II) ($|x - lT| < a$ and $0 < y < d$), which is a slit interior, the H-field is

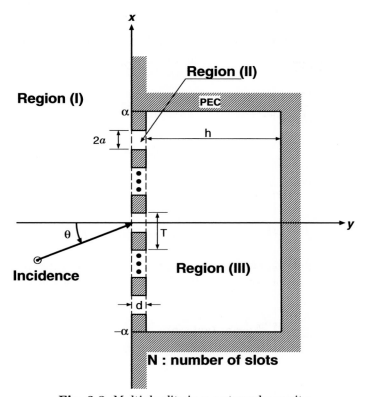

Fig. 3.8. Multiple slits in a rectangular cavity

$$H_z^{II}(x,y) = \sum_{m=0}^{\infty} \left[c_m^l \cos \xi_m (y - d) + d_m^l \sin \xi_m (y - d) \right]$$
$$\cdot \cos a_m (x - lT + a) \tag{4}$$

where $a_m = m\pi/(2a)$ and $\xi_m = \sqrt{k_0^2 - a_m^2}$. In region (III) ($|x| < \alpha$ and $d < y < d + h$), which is a cavity interior, the H-field is

$$H_z^{III}(x,y) = \sum_{q=0}^{\infty} e_q \cos \gamma_q (y - d - h) \cos \alpha_q (x + \alpha) \tag{5}$$

where $\alpha_q = q\pi/(2\alpha)$ and $\gamma_q = \sqrt{k_0^2 - \alpha_q^2}$.

The boundary conditions on E_x and H_z continuities at $y = 0$ require, respectively

$$E_x^i(x,0) + E_x^r(x,0) + E_x^s(x,0) = \begin{cases} E_x^{II}(x,0), & |x - lT| < a \\ 0, & \text{otherwise} \end{cases} \tag{6}$$

and

$$H_z^i(x,0) + H_z^r(x,0) + H_z^s(x,0) = H_z^{II}(x,0), \qquad |x - lT| < a . \qquad (7)$$

Applying the Fourier transform to (6) yields a representation for $\widetilde{H}_z^s(\zeta)$ in terms of the modal coefficients c_m^l and d_m^l. We then substitute $\widetilde{H}_z^s(\zeta)$ into (7), multiply (7) by $\cos a_n(x - rT + a)$, and perform integration from $(rT - a)$ to $(rT + a)$ to get

$$
\boxed{
\begin{aligned}
&2aik_x F_n(k_x a)e^{ik_x rT} \\
&- \frac{ia}{2\pi} \sum_{l=-L_1}^{L_2} \sum_{m=0}^{\infty} \xi_m \left[c_m^l \sin(\xi_m d) + d_m^l \cos(\xi_m d) \right] \Omega_2(k_0) \\
&= \varepsilon_n \left[c_n^r \cos(\xi_n d) - d_n^r \sin(\xi_n d) \right] , \qquad n = 0, 1, 2, \dots
\end{aligned}
}
\qquad (8)
$$

where

$$\Omega_2(k_0) = \int_{-\infty}^{\infty} a^2 \frac{\zeta^2}{\kappa_0} F_m(\zeta a) F_n(-\zeta a) \exp[i\zeta(l-r)T] \, d\zeta . \qquad (9)$$

The analytic evaluation of $\Omega_2(k_0)$ is available in Subsect. 1.2.3. In addition, the boundary conditions of E- and H-field continuities at $y = d$ yield

$$
\boxed{
\frac{1}{a\alpha} \sum_{l=-L_1}^{L_2} \sum_{m=0}^{\infty} \xi_m d_m^l \sum_{q=0}^{\infty} \frac{\Upsilon_{mq}^l \Upsilon_{nq}^r}{\gamma_q \tan(\gamma_q h)} = \varepsilon_n c_n^r
}
\qquad (10)
$$

where

$$
\Upsilon_{mq}^l = \begin{cases}
\dfrac{\alpha_q}{\alpha_q^2 - a_m^2} \\
\quad \cdot \{\sin[\alpha_q(a - \alpha - lT)] + (-1)^m \sin[\alpha_q(a + \alpha + lT)]\} \\
2a, \qquad\qquad\qquad\qquad m = q = 0 .
\end{cases}
\qquad (11)
$$

3.7.2 TE Scattering [30]

Assume that an electromagnetic TE wave is obliquely incident on thick multiple slits in a rectangular cavity. In region (I) the incident and reflected E-fields are represented as

$$E_z^i(x,y) = -Z_0 \exp(ik_x x + ik_y y) \qquad (12)$$
$$E_z^r(x,y) = Z_0 \exp(ik_x x - ik_y y) \qquad (13)$$

where $k_x = k_0 \sin\theta$, $k_y = k_0 \cos\theta$, $k_0 = \omega\sqrt{\mu_0 \epsilon_0} = 2\pi/\lambda$ is the free-space wavenumber, and Z_0 is the intrinsic impedance in free space. The scattered E-field in region (I) is given by

$$E_z^s(x,y) = \frac{1}{2\pi} \int_{-\infty}^{\infty} \widetilde{E}_z^s(\zeta) \exp(-i\zeta x + i\kappa_0 y) d\zeta \qquad (14)$$

where $\kappa_0 = \sqrt{k_0^2 - \zeta^2}$. In region (II) ($|x - lT| < a$ and $0 < y < d$) the E-field inside the slit is

$$E_z^{II}(x,y) = \sum_{m=1}^{\infty} \left[c_m^l \cos \xi_m (y - d) + d_m^l \sin \xi_m (y - d) \right]$$
$$\cdot \sin a_m (x - lT + a) \qquad (15)$$

where $a_m = m\pi/(2a)$ and $\xi_m = \sqrt{k_0^2 - a_m^2}$. In region (III) ($|x| < \alpha$ and $d < y < d + h$) the E-field is

$$E_z^{III}(x,y) = \sum_{q=1}^{\infty} e_q \sin \gamma_q (y - d - h) \sin \alpha_q (x + \alpha) \qquad (16)$$

where $\alpha_q = q\pi/(2\alpha)$ and $\gamma_q = \sqrt{k_0^2 - \alpha_q^2}$.

The boundary conditions on E_z and H_x at $y = 0$ require

$$E_z^i(x,0) + E_z^r(x,0) + E_z^s(x,0) = \begin{cases} E_z^{II}(x,0), & |x - lT| < a \\ 0, & \text{otherwise} \end{cases} \qquad (17)$$

and

$$H_x^i(x,0) + H_x^r(x,0) + H_x^s(x,0) = H_x^{II}(x,0), \qquad |x - lT| < a . \qquad (18)$$

Applying the Fourier transform to (17), substituting $\widetilde{E}_z^s(\zeta)$ into (18), multiplying (18) by $\sin a_n(x - rT + a)$, and performing integration from $(rT - a)$ to $(rT + a)$, we get

$$\frac{aa_n}{2\pi\omega\mu_0} \sum_{l=-L_1}^{L_2} \sum_{m=1}^{\infty} a_m [c_m^l \cos(\xi_m d) - d_m^l \sin(\xi_m d)] \Lambda_2(k_0)$$
$$= \frac{\xi_n}{i\omega\mu_0} \left[c_n^r \sin(\xi_n d) + d_n^r \cos(\xi_n d) \right] - 2a \cos\theta a_n F_n(k_x a) e^{ik_x rT} \qquad (19)$$

where

$$\Lambda_2(k_0) = \int_{-\infty}^{\infty} a^2 \kappa_0 F_m(\zeta a) F_n(-\zeta a) \exp[i\zeta(l - r)T] d\zeta . \qquad (20)$$

The explicit evaluation of $\Lambda_2(k_0)$ is given in Subsect. 1.2.3. Similarly from the boundary conditions at $y = d$, we obtain

$$\frac{1}{a\alpha} \sum_{l=-L_1}^{L_2} \sum_{m=1}^{\infty} c_m^l \sum_{q=1}^{\infty} \frac{\gamma_q \Upsilon_{mq}^l \Upsilon_{nq}^r}{\tan(\gamma_q h)} = -\xi_n d_n^r \qquad (21)$$

where

$$\Upsilon_{mq}^l = \frac{a_m}{\alpha_q^2 - a_m^2}$$
$$\cdot \{\sin[\alpha_q(a - \alpha - lT)] + (-1)^m \sin[\alpha_q(a + \alpha + lT)]\} . \qquad (22)$$

3.8 EM Scattering from Slits in Parallel-Conducting Planes [31]

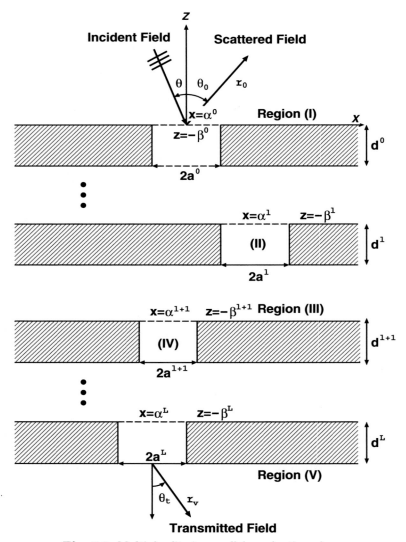

Fig. 3.9. Multiple slits in parallel-conducting planes

Electromagnetic wave mutual coupling between slits in two parallel conducting screens was studied in [32-34]. In this section we will investigate electromagnetic wave scattering from multiple slits in parallel conducting planes.

The present section is an extension of Sec. 3.3 where electromagnetic scattering from a single slit was discussed. In region (I) (air, $z > \beta^0 = 0$) a TM wave $H_y^i(x, z)$ is incident on a slit (width: $2a^0$ and depth: d^0) in a thick perfectly conducting plane. Region (II) ($-\beta^l - d^l < z < -\beta^l$, $\alpha^l - a^l < x < \alpha^l + a^l$, and relative permittivity ϵ_{rII}) and region (IV) ($-\beta^{l+1} - d^{l+1} < z < -\beta^{l+1}$, $\alpha^{l+1} - a^{l+1} < x < \alpha^{l+1} + a^{l+1}$, and relative permittivity ϵ_{rIV}) denote the lth and $(l+1)$th slits, respectively. Region (III) ($-\beta^{l+1} < z < -\beta^l - d^l$ and relative permittivity ϵ_{rIII}) denotes a dielectric slab bounded by the parallel conducting planes. Regions (I) and (V) are half-spaces that situate above and below the parallel-conducting planes, respectively. In region (I) (wavenumber $k = \omega\sqrt{\mu\epsilon_0}$) the total H-field is assumed to have the incident, reflected, and scattered components

$$H_y^i(x, z) = \exp(ik_x x - ik_z z) \tag{1}$$

$$H_y^r(x, z) = \exp(ik_x x + ik_z z) \tag{2}$$

$$H_y^s(x, z) = \frac{1}{2\pi} \int_{-\infty}^{\infty} \tilde{H}_y^s(\zeta) \exp(-i\zeta x + i\kappa z) \, d\zeta \tag{3}$$

where $k_x = k \sin\theta$, $k_z = k \cos\theta$, and $\kappa = \sqrt{k^2 - \zeta^2}$. In regions (II), (III), (IV), and (V) the total H-fields are given as

$$H_y^{II}(x, z) = \sum_{m=0}^{\infty} \left[b_m^l \cos\xi_m^l(z + \beta^l) + c_m^l \sin\xi_m^l(z + \beta^l) \right]$$
$$\cdot \cos a_m^l(x + a^l - \alpha^l) \tag{4}$$

$$H_y^{III}(x, z) = \frac{1}{2\pi} \int_{-\infty}^{\infty} \left\{ \tilde{H}_y^{III+}(\zeta) \exp[i\kappa(z + \beta^l + d^l)] \right.$$
$$\left. + \tilde{H}_y^{III-}(\zeta) \exp[-i\kappa(z + \beta^l + d^l)] \right\} e^{-i\zeta x} \, d\zeta \tag{5}$$

$$H_y^{IV}(x, z) = \sum_{m=0}^{\infty} \left[b_m^{l+1} \cos\xi_m^{l+1}(z + \beta^{l+1}) + c_m^{l+1} \sin\xi_m^{l+1}(z + \beta^{l+1}) \right]$$
$$\cdot \cos a_m^{l+1}(x + a^{l+1} - \alpha^{l+1}) \tag{6}$$

$$H_y^V(x, z) = \frac{1}{2\pi} \int_{-\infty}^{\infty} \tilde{H}^V(\zeta) \exp\left[-i\zeta x - i\kappa(z + \beta^L + d^L)\right] d\zeta \tag{7}$$

where $a_m^l = m\pi/(2a^l)$, $\xi_m^l = \sqrt{k^2 - (a_m^l)^2}$, and $\kappa = \sqrt{k^2 - \zeta^2}$.

The boundary condition on the tangential E- and H-field continuities at $z = 0$ yields

$$\frac{2ik_x}{[k_x^2 - (a_n^l)^2]} \left[-(-1)^n e^{ik_x a^0} + e^{-ik_x a^0} \right] e^{ik_x \alpha^0}$$
$$= \frac{i}{2\pi\epsilon_{rII}} \sum_{m=0}^{\infty} \xi_m^0 c_m^0 (a^0)^2 \left[\Omega_2(k) \right]_{a \to a^0} + b_n^0 a^0 \varepsilon_n \tag{8}$$

where

$$[\Omega_2(k)]_{a \to a^0} = \int_{-\infty}^{\infty} (a^0)^2 F_m(\zeta a^0) F_n(-\zeta a^0) \zeta^2 (\kappa)^{-1} \, d\zeta \,. \tag{9}$$

Note that a in $\Omega_2(k)$ needs to be replaced by a^0 that is a half of the slit width in region (I). The evaluation of $\Omega_2(k)$ is available in Subsect. 1.2.3 as

$$[\Omega_2(k)]_{a \to a^0} = \frac{2\pi\varepsilon_n}{a^0 \sqrt{k^2 - (a_m^0)^2}} \delta_{mn} - [\overline{\Omega_2(k)}]_{a \to a^0} \tag{10}$$

where δ_{mn} is the Kronecker delta, $\varepsilon_0 = 2$, and $\varepsilon_n = 1$ $(n = 1, 2, 3, \dots)$. Similarly the boundary conditions at $z = -\beta^l - d^l$ and $-\beta^{l+1}$ give, respectively,

$$
\begin{aligned}
& 2\pi(b_n^l \cos \xi_n^l d^l - c_n^l \sin \xi_n^l d^l) a^l \varepsilon_n \\
& = \sum_{m=0}^{\infty} \left[\frac{\epsilon_{rIII}}{\epsilon_{rII}} \xi_m^l (b_m^l \sin \xi_m^l d^l + c_m^l \cos \xi_m^l d^l) I_1 \right. \\
& \quad \left. - \frac{\epsilon_{rIII}}{\epsilon_{rIV}} \xi_m^{l+1} c_m^{l+1} I_2 \right]
\end{aligned}
\tag{11}
$$

$$
\begin{aligned}
& 2\pi b_n^{l+1} a^{l+1} \varepsilon_n \\
& = \sum_{m=0}^{\infty} \left[\frac{\epsilon_{rIII}}{\epsilon_{rII}} \xi_m^l (b_m^l \sin \xi_m^l d^l + c_m^l \cos \xi_m^l d^l) I_3 \right. \\
& \quad \left. - \frac{\epsilon_{rIII}}{\epsilon_{rIV}} \xi_m^{l+1} c_m^{l+1} I_4 \right]
\end{aligned}
\tag{12}
$$

where

$$I_1 = \int_{-\infty}^{\infty} \frac{\zeta^2 (a^l)^2 F_m(\zeta a^l)(a^l)^2 F_n(-\zeta a^l)}{\kappa \tan \kappa (\beta^l + d^l - \beta^{l+1})} \, d\zeta \tag{13}$$

$$I_2 = \int_{-\infty}^{\infty} \frac{\zeta^2 (a^{l+1})^2 F_m(\zeta a^{l+1})(a^l)^2 F_n(-\zeta a^l) \exp[i\zeta(a^{l+1} - a^l)]}{\kappa \sin \kappa (\beta^l + d^l - \beta^{l+1})} \, d\zeta \tag{14}$$

$$I_3 = \int_{-\infty}^{\infty} \frac{\zeta^2 (a^l)^2 F_m(\zeta a^l)(a^{l+1})^2 F_n(-\zeta a^{l+1}) \exp[i\zeta(a^l - a^{l+1})]}{\kappa \sin \kappa (\beta^l + d^l - \beta^{l+1})} \, d\zeta \tag{15}$$

$$I_4 = \int_{-\infty}^{\infty} \frac{\zeta^2 (a^{l+1})^2 F_m(\zeta a^{l+1})(a^{l+1})^2 F_n(-\zeta a^{l+1})}{\kappa \tan \kappa (\beta^l + d^l - \beta^{l+1})} \, d\zeta \,. \tag{16}$$

It is possible to transform (13) through (16) into numerically-efficient series based on the residue calculus. The boundary conditions on the tangential E- and H-field continuities at $z = -\beta^{l+1} - d^{l+1}$ yield

$$
\begin{aligned}
& 2\pi(b_n^L \cos \xi_n^L d^L - c_n^L \sin \xi_n^L d^L) a^L \varepsilon_n \\
& = \frac{i\epsilon_{rV}}{\epsilon_{rIV}} \sum_{m=0}^{\infty} \xi_m^L (b_m^L \sin \xi_m^L d^L + c_m^L \cos \xi_m^L d^L)(a^L)^2 [\Omega_2(k)]_{a \to a^L}
\end{aligned}
\tag{17}
$$

where

$$[\Omega_2(k)]_{a \to a^L} = \int_{-\infty}^{\infty} (a^L)^2 F_m(\zeta a^L) F_n(-\zeta a^L) \zeta^2 (\kappa)^{-1} \, d\zeta \ . \tag{18}$$

References for Chapter 3

1. Y. S. Kim and H. J. Eom, "Fourier-transform analysis of electrostatic potential distribution through a thick slit," *IEEE Trans. Electromagn. Compat.*, vol. 38, no. 1, pp. 77-79, Feb. 1996. Corrections to "Fourier-transform analysis of electrostatic potential distribution through a thick slit," *IEEE Trans. Electromagn. Compat.*, vol. 39, no. 1, page 66, Feb. 1997.
2. L. K. Warne and K. C. Chen, "Relation between equivalent antenna radius and transverse line dipole moments of a narrow slot aperture having depth," *IEEE Trans. Electromagn. Compat.*, vol. 30, no. 3, pp. 364-370, Aug. 1988.
3. Y. C. Noh and H. J. Eom, "Electrostatic potential due to a potential drop across a slit," *IEEE Trans. Microwave Theory Tech.*, vol. 46, no. 4, pp. 428-430, April 1998.
4. S. C. Kashyap and M. A. K. Hamid, "Diffraction Characteristics of a slit in a thick conducting screen," *IEEE Trans. Antennas Propagat.*, vol. 19, no. 4, pp. 499-507, July 1971.
5. K. Hongo and G. Ishii, "Diffraction of an electromagnetic plane wave by a thick slit," *IEEE Trans. Antennas Propagat.*, vol. 26, no. 3, pp. 494-499, May 1978.
6. D. T. Auckland and R. F. Harrington, "Electromagnetic transmission through a filled slit in a conducting plane of finite thickness, TE case," *IEEE Trans. Microwave Theory Tech.*, vol. 26, no. 7, pp. 499-505, July 1978.
7. O. M. Mendez, M. Cadilhac, and R. Petit, "Diffraction of a two-dimensional electromagnetic beam wave by a thick slit pierced in a perfectly conducting screen," *J. Opt. Soc. Am.*, vol. 73, no. 3, pp. 328-331, Mar. 1983.
8. J. M. Jin and J. L. Volakis, "TM scattering by an inhomogeneously filled aperture in a thick conducting plane," *IEE Proceedings*, vol. 137, pt. H, no. 3, pp. 153-159, June 1990.
9. J. B. Keller, "Diffraction by an aperture," *J. Appl. Phys.*, vol. 28, no. 4, pp. 426-444, April 1957.
10. S. H. Kang, H. J. Eom, and T. J. Park, "TM-scattering from a slit in a thick conducting screen: revisited," *IEEE Trans. Microwave Theory Tech.*, vol. 41, no. 5, pp. 895-899, May 1993.
11. T. J. Park, S. H. Kang, and H. J. Eom, "TE-scattering from a slit in a thick conducting screen: revisited," *IEEE Trans. Antennas Propagat.*, vol. 42, no. 1, pp. 112-114, Jan. 1994.
12. J. A. Stratton, *Electromagnetic Theory*, pp. 227-228, McGraw-Hill, 1941
13. Y. S. Kim, H. J. Eom, J. W. Lee, and K. Yoshitomi, "Scattering from multiple slits in a thick conducting plane," *Radio Sci.*, vol. 30, no. 5, pp. 1341-1347, Sept.-Oct. 1995.
14. R. Petit and G. Tayeb, "Numerical study of the symmetrical strip-grating-loaded slab," *J. Opt. Soc. Am. A.*, vol. 7, no. 3, pp. 373-378, March 1990.
15. K. Kobayashi and K. Miura, "Diffraction of a plane wave by a thick strip grating," *IEEE Trans. Antennas Propagat.*, vol. 37, no. 4, pp. 459-470, Apr. 1989.
16. Ya. N. Feld, G. A. Svistunov, A. G. Kyurkchan, and A. S. Leontev, "The diffraction of an electromagnetic wave by a system of plane parallel waveguides of finite length," *Radio Eng. Electron. Phys.*, vol. 18, no. 5, pp. 655-663, 1973.

17. T. Otsuki, "Diffraction by multiple slits," *J. Opt. Soc. Am. A*, vol. 7, no. 4, pp. 646-652, April 1990.
18. K. Kobayashi, "Diffraction of a plane wave by the parallel plate grating with dielectric loading," *Trans. of Inst. of Electron. Commun. Eng. Jap.*, vol. J64-B, no. 10, pp. 1091-1098, Oct. 1981.
19. Y. K. Cho, "Analysis of a narrow slit in a parallel-plate transmission line: E-polarization case," *Electron. Lett.*, vol. 23, no. 21, pp. 1105-1106, 8th Oct. 1987.
20. H. A. Auda, "Quasistatic characteristics of a slotted parallel-plate waveguides," *IEE Proceedings*, vol. 135, pt. H, no. 4, pp. 256-262, Aug. 1988.
21. J. P. Quintenz and D. G. Dudley, "Slots in a parallel-plate waveguide," *Radio Sci.*, vol. 11, no. 8-9, pp. 713-724, Aug.-Sept. 1976.
22. A. M. Barbosa, A. F. dos Santos, and J. Figanier, "Radiation field of a periodic strip grating excited by an aperiodic line source," *Radio Sci.*, vol. 19, no. 3, pp. 829-839, May-June 1984.
23. C. W. Chuang, "Generalized admittance matrix for a slotted parallel-plate waveguide," *IEEE Trans. Antennas Propagat.*, vol. 36, no. 9, pp. 1227-1230, Sept. 1988.
24. E. M. T. Jones and J. K. Shimizu, "A wide-band transverse-slot flush-mounted array," *IRE Trans. Antennas Propagat.*, vol. 8, no. 4, pp. 401-407, July, 1960.
25. J. A. Encinar, "Mode-matching and point-matching techniques applied to the analysis of metal-strip-loaded dielectric antennas," *IEEE Trans. Antennas Propagat.*, vol. 38, no. 9, pp. 1405-1412, Sept. 1990.
26. J. H. Lee and H. J. Eom, "Scattering and radiation from finite thick slits in parallel-plate waveguide," *IEEE Trans. Antennas Propagat.*, vol. 44, no.2, pp. 212-216, 1996.
27. J. H. Lee, H. J. Eom, Y. K. Cho, and W. J. Chun, "TM-wave radiation from finite thick slits in parallel-plate," *IEICE Trans. Commun.*, vol. E79-B, no. 6, pp. 875-878, June 1996.
28. J. V. Tejedor, L. Nuno, and M. F. Bataller, "Susceptibility analysis of arbitrarily shaped 2-D slotted screen using a hybrid generalized matrix finite-element technique," *IEEE Trans. Electromagn. Compat.*, vol. 40, no. 1, pp. 47-54, Feb. 1998.
29. H. H. Park and H. J. Eom, "Electromagnetic penetration into 2D multiple slotted rectangular cavity: TE-wave," *Electron. Lett.*, vol. 35, no. 1, pp. 31-32, 1999.
30. H. H. Park and H. J. Eom, "Electromagnetic penetration into 2D multiple slotted rectangular cavity: TM-wave," *IEEE Trans. Antennas Propagat.*, vol. 48, no. 2, pp. 331-333, Feb. 2000.
31. J. W. Lee and H. J. Eom, "TM scattering from slits in thick parallel conducting screens," *IEEE Trans. Antennas Propagat.*, vol. 46, no. 7, pp. 1117-1119, July 1998.
32. L. R. Alldredge, "Diffraction of microwaves by tandem slits," *IRE Trans. Antennas Propagat.*, pp. 640-649, Oct. 1956.
33. Y. E. Elmoazzen and L. Shafai, "Mutual coupling between parallel-plate waveguides," *IEEE Trans. Microwave Theory Tech.*, vol. 21, no. 12, pp. 825-833, Dec. 1973.
34. Y. Leviatan, "Electromagnetic coupling between two half-space regions separated by two slot-perforated parallel conducting screens," *IEEE Trans. Microwave Theory Tech.*, vol. 36, no. 1, pp. 44-51, Jan. 1978.

4. Waveguides and Couplers

4.1 Inset Dielectric Guide

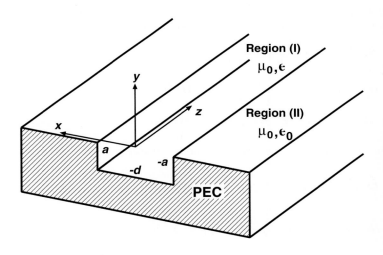

Fig. 4.1. Inset dielectric guide

The inset dielectric guide (IDG) is a dielectric-filled rectangular groove and its waveguiding characteristics have been extensively analyzed in [1-3]. In this section we will analyze an electromagnetic wave guided by the inset dielectric guide and obtain its dispersion relation. Region (I) ($-d < y < 0$ and $|x| < a$) is filled with a dielectric material of permittivity ϵ and region (II) ($y > 0$) is the air of ϵ_0. Due to a dielectric discontinuity at $y = 0$ between regions (I) and (II), a hybrid wave, which is a combination of the TE and TM waves, is assumed to propagate along the z-direction. The assumption of a hybrid wave propagation is necessary to satisfy the boundary conditions on the field continuities at $y = 0$. In region (I) the total E- and H-field z-components are

$$E_z^I(x,y,z) = \sum_{k=1}^{\infty} p_k \sin a_k(x+a) \sin \sigma_k(y+d) e^{i\beta z} \qquad (1)$$

$$H_z^I(x,y,z) = \sum_{m=0}^{\infty} q_m \cos a_m(x+a) \cos \sigma_m(y+d) e^{i\beta z} \qquad (2)$$

where $a_m = m\pi/(2a)$, $\sigma_m = \sqrt{k_1^2 - a_m^2 - \beta^2}$, and $k_1 = \omega\sqrt{\mu_0\epsilon} = \omega\sqrt{\mu_0\epsilon_0\epsilon_r}$.
In region (II) the fields are given as

$$E_z^{II}(x,y,z) = \frac{1}{2\pi} \int_{-\infty}^{\infty} \widetilde{E}_z(\zeta) \exp\{i[-\zeta x + \eta(\zeta)y + \beta z]\} \, d\zeta \qquad (3)$$

$$H_z^{II}(x,y,z) = \frac{1}{2\pi} \int_{-\infty}^{\infty} \widetilde{H}_z(\zeta) \exp\{i[-\zeta x + \eta(\zeta)y + \beta z]\} \, d\zeta \qquad (4)$$

where $\eta(\zeta) = \sqrt{k_2^2 - \zeta^2 - \beta^2}$ and $k_2 = \omega\sqrt{\mu_0\epsilon_0} = 2\pi/\lambda$. The transverse components of E and H fields are immediate from

$$E_t^{I,II}(x,y,z) = \frac{i}{k_{1,2}^2 - \beta^2} \left(\beta \nabla_t E_z^{I,II} + \omega\mu \nabla_t \times H_z^{I,II}\right) \qquad (5)$$

$$H_t^{I,II}(x,y,z) = \frac{i}{k_{1,2}^2 - \beta^2} \left(\beta \nabla_t H_z^{I,II} - \omega\epsilon_{I,II} \nabla_t \times E_z^{I,II}\right) \qquad (6)$$

where the subscript t means taking the component transverse to the z-direction.

The E_z and E_x field continuities at $y = 0$ yield

$$\widetilde{E}_z(\zeta) = \sum_{k=1}^{\infty} p_k \sin(\sigma_k d) a^2 a_k F_k(\zeta a) \qquad (7)$$

$$\widetilde{H}_z(\zeta) = \sum_{k=1}^{\infty} -i\zeta a^2 A_k \frac{F_k(\zeta a)}{\eta(\zeta)} p_k + \sum_{m=0}^{\infty} i\zeta a^2 B_m \frac{F_m(\zeta a)}{\eta(\zeta)} q_m \qquad (8)$$

where

$$A_m = \frac{\beta a_m(k_2^2 - k_1^2) \sin(\sigma_m d)}{i\omega\mu(k_1^2 - \beta^2)} \qquad (9)$$

$$B_m = \frac{\sigma_m(k_2^2 - \beta^2) \sin(\sigma_m d)}{i(k_1^2 - \beta^2)} \, . \qquad (10)$$

Multiplying the H_z and H_x field continuities at $|x| < a$ and $y = 0$ by $\cos a_n(x+a)$ and $\sin a_l(x+a)$, respectively, and integrating with respect to x from $-a$ to a, we obtain

$$\boxed{\sum_{k=1}^{\infty} \frac{A_k}{2\pi} J_{kn} p_k - \sum_{m=0}^{\infty} \frac{B_m}{2\pi} J_{mn} q_m = \cos(\sigma_n d)\varepsilon_n a q_n} \qquad (11)$$

$$\sum_{k=1}^{\infty} \left[\frac{-\beta a_l A_k}{2\pi} J_{kl} - \frac{i\omega\epsilon_0 \sin(\sigma_k d)}{2\pi} I_{kl} \right] p_k + \sum_{m=0}^{\infty} \frac{\beta a_l B_m}{2\pi} J_{ml} q_m$$

$$= \left(\frac{k_2^2 - \beta^2}{k_1^2 - \beta^2} \right) \cos(\sigma_l d) a(-\beta a_l q_l - \omega\epsilon_l p_l) \qquad (12)$$

where

$$J_{mn} = a^2 \int_{-\infty}^{\infty} \frac{(a\zeta)^2}{\sqrt{k_2^2 - \zeta^2 - \beta^2}} F_m(\zeta a) F_n(-\zeta a) \, d\zeta \qquad (13)$$

$$I_{mn} = a^2 a_m a_n \int_{-\infty}^{\infty} a^2 \sqrt{k_2^2 - \zeta^2 - \beta^2} F_m(\zeta a) F_n(-\zeta a) \, d\zeta . \qquad (14)$$

Note $J_{mn} = a^2 \Omega_1 \left(\sqrt{k_2^2 - \beta^2} \right)$ and $I_{mn} = a^2 a_m a_n \Lambda_1 \left(\sqrt{k_2^2 - \beta^2} \right)$ where the expressions for Ω_1 and Λ_1 are available in Subsect. 1.1.3.
For $m + n$ is odd, $\Omega_1 \left(\sqrt{k_2^2 - \beta^2} \right) = \Lambda_1 \left(\sqrt{k_2^2 - \beta^2} \right) = 0$.
For $m + n$ is even,

$$\Lambda_1 \left(\sqrt{k_2^2 - \beta^2} \right) = \frac{2\pi \sqrt{k_2^2 - \beta^2 - a_m^2}}{a a_m^2} \delta_{mn} - \bar{\Lambda}_1 \left(\sqrt{k_2^2 - \beta^2} \right) \qquad (15)$$

$$\Omega_1 \left(\sqrt{k_2^2 - \beta^2} \right) = \frac{2\pi\varepsilon_n}{a\sqrt{k_2^2 - \beta^2 - a_m^2}} \delta_{mn} - \bar{\Omega}_1 \left(\sqrt{k_2^2 - \beta^2} \right) . \qquad (16)$$

In order to determine β we form a determinant

$$\begin{vmatrix} \Psi_1 & \Psi_2 \\ \Psi_3 & \Psi_4 \end{vmatrix} = 0 . \qquad (17)$$

The dispersion relation (17) is given by a product of even and odd mode determinants where Ψ_1, Ψ_2, Ψ_3, and Ψ_4 elements are

$$\psi_{1,nk} = \frac{A_k a\varepsilon_k}{\eta(a_k)} \delta_{nk} - \frac{a^2 [\bar{\Omega}_1]_{mn \to nk} A_k}{2\pi} \qquad (18)$$

$$\psi_{2,nm} = - \left[\frac{B_m a\varepsilon_m}{\eta(a_m)} + \cos(\sigma_m d)\varepsilon_m a \right] \delta_{nm} + \frac{a^2 [\bar{\Omega}_1]_{mn \to nm} B_m}{2\pi} \qquad (19)$$

$$\psi_{3,lk} = \left[\frac{-\beta A_k a_l a\varepsilon_k}{\eta(a_k)} - i\omega\epsilon_0 a \sin(\sigma_k d)\eta(a_k) \right.$$

$$\left. + \left(\frac{k_2^2 - \beta^2}{k_1^2 - \beta^2} \right) \omega\epsilon_k \cos(\sigma_k d)a \right] \delta_{lk}$$

$$+ \frac{a^2 [\bar{\Omega}_1]_{mn \to lk} \beta A_k a_l}{2\pi} + \frac{i a^2 a_l a_k [\bar{\Lambda}_1]_{mn \to lk} \omega\epsilon_0 \sin(\sigma_k d)}{2\pi} \qquad (20)$$

$$\psi_{4,lm} = \left[\frac{\beta B_m a_l a \varepsilon_m}{\eta(a_m)} + \left(\frac{k_2^2 - \beta^2}{k_1^2 - \beta^2}\right)\beta a_m \cos(\sigma_m d)a\right]\delta_{lm}$$
$$- \frac{a^2[\bar{\Omega}_1]_{mn\to lm}\beta B_m a_l}{2\pi} . \tag{21}$$

Note that the indices m and n associated with $\bar{\Omega}_1$ and $\bar{\Lambda}_1$ in (18) through (21) need to be changed according to the indices for $\psi_{1,2,3,4}$.

A dominant-mode approximate solution, $\psi_{2,00} = 0$, is given by

$$\sqrt{k_2^2 - \beta^2} + i\sqrt{k_1^2 - \beta^2}\cot(\sigma_0 d) - \frac{i\sqrt{k_2^2 - \beta^2}}{4\pi a}\bar{J} = 0 \tag{22}$$

where

$$\bar{J} = \frac{8i}{\sqrt{k_2^2 - \beta^2}}$$
$$- 4i\int_0^\infty \left[\frac{\exp(i2\sqrt{k_2^2 - \beta^2}a)\exp(-2\sqrt{k_2^2 - \beta^2}av)}{\sqrt{k_2^2 - \beta^2}(1 + iv)^2\sqrt{v(v - 2i)}}\right]dv . \tag{23}$$

4.2 Groove Guide [4]

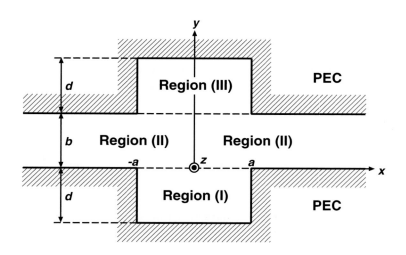

Fig. 4.2. Groove guide

The groove guide is known to have a low-loss high-power capacity above 100 GHz and experimental and theoretical studies were performed in [5-7]. In this section we will analyze its guiding characteristics and derive the dispersion relations for the TE and TM waves. The groove guide interior consisting of regions (I), (II), and (III) is filled with a homogeneous dielectric medium with permittivity ϵ. This means that the groove guide can support either a TE or TM wave propagation. The E-field, $\boldsymbol{E}(x, y, z)$, propagating along the z-direction may be represented as $\boldsymbol{E}(x, y)\mathrm{e}^{\mathrm{i}\beta z}$. In regions (I), (II), and (III), their respective E_z and H_z fields are

$$E_z^I(x, y) = \sum_{m=1}^{\infty} p_m \sin a_m(x + a) \sin \sigma_m(y + d) \tag{1}$$

$$H_z^I(x, y) = \sum_{m=0}^{\infty} q_m \cos a_m(x + a) \cos \sigma_m(y + d) \tag{2}$$

$$E_z^{II}(x, y) = \frac{1}{2\pi} \int_{-\infty}^{\infty} \left(\widetilde{E}_z^+ \mathrm{e}^{\mathrm{i}\kappa y} + \widetilde{E}_z^- \mathrm{e}^{-\mathrm{i}\kappa y} \right) \mathrm{e}^{-\mathrm{i}\zeta x} \, \mathrm{d}\zeta \tag{3}$$

$$H_z^{II}(x, y) = \frac{1}{2\pi} \int_{-\infty}^{\infty} \left(\widetilde{H}_z^+ \mathrm{e}^{\mathrm{i}\kappa y} + \widetilde{H}_z^- \mathrm{e}^{-\mathrm{i}\kappa y} \right) \mathrm{e}^{-\mathrm{i}\zeta x} \, \mathrm{d}\zeta \tag{4}$$

$$E_z^{III}(x, y) = \sum_{m=1}^{\infty} r_m \sin a_m(x + a) \sin \sigma_m(y - b - d) \tag{5}$$

$$H_z^{III}(x, y) = \sum_{m=0}^{\infty} s_m \cos a_m(x + a) \cos \sigma_m(y - b - d) \tag{6}$$

where $a_m = m\pi/(2a)$, $\sigma_m = \sqrt{k^2 - a_m^2 - \beta^2}$, $\kappa = \sqrt{k^2 - \beta^2 - \zeta^2}$, and the wavenumber is $k = \omega\sqrt{\mu\epsilon} = 2\pi/\lambda$. The remaining field components are immediately available from (5) and (6) in Sect. 4.1.

4.2.1 TM Propagation

Consider a TM wave propagation with $E_z \neq 0$ and $H_z = 0$. The boundary condition $E_z^{II}(x, 0) = E_z^I(x, 0)$ is written as

$$\frac{1}{2\pi} \int_{-\infty}^{\infty} \left(\widetilde{E}_z^+ + \widetilde{E}_z^- \right) \mathrm{e}^{-\mathrm{i}\zeta x} \, \mathrm{d}\zeta$$
$$= \begin{cases} \displaystyle\sum_{m=1}^{\infty} p_m \sin a_m(x + a) \sin(\sigma_m d), & |x| < a \\ 0, & |x| > a \ . \end{cases} \tag{7}$$

Applying the Fourier transform to (7) yields

$$\widetilde{E}_z^+ + \widetilde{E}_z^- = \sum_{m=1}^{\infty} p_m a_m \sin(\sigma_m d) a^2 F_m(\zeta a) \ . \tag{8}$$

Similarly the condition $E_z^{III}(x, b) = E_z^{II}(x, b)$ gives

$$\widetilde{E}_z^+ e^{i\kappa b} + \widetilde{E}_z^- e^{-i\kappa b} = -\sum_{m=1}^{\infty} r_m a_m \sin(\sigma_m d) a^2 F_m(\zeta a) \ . \tag{9}$$

The condition $H_x^{II}(x,0) = H_x^{I}(x,0)$ for $|x| < a$ and the condition $H_x^{III}(x,b) = H_x^{II}(x,b)$ for $|x| < a$ give, respectively

$$\sum_{m=1}^{\infty} p_m \sigma_m \cos(\sigma_m d) \sin a_m (x+a)$$

$$= \frac{1}{2\pi} \int_{-\infty}^{\infty} \kappa \left[-A\cot(\kappa b) + B\csc(\kappa b) \right] e^{-i\zeta x}\, d\zeta \tag{10}$$

$$\sum_{m=1}^{\infty} r_m \sigma_m \cos(\sigma_m d) \sin a_m (x+a)$$

$$= \frac{1}{2\pi} \int_{-\infty}^{\infty} \kappa \left[-A\csc(\kappa b) + B\cot(\kappa b) \right] e^{-i\zeta x}\, d\zeta \tag{11}$$

where

$$A = \sum_{m=1}^{\infty} p_m a_m \sin(\sigma_m d) a^2 F_m(\zeta a) \tag{12}$$

$$B = -\sum_{m=1}^{\infty} r_m a_m \sin(\sigma_m d) a^2 F_m(\zeta a) \ . \tag{13}$$

We multiply (10) and (11) by $\sin a_n(x+a)$ and integrate from $-a$ to a to obtain

$$\sum_{m=1}^{\infty} \left\{ p_m \left[\frac{4a\sigma_m}{(m\pi)^2} \cos(\sigma_m d)\delta_{mn} + \sin(\sigma_m d)I_1 \right] \right.$$

$$\left. + r_m \sin(\sigma_m d) I_2 \right\} = 0 \tag{14}$$

$$\sum_{m=1}^{\infty} \left\{ p_m \sin(\sigma_m d) I_2 \right.$$

$$\left. + r_m \left[\frac{4a\sigma_m}{(m\pi)^2} \cos(\sigma_m d)\delta_{mn} + \sin(\sigma_m d)I_1 \right] \right\} = 0 \tag{15}$$

where

$$I_1 = \frac{1}{2\pi} \int_{-\infty}^{\infty} \cot(\kappa b) a^2 F_m(\zeta a) F_n(-\zeta a)\kappa\, d\zeta \tag{16}$$

$$I_2 = \frac{1}{2\pi} \int_{-\infty}^{\infty} \csc(\kappa b) a^2 F_m(\zeta a) F_n(-\zeta a)\kappa\, d\zeta \ . \tag{17}$$

Using the residue calculus, we transform I_1 and I_2 into fast convergent series

$$I_1 = \begin{cases} \dfrac{4a\sigma_m \cot(\sigma_m b)}{(m\pi)^2} \delta_{mn} - i \sum_{v=1}^{\infty} U_{mn}(v\pi/b), & m+n = \text{even} \\ 0, & m+n = \text{odd} \end{cases} \tag{18}$$

$$I_2 = \begin{cases} \dfrac{4a\sigma_m \csc(\sigma_m b)}{(m\pi)^2} \delta_{mn} - i \sum_{v=1}^{\infty} (-1)^v U_{mn}(v\pi/b), & m+n = \text{even} \\ 0, & m+n = \text{odd} \end{cases} \tag{19}$$

where

$$U_{mn}(w) = \frac{2w^2}{a^2 b \sqrt{k^2 - \beta^2 - w^2}}$$
$$\cdot \left[\frac{1 - (-1)^m \exp\left(i2\sqrt{k^2 - \beta^2 - w^2} a\right)}{(\sigma_m^2 - w^2)(\sigma_n^2 - w^2)} \right] . \tag{20}$$

A dispersion relation is obtained by setting a determinant of simultaneous equations for p_m and r_m to zero

$$\begin{vmatrix} \Psi_1 & \Psi_2 \\ \Psi_2 & \Psi_1 \end{vmatrix} = 0 \tag{21}$$

where the elements of Ψ_1 and Ψ_2 for $m+n = \text{even}$ are

$$\psi_{1,mn} = \frac{4a\sigma_m}{(m\pi)^2} [\cos(\sigma_m d) + \sin(\sigma_m d) \cot(\sigma_m b)] \delta_{mn}$$
$$-i \sin(\sigma_m d) \sum_{v=1}^{\infty} U_{mn}(v\pi/b) \tag{22}$$

$$\psi_{2,mn} = \frac{4a\sigma_m}{(m\pi)^2} \sin(\sigma_m d) \csc(\sigma_m b) \delta_{mn}$$
$$-i \sin(\sigma_m d) \sum_{v=1}^{\infty} (-1)^v U_{mn}(v\pi/b) . \tag{23}$$

In low-frequency limit, a dominant-mode ($m = 1$) approximate solution is

$$\psi_{1,11} - \psi_{2,11} = 0 . \tag{24}$$

4.2.2 TE Propagation

We consider a TE wave propagation by assuming $E_z = 0$ and $H_z \neq 0$. By applying the Fourier transform to the condition $E_x^{II}(x,0) = E_x^I(x,0)$, we get

$$\tilde{H}_z^+ - \tilde{H}_z^- = \frac{\zeta}{\kappa} \sum_{m=0}^{\infty} q_m \sigma_m \sin(\sigma_m d) a^2 F_m(\zeta a) . \tag{25}$$

Similarly from the condition $E_x^{III}(x,b) = E_x^{II}(x,b)$, we get

$$\widetilde{H}_z^+ e^{i\kappa b} - \widetilde{H}_z^- e^{-i\kappa b} = -\frac{\zeta}{\kappa} \sum_{m=0}^{\infty} s_m \sigma_m \sin(\sigma_m d) a^2 F_m(\zeta a) \ . \tag{26}$$

From the condition $H_z^{II}(x,0) = H_z^{I}(x,0)$ for $|x| < a$ and the condition $H_z^{III}(x,b) = H_z^{II}(x,b)$ for $|x| < a$, we obtain

$$\sum_{m=0}^{\infty} \left\{ q_m \left[\frac{\varepsilon_m}{a\sigma_m} \cos(\sigma_m d)\delta_{mn} + \sin(\sigma_m d) I_3 \right] \right.$$
$$\left. + s_m \sin(\sigma_m d) I_4 \right\} = 0 \tag{27}$$

$$\sum_{m=0}^{\infty} \left\{ q_m \sin(\sigma_m d) I_4 \right.$$
$$\left. + s_m \left[\frac{\varepsilon_m}{a\sigma_m} \cos(\sigma_m d)\delta_{mn} + \sin(\sigma_m d) I_3 \right] \right\} = 0 \tag{28}$$

where

$$I_3 = \frac{1}{2\pi} \int_{-\infty}^{\infty} \cot(\kappa b) \frac{(\zeta a)^2}{\kappa} F_m(\zeta a) F_n(-\zeta a) \, d\zeta \tag{29}$$

$$I_4 = \frac{1}{2\pi} \int_{-\infty}^{\infty} \csc(\kappa b) \frac{(\zeta a)^2}{\kappa} F_m(\zeta a) F_n(-\zeta a) \, d\zeta \ . \tag{30}$$

Note that I_3 and I_4 are represented in fast convergent series

$$I_3 = \begin{cases} \dfrac{\varepsilon_m \cot(\sigma_m b)}{a\sigma_m} \delta_{mn} - i \sum_{v=0}^{\infty} \dfrac{1}{\varepsilon_v} V_{mn}(v\pi/b), & m+n = \text{even} \\ 0, & m+n = \text{odd} \end{cases} \tag{31}$$

$$I_4 = \begin{cases} \dfrac{\varepsilon_m \csc(\sigma_m b)}{a\sigma_m} \delta_{mn} - i \sum_{v=0}^{\infty} \dfrac{(-1)^v}{\varepsilon_v} V_{mn}(v\pi/b), & m+n = \text{even} \\ 0, & m+n = \text{odd} \end{cases} \tag{32}$$

where

$$V_{mn}(w) = \frac{2}{a^2 b} \sqrt{k^2 - \beta^2 - w^2}$$
$$\cdot \left[\frac{1 - (-1)^m \exp\left(i2\sqrt{k^2 - \beta^2 - w^2}a\right)}{(\sigma_m^2 - w^2)(\sigma_n^2 - w^2)} \right] \ . \tag{33}$$

To determine β, we form a determinant for $m+n = \text{even}$

$$\begin{vmatrix} \Psi_3 & \Psi_4 \\ \Psi_4 & \Psi_3 \end{vmatrix} = 0 \tag{34}$$

where

$$\psi_{3,mn} = \frac{\varepsilon_m}{a\sigma_m}[\cos(\sigma_m d) + \sin(\sigma_m d)\cot(\sigma_m b)]\delta_{mn}$$

$$-\mathrm{i}\sin(\sigma_m d)\sum_{v=0}^{\infty}\frac{1}{\varepsilon_v}V_{mn}(v\pi/b) \tag{35}$$

$$\psi_{4,mn} = \frac{\varepsilon_m}{a\sigma_m}\sin(\sigma_m d)\csc(\sigma_m b)\delta_{mn}$$

$$-\mathrm{i}\sin(\sigma_m d)\sum_{v=0}^{\infty}\frac{(-1)^v}{\varepsilon_v}V_{mn}(v\pi/b)\ . \tag{36}$$

A dominant-mode solution is approximately given by

$$\psi_{3,00} - \psi_{4,00} = 0\ . \tag{37}$$

4.3 Multiple Groove Guide [8]

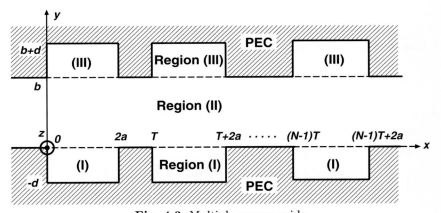

Fig. 4.3. Multiple groove guide

A multiple groove guide is a potential high-power coupling structure that consists of a finite number of parallel rectangular groove guides. A double-groove guide has been analyzed in [9] to assess its utility as a waveguide or a power coupler. In this section we will investigate the guiding and coupling characteristics of a multiple groove guide by deriving its dispersion relation. A theoretical analysis given in this section is an extension of the single groove case discussed in Sect. 4.2. Consider a multiple rectangular groove guide (N: number of groove guides) where the TE wave propagates along the z-direction with $\boldsymbol{H}(x,y,z) = \boldsymbol{H}(x,y)\mathrm{e}^{\mathrm{i}\beta z}$. In regions (I) ($-d < y < 0$), (II) ($0 < y < b$), and (III) ($b < y < b+d$), the H_z components are

$$H_z^I(x,y) = \sum_{n=0}^{N-1} \sum_{m=0}^{\infty} q_m^n \cos a_m (x - nT) \cos \xi_m (y + d)$$
$$\cdot [u(x - nT) - u(x - nT - 2a)] \tag{1}$$

$$H_z^{II}(x,y) = \frac{1}{2\pi} \int_{-\infty}^{\infty} \left(\widetilde{H}_z^+ e^{i\eta y} + \widetilde{H}_z^- e^{-i\eta y} \right) e^{-i\zeta x} d\zeta \tag{2}$$

$$H_z^{III}(x,y) = \sum_{n=0}^{N-1} \sum_{m=0}^{\infty} s_m^n \cos a_m (x - nT) \cos \xi_m (y - b - d)$$
$$\cdot [u(x - nT) - u(x - nT - 2a)] \tag{3}$$

where $a_m = m\pi/(2a)$, $\xi_m = \sqrt{k^2 - a_m^2 - \beta^2}$, $\eta = \sqrt{k^2 - \zeta^2 - \beta^2}$, $k = 2\pi/\lambda_0$, and $u(\cdot)$ is a unit step function.

Applying the Fourier transform to the E_x continuity at $y = 0$ yields

$$\widetilde{H}_z^+ - \widetilde{H}_z^- = - \sum_{n=0}^{N-1} \sum_{m=0}^{\infty} \frac{1}{i\eta} q_m^n \xi_m \sin(\xi_m d) G_m^n(\zeta) \tag{4}$$

where

$$G_m^n(\zeta) = \frac{i\zeta[1 - (-1)^m e^{i\zeta 2a}]}{\zeta^2 - a_m^2} e^{i\zeta nT} . \tag{5}$$

Similarly the E_x continuity at $y = b$ gives

$$\widetilde{H}_z^+ e^{i\eta b} - \widetilde{H}_z^- e^{-i\eta b} = \sum_{n=0}^{N-1} \sum_{m=0}^{\infty} \frac{1}{i\eta} s_m^n \xi_m \sin(\xi_m d) G_m^n(\zeta) . \tag{6}$$

We multiply the H_z continuity at $y = 0$ by $\cos a_l(x - pT)$, $(p = 0, \dots, N-1)$, and integrate over $(pT < x < pT + 2a)$ to get

$$\boxed{\begin{aligned} &\sum_{n=0}^{N-1} \sum_{m=0}^{\infty} \left\{ q_m^n [\xi_m \sin(\xi_m d) I_1 + a \cos(\xi_m d) \delta_{ml} \delta_{np} \varepsilon_m] \right. \\ &\left. + s_m^n \xi_m \sin(\xi_m d) I_2 \right\} = 0 \end{aligned}} \tag{7}$$

where

$$I_1 = \frac{1}{2\pi} \int_{-\infty}^{\infty} \frac{\cot(\eta b)}{\eta} G_m^n(\zeta) G_l^p(-\zeta) d\zeta \tag{8}$$

$$I_2 = \frac{1}{2\pi} \int_{-\infty}^{\infty} \frac{\csc(\eta b)}{\eta} G_m^n(\zeta) G_l^p(-\zeta) d\zeta . \tag{9}$$

Note that I_1 and I_2 are transformed into rapidly-convergent series

$$I_1 = a \frac{\varepsilon_m \delta_{ml} \delta_{np}}{\xi_m \tan(\xi_m b)} - \frac{\mathrm{i}}{b} \sum_{v=0}^{\infty} \frac{\zeta_v A_1}{\varepsilon_v (\zeta_v^2 - a_m^2)(\zeta_v^2 - a_l^2)} \tag{10}$$

$$I_2 = a \frac{\varepsilon_m \delta_{ml} \delta_{np}}{\xi_m \sin(\xi_m b)} - \frac{\mathrm{i}}{b} \sum_{v=0}^{\infty} (-1)^v \frac{\zeta_v A_1}{\varepsilon_v (\zeta_v^2 - a_m^2)(\zeta_v^2 - a_l^2)} \tag{11}$$

$$\begin{aligned} A_1 = {} & [(-1)^{m+l} + 1] \exp(\mathrm{i}\zeta_v |n - p|T) \\ & - (-1)^m \exp[\mathrm{i}\zeta_v |(n-p)T + 2a|] \\ & - (-1)^l \exp[\mathrm{i}\zeta_v |(n-p)T - 2a|] \end{aligned} \tag{12}$$

where $\zeta_v = \sqrt{k^2 - (v\pi/b)^2 - \beta^2}$. Similarly the E_x and H_z continuities at $y = b$ between regions (II) and (III) yield

$$\begin{aligned} \sum_{n=0}^{N-1} \sum_{m=0}^{\infty} & \Big\{ q_m^n \xi_m \sin(\xi_m d) I_2 \\ & + s_m^n [\xi_m \sin(\xi_m d) I_1 + a \cos(\xi_m d) \delta_{ml} \delta_{np} \varepsilon_m] \Big\} = 0 \, . \end{aligned} \tag{13}$$

A dispersion relation is formed from (7) and (13) as

$$\begin{vmatrix} \Psi_1 & \Psi_2 \\ \Psi_2 & \Psi_1 \end{vmatrix} = 0 \tag{14}$$

where the elements of Ψ_1 and Ψ_2 are

$$\psi_{1,ml}^{np} = \xi_m \sin(\xi_m d) I_1 + a \cos(\xi_m d) \delta_{ml} \delta_{np} \varepsilon_m \tag{15}$$

$$\psi_{2,ml}^{np} = \xi_m \sin(\xi_m d) I_2 \, . \tag{16}$$

When $N = 1$ (single-groove case), (14) reduces to (34) in Subsect. 4.2.2. When $N = 2$ (double-groove case), (14) reduces to a dominant-mode solution $(m=0)$

$$\psi_{1,00}^{00} - \psi_{2,00}^{00} = \pm(\psi_{1,00}^{10} - \psi_{2,00}^{10}) \tag{17}$$

where $\pm$ sign corresponds to the TE_{12} and TE_{11} waves, respectively.

4.4 Corrugated Coaxial Line [10]

A study of electromagnetic wave scattering from a corrugated coaxial line is of interest due to practical applications for microwave filter design [11,12]. In this section we will analyze scattering from multiple grooves in the inner conductor of a coaxial line. An incident TEM wave propagates along a coaxial line whose inner conductor has an N number of grooves. In region (I) ($b < r < a$) the H-field consists of the incident and scattered fields

$$H_{\phi I}^i(r, z) = \frac{\mathrm{e}^{\mathrm{i}k_1 z}}{\eta_1 r} \tag{1}$$

$$H_{\phi I}(r, z) = \frac{\mathrm{i}\omega\epsilon_1}{2\pi} \int_{-\infty}^{\infty} \frac{1}{\kappa} \widetilde{E}(\zeta) R'(\kappa r) \mathrm{e}^{-\mathrm{i}\zeta z} \mathrm{d}\zeta \tag{2}$$

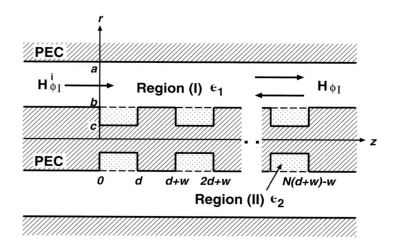

Fig. 4.4. Corrugated coaxial line

where $\kappa = \sqrt{k_1^2 - \zeta^2}$, $\eta_1 = \sqrt{\mu_0/\epsilon_1}$, $R(\kappa r) = J_0(\kappa r)N_0(\kappa a) - N_0(\kappa r)J_0(\kappa a)$, $k_1 = \omega\sqrt{\mu_0\epsilon_1} = \omega\sqrt{\mu_0\epsilon_{r1}\epsilon_0}$, and $R'(\kappa r) = \dfrac{dR(\kappa r)}{d(\kappa r)}$. Note that (r, ϕ, z) are circular cylindrical coordinates, and $J_0(\cdot)$ and $N_0(\cdot)$ represent the 0th order Bessel functions of the first and second kinds, respectively. In region (II) $(c < r < b)$ with permittivity ϵ_2 the H-field is

$$H_{\phi II}(r, z) = i\omega\epsilon_2 \sum_{n=0}^{N-1} \sum_{m=0}^{\infty} \frac{p_m^n}{\kappa_m} R_0'(\kappa_m r) \cos a_m[z - s(n)]$$
$$\cdot \{u[z - s(n)] - u[z - d - s(n)]\} \qquad (3)$$

where

$$R_0(\kappa_m r) = \begin{cases} J_0(\kappa_m r)N_0(\kappa_m c) - N_0(\kappa_m r)J_0(\kappa_m c), & c \neq 0 \\ J_0(\kappa_m r), & c = 0 \end{cases} \qquad (4)$$

$$s(n) = n(d + w) \qquad (5)$$

$\kappa_m = \sqrt{k_2^2 - a_m^2}$, $k_2 = \omega\sqrt{\mu_0\epsilon_{r2}\epsilon_0}$, $a_m = m\pi/d$, $u(\cdot)$ is a unit step function, N is the number of grooves, and p_m^n is an unknown coefficient associated with the mth mode in the $(n + 1)$th groove.

The tangential E-field continuity is

$$E_{zI}(b, z) = \begin{cases} E_{zII}(b, z), & s(n) < z < d + s(n), \quad n = 0, 1, \dots, N - 1 \\ 0, & \text{otherwise} . \end{cases} \qquad (6)$$

Applying the Fourier transform to (6) yields

$$\widetilde{E}(\zeta) = \sum_{n=0}^{N-1} \sum_{m=0}^{\infty} \frac{p_m^n R_0(\kappa_m b)}{R(\kappa b)} G_m^n(\zeta) \tag{7}$$

where

$$G_m^n(\zeta) = \frac{-\mathrm{i}\zeta[(-1)^m \mathrm{e}^{\mathrm{i}\zeta d} - 1]}{\zeta^2 - a_m^2} \exp[\mathrm{i}\zeta s(n)] \ . \tag{8}$$

The H_ϕ field continuity at $r = b$ over the groove apertures requires

$$H_{\phi I}^i(b, z) + H_{\phi I}(b, z) = H_{\phi II}(b, z) \ . \tag{9}$$

We multiply (9) by $\cos a_l[z - s(p)]$ (where $l = 0, 1, 2, \ldots$ and $p = 0, 1, 2, \ldots, N-1$) and integrate from $s(p)$ to $s(p) + d$ to obtain

$$\boxed{\begin{aligned} &\sum_{n=0}^{N-1} \sum_{m=0}^{\infty} \left[R_0(\kappa_m b) I - \frac{d\epsilon_{r2}\varepsilon_m}{2\epsilon_{r1}\kappa_m} R_0'(\kappa_m b) \delta_{ml}\delta_{np} \right] p_m^n \\ &= \frac{\mathrm{i}}{k_1 b} G_l^p(k_1) \end{aligned}} \tag{10}$$

where

$$I = \frac{1}{2\pi} \int_{-\infty}^{\infty} \frac{R'(\kappa b)}{\kappa R(\kappa b)} G_m^n(\zeta) G_l^p(-\zeta) \mathrm{d}\zeta \ . \tag{11}$$

Using the residue calculus, it is possible to transform I into rapidly-convergent series that are efficient for numerical computation. The result is

$$\begin{aligned} I = &\ \frac{\varepsilon_m d}{2} \frac{R'(\kappa b)}{\kappa R(\kappa b)} \delta_{ml}\delta_{np} \Big|_{\zeta=a_m} \\ &- \sum_{j=1}^{\infty} \frac{2\mathrm{i}\zeta[1 - (-1)^m \mathrm{e}^{\mathrm{i}\zeta d}]}{b[1 - J_0^2(\kappa b)/J_0^2(\kappa a)](\zeta^2 - a_m^2)(\zeta^2 - a_l^2)} \Big|_{\zeta=\zeta_j} \\ &- \frac{\mathrm{i}k_1[1 - (-1)^m \mathrm{e}^{\mathrm{i}k_1 d}]}{b \ln(b/a)(k_1^2 - a_m^2)(k_1^2 - a_l^2)}, \\ &\qquad n = p \text{ and } m + l \text{ is even} \end{aligned} \tag{12}$$

$$= 0, \qquad\qquad n = p \text{ and } m + l \text{ is odd} \tag{13}$$

$$\begin{aligned} = &-\sum_{j=1}^{\infty} \frac{\mathrm{i}\zeta[(-1)^m \mathrm{e}^{\mathrm{i}\zeta d} - 1][(-1)^l \mathrm{e}^{-\mathrm{i}\zeta d} - 1] \exp[\mathrm{i}\zeta[s(n) - s(p)]]}{b[1 - J_0^2(\kappa b)/J_0^2(\kappa a)](\zeta^2 - a_m^2)(\zeta^2 - a_l^2)} \Big|_{\zeta=\zeta_j} \\ &- \frac{\mathrm{i}k_1[(-1)^m \mathrm{e}^{\mathrm{i}k_1 d} - 1][(-1)^l \mathrm{e}^{-\mathrm{i}k_1 d} - 1] \exp[\mathrm{i}k_1[s(n) - s(p)]]}{2b \ln(b/a)(k_1^2 - a_m^2)(k_1^2 - a_l^2)}, \\ &\qquad n > p \end{aligned} \tag{14}$$

$$= \sum_{j=1}^{\infty} \frac{\mathrm{i}\zeta[(-1)^m \mathrm{e}^{\mathrm{i}\zeta d} - 1][(-1)^l \mathrm{e}^{-\mathrm{i}\zeta d} - 1] \exp[\mathrm{i}\zeta[s(n) - s(p)]]}{b[1 - J_0^2(\kappa b)/J_0^2(\kappa a)](\zeta^2 - a_m^2)(\zeta^2 - a_l^2)} \Bigg|_{\zeta = -\zeta_j}$$

$$- \frac{\mathrm{i}k_1[(-1)^m \mathrm{e}^{-\mathrm{i}k_1 d} - 1][(-1)^l \mathrm{e}^{\mathrm{i}k_1 d} - 1] \exp[-\mathrm{i}k_1[s(n) - s(p)]]}{2b \ln (b/a)(k_1^2 - a_m^2)(k_1^2 - a_l^2)}, \quad (15)$$

$$n < p \,.$$

Note that ζ_j is a root of the characteristic equation $R(\kappa b) = 0$. The first term in (12) represents a contribution of mode-scattering in region (II), and the second and third terms in (12) account for the effects of higher and TEM mode-scattering in a coaxial line, respectively. The field reflection coefficient Γ at $z = 0$ and the transmission coefficient T at $z = (N - 1)(d + w) + d = N(d + w) - w$ are

$$\Gamma = -\frac{H_{\phi I}}{H_{\phi I}^i} = -L_0^- \tag{16}$$

$$T = \frac{H_{\phi I}^i + H_{\phi I}}{H_{\phi I}^i} = 1 + L_0^+ \tag{17}$$

$$L_0^{\pm} = \mp \sum_{n=0}^{N-1} \sum_{m=0}^{\infty} \frac{p_m^n R_0(\kappa_m b) \mathrm{i}k_1 [1 - (-1)^m \mathrm{e}^{\mp \mathrm{i}k_1 d}]}{2 \ln (b/a)(k_1^2 - a_m^2)} \exp[\mp \mathrm{i}k_1 s(n)] \,. \tag{18}$$

4.5 Coaxial Line with a Gap [13]

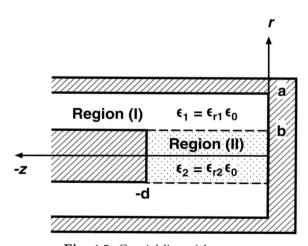

Fig. 4.5. Coaxial line with a gap

A study of TEM wave scattering from a coaxial line terminated by a gap is of practical interest in microwave circuit design [14,15]. In this section we will analyze scattering from a coaxial line terminated by a gap, where the inner conductor of a coaxial line is partially removed and replaced by a dielectric medium with permittivity $\epsilon_2 = \epsilon_0 \epsilon_{r2}$. A theoretical analysis shown in this section is similar to that discussed in Sect. 4.4. An incident TEM wave, $H_\phi^i = \dfrac{e^{ik_1 z}}{\eta_1 r}$, propagates from left inside a coaxial line. In region (I) the scattered H-field is

$$H_{\phi I}(r, z) = \frac{e^{-ik_1 z}}{\eta_1 r} + i\omega\epsilon_1 \frac{2}{\pi} \int_0^\infty \frac{1}{\kappa} \widetilde{E}_I(\zeta) R'(\kappa r) \cos(\zeta z) d\zeta \tag{1}$$

where $\kappa = \sqrt{k_1^2 - \zeta^2}$, $k_1 = \omega\sqrt{\mu\epsilon_1} = 2\pi/\lambda_1$, $\eta_1 = \sqrt{\mu/\epsilon_1}$, $R(\kappa r) = J_0(\kappa r)N_0(\kappa a) - N_0(\kappa r)J_0(\kappa a)$, and $R'(\cdot) = dR(\cdot)/d(\cdot)$. Note that $J_0(\cdot)$ and $N_0(\cdot)$ are the 0th order Bessel and Neumann functions, respectively. In region (II) the scattered H-field is

$$H_{\phi II}(r, z) = i\omega\epsilon_2 \sum_{m=0}^\infty \frac{1}{\kappa_m} R_0'(\kappa_m r) \cos(a_m z) \tag{2}$$

where $R_0(\kappa_m r) = p_m J_0(\kappa_m r)$, $R_0'(\cdot) = dR_0(\cdot)/d(\cdot)$, $\kappa_m = \sqrt{k_2^2 - a_m^2}$, $k_2 = \omega\sqrt{\mu\epsilon_2} = 2\pi/\lambda_2$, and $a_m = m\pi/d$.

The E_z-field continuity at $r = b$ is

$$E_{zI}(b, z) = \begin{cases} E_{zII}(b, z), & -d < z < 0 \\ 0, & \text{otherwise} . \end{cases} \tag{3}$$

Applying the Fourier cosine transform to (3) gives

$$\widetilde{E}_I(\zeta) = \sum_{m=0}^\infty \frac{R_0(\kappa_m b)}{R(\kappa b)} \Xi_m(\zeta) \tag{4}$$

where

$$\Xi_m(\zeta) = \frac{(-1)^m \zeta \sin(\zeta d)}{(\zeta^2 - a_m^2)} . \tag{5}$$

The H_ϕ-field continuity at $r = b$ for $(-d < z < 0)$ is

$$H_{\phi I}(b, z) + \frac{e^{ik_1 z}}{\eta_1 b} = H_{\phi II}(b, z) . \tag{6}$$

Multiplying (6) by $\cos(a_s z)$ (where $s = 0, 1, 2, \dots$) and integrating with respect to z from $-d$ to 0, we obtain

$$\boxed{\sum_{m=0}^\infty \left[J_0(\kappa_m b) I_{ms} + \frac{\varepsilon_m d\epsilon_{r2}}{2\kappa_m \epsilon_{r1}} J_1(\kappa_m b)\delta_{ms} \right] p_m = \frac{2i}{k_1 b} \Xi_s(k_1)} \tag{7}$$

where δ_{ms} is the Kronecker delta and

$$I_{ms} = \frac{2}{\pi} \int_0^\infty \frac{R'(\kappa b)}{\kappa R(\kappa b)} \Xi_m(\zeta)\Xi_s(\zeta)\,\mathrm{d}\zeta \ . \tag{8}$$

We transform I_{ms} into a fast-convergent series

$$
\begin{aligned}
I_{ms} = {}& \frac{\varepsilon_m d R'(\kappa b)}{2\kappa R(\kappa b)} \delta_{ms}\Bigg|_{\zeta=a_m} \\
& - \frac{\mathrm{i}(-1)^{m+s} k_1 (1 - \mathrm{e}^{2\mathrm{i}k_1 d})}{2b \ln(b/a)(k_1^2 - a_m^2)(k_1^2 - a_s^2)} \\
& - \sum_{n=1}^\infty \frac{\mathrm{i}(-1)^{m+s}\zeta(1 - \mathrm{e}^{2\mathrm{i}\zeta d})}{b[1 - J_0^2(\kappa b)/J_0^2(\kappa a)](\zeta^2 - a_m^2)(\zeta^2 - a_s^2)}\Bigg|_{\zeta=\zeta_n}
\end{aligned}
\tag{9}
$$

where ζ_n is given by $R(\kappa b)\,|_{\zeta=\zeta_n} = 0$.

The scattered field for $(z < -d)$ in region (I) is

$$H_{\phi I}(r, z) = \frac{(1 + L_0)\mathrm{e}^{-\mathrm{i}k_1 z}}{\eta_1 r} - \sum_{n=1}^\infty L_n(\zeta) R'(\kappa r)\mathrm{e}^{\mathrm{i}\zeta z}\Bigg|_{\zeta=-\zeta_n} \tag{10}$$

where

$$L_0 = \frac{k_1 \sin(k_1 d)}{\ln(b/a)} \sum_{m=0}^\infty \frac{(-1)^m R_0(\kappa_m b)}{k_1^2 - a_m^2} \tag{11}$$

$$L_n(\zeta) = \frac{2\omega\epsilon_1 \sin(\zeta d)}{bR'(\kappa b)[1 - J_0^2(\kappa b)/J_0^2(\kappa a)]} \sum_{m=0}^\infty \frac{(-1)^m R_0(\kappa_m b)}{\zeta^2 - a_m^2} \ . \tag{12}$$

The field reflection coefficient Γ at $z = -d$ is

$$\Gamma = -\frac{H_{\phi I}}{H_\phi^i}\Bigg|_{z=-d} = -(1 + L_0)\mathrm{e}^{\mathrm{i}2k_1 d} \ . \tag{13}$$

4.6 Coaxial Line with a Cavity [16]

A coaxial line-radial line junction has been widely used as an antenna feeder or a power combiner and was studied in [17] to obtain its equivalent circuit representation. A coaxial line with a material-filled cavity is useful as a microwave sensor to estimate the material permittivity but its theoretical investigation is very little. In this section we will investigate TEM-wave scattering from an infinitely-long coaxial line with a dielectric-filled circular cylindrical cavity. The geometry of a scattering problem is shown in Fig. 4.6. A theoretical analysis given in this section is similar to the discussions in Sects. 4.4 and 4.5. In region (I) $(a < r < b$ and permittivity: $\epsilon_0\epsilon_{r1})$ the total E-field is assumed to have incident and scattered components

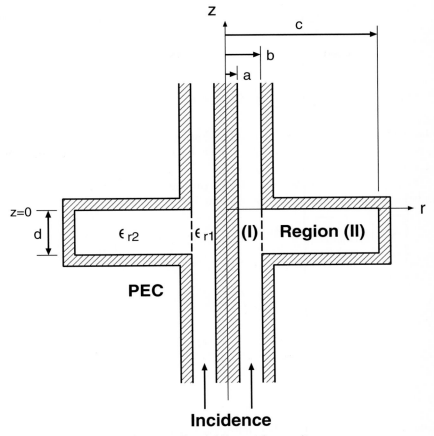

Fig. 4.6. Coaxial line with a cavity

$$E_{rI}^{i}(r,z) = \frac{e^{ik_1 z}}{r} \tag{1}$$

$$E_{zI}(r,z) = \frac{1}{2\pi} \int_{-\infty}^{\infty} \widetilde{E}_I(\zeta) R(\kappa r) e^{-i\zeta z} d\zeta \tag{2}$$

$$E_{rI}(r,z) = \frac{-1}{2\pi} \int_{-\infty}^{\infty} \widetilde{E}_I(\zeta) i\zeta R'(\kappa r)\kappa^{-1} e^{-i\zeta z} d\zeta \tag{3}$$

$$H_{\phi I}(r,z) = i\omega\epsilon_0\epsilon_{r1} \frac{1}{2\pi} \int_{-\infty}^{\infty} \widetilde{E}_I(\zeta) R'(\kappa r)\kappa^{-1} e^{-i\zeta z} d\zeta \tag{4}$$

where $R(\kappa r) = J_0(\kappa r)N_0(\kappa a) - N_0(\kappa r)J_0(\kappa a)$, $R'(\cdot) = dR(\cdot)/d(\cdot)$, $k_1 = \omega\sqrt{\mu_0\epsilon_0\epsilon_{r1}} = 2\pi/\lambda_1$: wavenumber, $\kappa = \sqrt{k_1^2 - \zeta^2}$, and $J_0(\cdot)$ and $N_0(\cdot)$ denote the 0th order Bessel and Neumann functions, respectively. In region (II) ($b < r < c$, $-d < z < 0$, and permittivity: $\epsilon_0\epsilon_{r2}$) the scattered fields are

$$E_{zII}(r,z) = \sum_{m=0}^{\infty} p_m Q(\kappa_m r)\cos(a_m z) \tag{5}$$

$$E_{rII}(r,z) = -\sum_{m=0}^{\infty} p_m a_m Q'(\kappa_m r)\kappa_m^{-1}\sin(a_m z) \tag{6}$$

$$H_{\phi II}(r,z) = i\omega\epsilon_0\epsilon_{r2}\sum_{m=0}^{\infty} p_m Q'(\kappa_m r)\kappa_m^{-1}\cos(a_m z) \tag{7}$$

where $Q(\kappa_m r) = J_0(\kappa_m r)N_0(\kappa_m c) - N_0(\kappa_m r)J_0(\kappa_m c)$, $\kappa_m = \sqrt{k_2^2 - a_m^2}$, $k_2 = \omega\sqrt{\mu_0\epsilon_0\epsilon_{r2}}$, and $a_m = m\pi/d$. If a cavity radius c is chosen to be infinite (∞), $Q(\kappa_m r)$ is given by $H_0^{(1)}(\kappa_m r)$.

Enforcing the boundary conditions on the E_z-field continuity at $r = b$ and the H_ϕ-field continuity at $r = b$ and $(-d < z < 0)$, we get

$$\sum_{m=0}^{\infty}\left[Q(\kappa_m b)I_{ms} - \frac{d\epsilon_{r2}\varepsilon_m}{2\epsilon_{r1}\kappa_m}Q'(\kappa_m b)\delta_{ms}\right]p_m = \frac{i}{k_1 b}G_s(k_1) \tag{8}$$

where

$$G_s(k_1) = \frac{-ik_1\left[1 - (-1)^s e^{-ik_1 d}\right]}{(k_1^2 - a_s^2)} \tag{9}$$

$$I_{ms} = \frac{d\varepsilon_m R'(\kappa b)}{2\kappa R(\kappa b)}\delta_{ms}\bigg|_{\zeta=a_m} - \frac{G_m(-k_1)}{b\ln(b/a)(k_1^2 - a_s^2)}$$
$$-\sum_{n=1}^{\infty}\frac{2G_m(-\zeta)}{b[1 - J_0^2(\kappa b)/J_0^2(\kappa a)](\zeta^2 - a_s^2)}\bigg|_{\zeta=\zeta_n}. \tag{10}$$

Note that δ_{ms} is the Kronecker delta, $\varepsilon_m = 2$ $(m{=}0)$, and 1 $(m{=}1,2,\dots)$.

The field reflection (Γ) and transmission (T) coefficients in region (I) are shown to be

$$\Gamma = \frac{E_{rI}^r(r,z)}{E_{rI}^i(r,z)} = -\sum_{m=0}^{\infty}\frac{p_m Q(\kappa_m b)}{2\ln(b/a)}G_m(k_1)e^{-2ik_1 z} \tag{11}$$

$$T = \frac{E_{rI}^t(r,z)}{E_{rI}^i(r,z)} = 1 + \sum_{m=0}^{\infty}\frac{p_m Q(\kappa_m b)}{2\ln(b/a)}G_m(-k_1). \tag{12}$$

It is also interesting to consider scattering from a shorted coaxial line with a cavity as shown in Fig. 4.7. When the coaxial line is shorted at $z = 0$, the field reflection coefficient Γ is given as

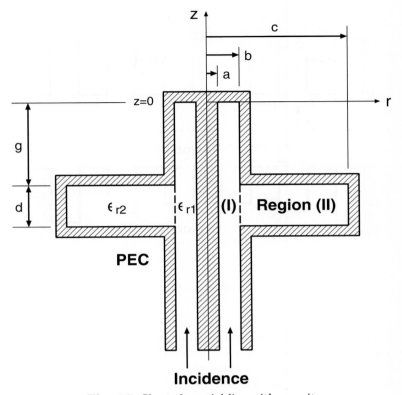

Fig. 4.7. Shorted coaxial line with a cavity

$$
\Gamma = \frac{E^r_{rI}(r,z)}{E^i_{rI}(r,z)}
$$

$$
= -\left\{1 + \sum_{m=0}^{\infty} \frac{p_m Q(\kappa_m b) k_1}{\ln(b/a)(k_1^2 - a_m^2)}\right.
$$

$$
\left. \cdot [(-1)^m \sin(k_1(d+g)) - \sin(k_1 g)]\right\} e^{-2ik_1 z} . \tag{13}
$$

4.7 Corrugated Circular Cylinder [18]

A study of electromagnetic wave propagation within a conducting corrugated circular cylinder is important for the design of antenna feed horn and gyrotron. The radiation characteristics from an open-ended corrugated circular waveguide should have low sidelobes, low cross polarization levels, axial beam symmetry, and low attenuation [19-22] for antenna feed application. In

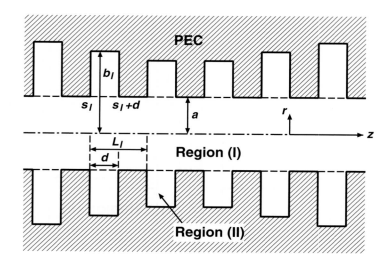

Fig. 4.8. Corrugated conducting circular cylinder

this section we will analyze scattering from a corrugated circular cylindrical waveguide that supports the propagation of a hybrid wave (a combination of the TE and TM waves). This implies that the boundary conditions need to be enforced in terms of the TE and TM waves simultaneously. An incident TE_{11} wave propagates within a perfectly conducting cylindrical waveguide that is filled with a dielectric medium with permittivity ϵ and permeability μ. For simplicity we introduce the normalized fields $\boldsymbol{E}(t')$ and $\boldsymbol{H}(t')$ that are given by $\boldsymbol{E}(t) = \sqrt{\mu}\boldsymbol{E}(t')$, $\boldsymbol{H}(t) = -i\sqrt{\epsilon}\boldsymbol{H}(t')$, and $t = \sqrt{\mu\epsilon}t'$. Maxwell's equations for the normalized fields $\boldsymbol{E}(t')$ and $\boldsymbol{H}(t')$ are therefore written in time domain as

$$\nabla \times \boldsymbol{E}(t') = i\frac{\partial \boldsymbol{H}(t')}{\partial t'} \tag{1}$$

$$\nabla \times \boldsymbol{H}(t') = i\frac{\partial \boldsymbol{E}(t')}{\partial t'} . \tag{2}$$

It is possible to represent the fields in terms of the Hertz vector potentials as

$$\boldsymbol{E} = \nabla \times \nabla \times \boldsymbol{\Pi}_{tm} + \omega\nabla \times \boldsymbol{\Pi}_{te} \equiv \boldsymbol{E}_{tm} + \boldsymbol{E}_{te} \tag{3}$$

$$\boldsymbol{H} = \nabla \times \nabla \times \boldsymbol{\Pi}_{te} + \omega\nabla \times \boldsymbol{\Pi}_{tm} \equiv \boldsymbol{H}_{te} + \boldsymbol{H}_{tm} \tag{4}$$

where $\boldsymbol{\Pi}_{tm}$ and $\boldsymbol{\Pi}_{te}$ are the electric (TM) and magnetic (TE) Hertz vector potentials, respectively. In region (I) $(r < a)$ the scattered field is represented by the Hertz vector potentials

$$\boldsymbol{\Pi}_{tm}^{s} = \hat{z}\cos\phi \int_{-\infty}^{\infty} \widetilde{\Pi}_{tm}(\zeta) J_1(\kappa r) \mathrm{e}^{\mathrm{i}\zeta z}\mathrm{d}\zeta \tag{5}$$

$$\boldsymbol{\Pi}_{te}^{s} = \hat{z}\sin\phi \int_{-\infty}^{\infty} \widetilde{\Pi}_{te}(\zeta) J_1(\kappa r) \mathrm{e}^{\mathrm{i}\zeta z}\mathrm{d}\zeta \tag{6}$$

where $\kappa = \sqrt{\omega^2 - \zeta^2}$ and $J_1(\cdot)$ is the first-order Bessel function. In region (II) ($r > a$ and $s_l < z < s_l + d$, corrugation region) the fields in the lth groove are

$$\boldsymbol{\Pi}_{tm}^{d} = -\hat{z}\sum_{n=0}^{\infty} E_{nl} P_{nl}(\kappa_n r)\cos\phi\cos\frac{n\pi}{d}(z - s_l) \tag{7}$$

$$\boldsymbol{\Pi}_{te}^{d} = \hat{z}\sum_{n=1}^{\infty} H_{nl} Q_{nl}(\kappa_n r)\sin\phi\sin\frac{n\pi}{d}(z - s_l) \tag{8}$$

where

$$P_{nl}(\kappa_n r) = J_1(\kappa_n r)N_1(\kappa_n b_l) - J_1(\kappa_n b_l)N_1(\kappa_n r) \tag{9}$$

$$Q_{nl}(\kappa_n r) = J_1(\kappa_n r)N_1'(\kappa_n b_l) - J_1'(\kappa_n b_l)N_1(\kappa_n r) \tag{10}$$

$\kappa_n = \sqrt{\omega^2 - (n\pi/d)^2}$, and b_l is a radius at the lth corrugation. The incident field takes the form of

$$\boldsymbol{\Pi}_{te}^{i} = \hat{z} J_1(\beta_c r)\sin\phi\, \mathrm{e}^{\mathrm{i}\beta z} \tag{11}$$

where $\beta = \sqrt{\omega^2 - \beta_c^2}$, $\beta_c = \chi_1'/a$, and χ_1' is the first root of derivative of the Bessel function $J_1(\cdot)$.

The tangential E_z continuity at $r = a$ requires

$$(\boldsymbol{E}_i + \boldsymbol{E}_{tm}^{\mathrm{I}} + \boldsymbol{E}_{te}^{\mathrm{I}})\cdot\hat{z}$$
$$= \begin{cases} (\boldsymbol{E}_{tm}^{\mathrm{II}} + \boldsymbol{E}_{te}^{\mathrm{II}})\cdot\hat{z}, & s_l < z < s_{l+1}\ (l = 1,2,3,\dots) \\ 0, & \text{otherwise} \end{cases} \tag{12}$$

where a symbol $(\cdot)$ denotes the dot product of two vectors. Applying the Fourier transform to (12) and solving for $\widetilde{\Pi}_{tm}(\zeta)$, we get

$$\widetilde{\Pi}_{tm}(\zeta) = \frac{1}{2\pi\kappa^2 J_1(\kappa a)}\sum_{n,l} E_{nl} P_{nl}(k_n a)\kappa_n^2 \exp(-\mathrm{i}\zeta s_l)\mathrm{i}\zeta \Xi_n(\zeta) \tag{13}$$

where

$$\Xi_n(\zeta) = \frac{[1 - (-1)^n \mathrm{e}^{-\mathrm{i}\zeta d}]}{\zeta^2 - (n\pi/d)^2}. \tag{14}$$

The tangential E_ϕ continuity at $r = a$ requires

$$(\boldsymbol{E}_i + \boldsymbol{E}_{tm}^{\mathrm{I}} + \boldsymbol{E}_{te}^{\mathrm{I}})\cdot\hat{\phi}$$
$$= \begin{cases} (\boldsymbol{E}_{tm}^{\mathrm{II}} + \boldsymbol{E}_{te}^{\mathrm{II}})\cdot\hat{\phi}, & s_l < z < s_{l+1}\ (l = 1,2,3,\dots) \\ 0, & \text{otherwise}\ . \end{cases} \tag{15}$$

Applying the Fourier transform to (15), substituting $\widetilde{\Pi}_{tm}(\zeta)$ into (15), and solving for $\widetilde{\Pi}_{te}(\zeta)$, we obtain

$$\widetilde{\Pi}_{te}(\zeta)\kappa\omega J_1'(\kappa a) =$$

$$-\frac{1}{2\pi}\sum_{n,l} E_{nl}\frac{P_{nl}(\kappa_n a)}{a}\left[\left(\frac{n\pi}{d}\right)^2 - \left(\frac{\zeta\kappa_n}{\kappa}\right)^2\right]\exp(-\mathrm{i}\zeta s_l)\varXi_n(\zeta)$$

$$-\frac{1}{2\pi}\sum_{n,l} H_{nl}\omega\kappa_n Q_{nl}'(\kappa_n a)\frac{n\pi}{d}\exp(-\mathrm{i}\zeta s_l)\varXi_n(\zeta) . \tag{16}$$

The tangential H_z continuity at $r = a$ in the qth groove

$$(\boldsymbol{H}_i + \boldsymbol{H}_{tm}^{\mathrm{I}} + \boldsymbol{H}_{te}^{\mathrm{I}})\cdot\hat{z} = (\boldsymbol{H}_{tm}^{\mathrm{II}} + \boldsymbol{H}_{te}^{\mathrm{II}})\cdot\hat{z} \tag{17}$$

is rewritten as

$$\beta_c^2 J_1(\beta_c a)\mathrm{e}^{\mathrm{i}\beta z} + \int_{-\infty}^{\infty}\widetilde{\Pi}_{te}(\zeta)J_1(\kappa a)\mathrm{e}^{\mathrm{i}\zeta z}\mathrm{d}\zeta$$

$$= \sum_{n,l} H_{nl}Q_{nl}(k_n a)\kappa_n^2\sin\frac{n\pi}{d}(z - s_l) . \tag{18}$$

Multiplying (18) by $\sin\frac{m\pi}{d}(z - s_q)$ and integrating from s_q to $s_q + d$ with respect to z gives

$$
\boxed{
\begin{aligned}
&-\beta_c^2 J_1(\beta_c a)\exp(\mathrm{i}\beta s_l)\left(\frac{m\pi}{d}\right)\varXi_m(-\beta) \\
&+\sum_{n,l} E_{nl}\frac{P_{nl}(\kappa_n a)}{\omega a}\left[\left(\frac{n\pi}{d}\right)^2 I_2 - \kappa_n^2 I_1\right]\frac{m\pi}{d} \\
&+\sum_{n,l} H_{nl}\kappa_n Q_{nl}'(\kappa_n a)\frac{nm\pi^2}{d^2}I_2 \\
&= H_{mq}Q_{mq}(\kappa_m a)\kappa_m^2\frac{d}{2}(1 - \delta_{m0})
\end{aligned}
}
\tag{19}
$$

where

$$I_1 = \frac{1}{2\pi}\int_{-\infty}^{\infty}\frac{\zeta^2 J_1(\kappa a)}{\kappa J_1'(\kappa a)}\exp[\mathrm{i}\zeta(s_q - s_l)]\varXi_m(-\zeta)\varXi_n(\zeta)\mathrm{d}\zeta \tag{20}$$

$$I_2 = \frac{1}{2\pi}\int_{-\infty}^{\infty}\frac{\kappa J_1(\kappa a)}{J_1'(\kappa a)}\exp[\mathrm{i}\zeta(s_q - s_l)]\varXi_m(-\zeta)\varXi_n(\zeta)\mathrm{d}\zeta . \tag{21}$$

Similarly the tangential H_ϕ continuity at $r = a$ in the lth groove

$$(\boldsymbol{H}_i + \boldsymbol{H}_{tm}^{\mathrm{I}} + \boldsymbol{H}_{te}^{\mathrm{I}})\cdot\hat{\phi} = (\boldsymbol{H}_{tm}^{\mathrm{II}} + \boldsymbol{H}_{te}^{\mathrm{II}})\cdot\hat{\phi} \tag{22}$$

is rewritten as

$$\frac{i\beta}{a} J_1(\beta_c a) e^{i\beta z} + \int_{-\infty}^{\infty} \left[\tilde{\Pi}_{te}(\zeta) i\zeta \frac{J_1(\kappa a)}{a} - \tilde{\Pi}_{tm}(\zeta) \kappa \omega J_1'(\kappa a) \right] e^{i\zeta z} d\zeta$$

$$= \sum_{n,l} \left[H_{nl} \frac{n\pi}{d} \frac{Q_{nl}(\kappa_n a)}{r} + E_{nl} \kappa_n \omega P_{nl}'(\kappa_n a) \right] \cos \frac{n\pi}{d}(z - s_l) . \quad (23)$$

Multiplying (23) by $\cos \dfrac{m\pi}{d}(z - s_q)$ and integrating from s_q to $s_q + d$ with respect to z, we get

$$-\frac{\beta^2}{a} J_1(\beta_c a) \exp(i\beta s_q) \Xi_m(-\beta) + \sum_{n,l} H_{nl} \frac{n\pi \kappa_n}{ad} Q_{nl}'(\kappa_n a) I_1$$

$$+ \sum_{n,l} E_{nl} \frac{P_{nl}(\kappa_n a)}{a^2 \omega} \left[\left(\frac{n\pi}{d} \right)^2 I_1 - \kappa_n^2 I_3 + (\omega \kappa_n a)^2 I_4 \right]$$

$$= \left[H_{mq} \frac{m\pi}{d} \frac{Q_{mq}(\kappa_m a)}{a} + E_{mq} \kappa_m \omega P_{mq}'(\kappa_m a) \right] \frac{d}{2}(1 + \delta_{m0}) \quad (24)$$

where

$$I_3 = \frac{1}{2\pi} \int_{-\infty}^{\infty} \frac{\zeta^4 J_1(\kappa a)}{\kappa^3 J_1'(\kappa a)} \exp[i\zeta(s_q - s_l)] \Xi_m(-\zeta) \Xi_n(\zeta) d\zeta \quad (25)$$

$$I_4 = \frac{1}{2\pi} \int_{-\infty}^{\infty} \frac{\zeta^2 J_1'(\kappa a)}{\kappa J_1(\kappa a)} \exp[i\zeta(s_q - s_l)] \Xi_m(-\zeta) \Xi_n(\zeta) d\zeta . \quad (26)$$

It is possible to evaluate the integrals, I_1, I_2, I_3, and I_4 as fast-convergent series by modifying them in the complex ζ-plane. In view of (24) in Subsect. 1.1.3, the rewritten integrals contain poles due to the Bessel functions and poles due to Ξ_m and Ξ_n when $m = n$ and $q = l$. Evaluating the pole contributions with a residue calculus, we obtain

$$I_1 = -i \sum_j \frac{\zeta_j' J_1(\chi_j')}{a J_1''(\chi_j')} A(\zeta_j')$$

$$+ \delta_{ql} \delta_{mn}(1 + 3\delta_{m0}) \frac{d J_1(\kappa_m a)}{2\kappa_m J_1'(\kappa_m a)} \quad (27)$$

$$I_2 = -i \sum_j \frac{\chi_j'^2 J_1(\chi_j')}{a^3 \zeta_j' J_1''(\chi_j')} A(\zeta_j')$$

$$+ \delta_{ql} \delta_{mn} \frac{\kappa_m J_1(\kappa_m a)}{J_1'(\kappa_m a)} \frac{d^3}{(2m^2\pi^2 - 3\delta_{m0})} \quad (28)$$

$$I_3 = -i \sum_j \frac{a\zeta_j'^3 J_1(\chi_j')}{\chi_j'^2 J_1''(\chi_j')} A(\zeta_j')$$

$$+ \delta_{ql} \delta_{mn} \frac{(m\pi)^2 J_1(\kappa_m a)}{2d\kappa_m^3 J_1'(\kappa_m a)} - i \frac{\omega^3 a}{2} A(\omega) \quad (29)$$

$$I_4 = -i \sum_j \frac{\zeta_j}{a} A(\zeta_j)$$

$$+\delta_{ql}\delta_{mn}(1+3\delta_{m0})\frac{dJ_1'(\kappa_m a)}{2\kappa_m J_1(\kappa_m a)} - i\frac{\omega}{2a}A(\omega) \tag{30}$$

where

$$A(\zeta) = \frac{A_1}{\left[\zeta^2 - (m\pi/d)^2\right]\left[\zeta^2 - (n\pi/d)^2\right]} \tag{31}$$

$$\begin{aligned} A_1 &= [1+(-1)^{m+n}]\exp(i\zeta|s_q - s_l|) \\ &\quad -(-1)^n \exp(i\zeta|s_q - s_l - d|) \\ &\quad -(-1)^m \exp(i\zeta|s_q - s_l + d|) \ . \end{aligned} \tag{32}$$

We note that $\zeta_j = \sqrt{\omega^2 - (\chi_j/a)^2}$, $\zeta_j' = \sqrt{\omega^2 - (\chi_j'/a)^2}$, and χ_j and χ_j' are the jth roots of $J_1(\cdot)$ and $J_1'(\cdot)$, respectively.

We evaluate the scattered E-field when $z \to \pm\infty$ as

$$\boldsymbol{E} = \sum_j C_j^e \boldsymbol{e}_j^e \exp(\pm i\zeta_j' z) + \sum_j C_j^m \boldsymbol{e}_j^m \exp(\pm i\zeta_j z) \tag{33}$$

where

$$\begin{aligned} \boldsymbol{e}_j^m &= \hat{r}\frac{i\zeta_j\chi_j}{a}J_1'\left(\frac{\chi_j r}{a}\right)\cos\phi - \hat{\phi}\frac{i\zeta_j}{r}J_1\left(\frac{\chi_j r}{a}\right)\sin\phi \\ &\quad \pm\hat{z}\left(\frac{\chi_j}{a}\right)^2 J_1\left(\frac{\chi_j r}{a}\right)\cos\phi \end{aligned} \tag{34}$$

$$\boldsymbol{e}_j^e = \hat{r}\frac{\omega}{r}J_1\left(\frac{\chi_j' r}{a}\right)\cos\phi - \hat{\phi}\frac{\omega\chi_j'}{a}J_1'\left(\frac{\chi_j' r}{a}\right)\sin\phi \tag{35}$$

$$C_j^m = \frac{1}{\chi_j J_1'(\chi_j)}\sum_{n,l} E_{nl}P_{nl}(\kappa_n a)\kappa_n^2 B_{\pm}(\zeta_j) \tag{36}$$

$$C_j^e = \frac{i\omega}{\zeta_j'\chi_j'^2 J_1''(\chi_j')}\sum_{n,l}(D_1 E_{nl} + D_2 H_{nl})B_{\pm}(\zeta_j') \tag{37}$$

$$D_1 = \left[\left(\frac{\chi_j'}{a}\right)^2 - \kappa_n^2\right]P_{nl}(\kappa_n a) \tag{38}$$

$$D_2 = \frac{\kappa_n n\pi\chi_j'^2}{ad\omega}Q_{nl}'(\kappa_n a) \tag{39}$$

$$B_{\pm}(\xi) = \frac{\exp(\mp i\xi s_l)[1-(-1)^n \exp(\mp i\xi d)]}{\xi^2 - (n\pi/d)^2} \ . \tag{40}$$

The reflected TE wave power at $z = -\infty$ is

$$P_r^e = \frac{\pi}{2}\sum_{j=1}^{j_1}|C_j^e|^2\omega\,\zeta_j'(\chi_j'^2 - 1)J_1^2(\chi_j') \tag{41}$$

where j_1 is the largest number that satisfies the condition $\omega^2 - (\chi_j'/a)^2 > 0$. The reflected TM wave power at $z = -\infty$ is similarly given as

$$P_r^m = \frac{\pi}{2} \sum_{j=1}^{j_2} |C_j^m|^2 \omega \, \zeta_j \chi_j^2 {J_1'}^2(\chi_j) \tag{42}$$

where j_2 is the largest number satisfying $\omega^2 - (\chi_j/a)^2 > 0$. The total reflected power is given by $(P_r^e + P_r^m)$.

4.8 Parallel-Plate Double Slit Directional Coupler [23]

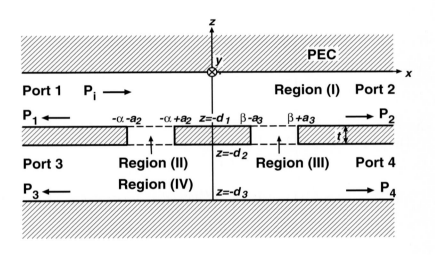

Fig. 4.9. Parallel-plate double slit directional coupler

Electromagnetic wave coupling by slits between rectangular waveguides has been considered in [24-28] for directional coupler applications. In this section we will examine TE (transverse electric to the x-axis) wave coupling through a double slit between two parallel-plate waveguides. The present study is directly applicable to a practical directional coupler that has coupling elements in the common narrow wall of rectangular waveguides. In region (I) $(-d_1 < z < 0)$ the total E-field consists of the incident and scattered components

$$E_y^i(x,z) = e^{ik_{xs}x} \sin(k_{zs}z) \tag{1}$$

$$E_y^I(x,z) = \frac{1}{2\pi} \int_{-\infty}^{\infty} \widetilde{E}_y^I(\zeta) \sin \kappa_1 z e^{-i\zeta x} \, d\zeta \tag{2}$$

where $0 < s < \dfrac{k_1 d_1}{\pi}$, $k_{zs} = \dfrac{s\pi}{d_1}$, $k_{xs} = \sqrt{k_1^2 - k_{zs}^2}$, and $\kappa = \sqrt{k_1^2 - \zeta^2}$. In region (II) $(-d_2 < z < -d_1,\ -\alpha - a_2 < x < -\alpha + a_2)$ the E-field is

$$E_y^{II}(x,z) = \sum_{m=1}^{\infty} [b_m \cos(\xi_{2m}z) + c_m \sin(\xi_{2m}z)] \sin a_{2m}(x + \alpha + a_2) \tag{3}$$

where $a_{2m} = \dfrac{m\pi}{2a_2}$ and $\xi_{2m} = \sqrt{k_2^2 - a_{2m}^2}$. In region (III) $(-d_2 < z < -d_1,\ \beta - a_3 < x < \beta + a_3)$ the E-field is

$$E_y^{III}(x,z) = \sum_{m=1}^{\infty} [e_m \cos(\xi_{3m}z) + f_m \sin(\xi_{3m}z)] \sin a_{3m}(x - \beta + a_3) \tag{4}$$

where $a_{3m} = \dfrac{m\pi}{2a_3}$ and $\xi_{3m} = \sqrt{k_3^2 - a_{3m}^2}$. In region (IV) $(-d_3 < z < -d_2$) the transmitted E-field is

$$E_y^{IV}(x,z) = \frac{1}{2\pi} \int_{-\infty}^{\infty} \widetilde{E}_y^{IV}(\zeta) \sin \kappa_4(z + d_3) e^{-i\zeta x} \, d\zeta \tag{5}$$

where $\kappa_4 = \sqrt{k_4^2 - \zeta^2}$.

The tangential E-field continuity at $z = -d_1$ is given by

$$E_y^i(x,-d_1) + E_y^I(x,-d_1) = \begin{cases} E_y^{II}(x,-d_1), & -\alpha - a_2 < x < -\alpha + a_2 \\[2mm] E_y^{III}(x,-d_1), & \beta - a_3 < x < \beta + a_3 \\[2mm] 0, & \text{otherwise .} \end{cases} \tag{6}$$

Taking the Fourier transform of (6) gives

$$-\sin(\kappa d_1)\widetilde{E}_y^I(\zeta)$$

$$= e^{-i\zeta\alpha} \sum_{m=1}^{\infty} [b_m \cos(\xi_{2m}d_1) - c_m \sin(\xi_{2m}d_1)] a_{2m} a_2^2 F_m(\zeta a_2)$$

$$+ e^{i\zeta\beta} \sum_{m=1}^{\infty} [e_m \cos(\xi_{3m}d_1) - f_m \sin(\xi_{3m}d_1)] a_{3m} a_3^2 F_m(\zeta a_3) . \tag{7}$$

The tangential H-field continuity between regions (I) and (II) for $-\alpha - a_2 < x < -\alpha + a_2$ is

$$H_x^i(x,-d_1) + H_x^I(x,-d_1) = H_x^{II}(x,-d_1) . \tag{8}$$

Multiplying (8) by $\sin a_{2n}(x+\alpha+a_2)$ and integrating from $-\alpha-a_2$ to $-\alpha+a_2$, we get

$$2\pi k_{zs}(-1)^s a_{2n}a_2^2 F(k_{xs}a_2)e^{-ik_{xs}\alpha}$$

$$= \sum_{m=1}^{\infty} [b_m \cos(\xi_{2m}d_1) - c_m \sin(\xi_{2m}d_1)]a_{2m}a_{2n}I_{1mn}$$

$$+ \sum_{m=1}^{\infty} [e_m \cos(\xi_{3m}d_1) - f_m \sin(\xi_{3m}d_1)]a_{3m}a_{2n}I_{2mn}$$

$$+ 2\pi[b_n\xi_{2n}\sin(\xi_{2n}d_1) + c_n\xi_{2n}\cos(\xi_{2n}d_1)]a_2 \tag{9}$$

where

$$I_{1mn} = \int_{-\infty}^{\infty} \kappa_1 \cot(\kappa_1 d_1)a_2^2 F_m(\zeta a_2)a_2^2 F_n(-\zeta a_2)\mathrm{d}\zeta \tag{10}$$

$$I_{2mn} = \int_{-\infty}^{\infty} \kappa_1 \cot(\kappa_1 d_1)a_3^2 F_m(\zeta a_3)a_2^2 F_n(-\zeta a_2)e^{i\zeta(\alpha+\beta)}\mathrm{d}\zeta \ . \tag{11}$$

It is possible to transform I_{1mn} and I_{2mn} into fast-convergent series based on the technique of contour integration. Similar evaluations are available in Sects. 1.5 and 5.3. The tangential E and H-field continuities between regions (I) and (III) give

$$2\pi k_{zs}(-1)^s a_{3n}a_3^2 F(k_{xs}a_3)e^{ik_{xs}\beta}$$

$$= \sum_{m=1}^{\infty} [b_m \cos(\xi_{2m}d_1) - c_m \sin(\xi_{2m}d_1)]a_{2m}a_{2n}I_{3mn}$$

$$+ \sum_{m=1}^{\infty} [e_m \cos(\xi_{3m}d_1) - f_m \sin(\xi_{3m}d_1)]a_{3m}a_{2n}I_{4mn}$$

$$+ 2\pi[e_n\xi_{3n}\sin(\xi_{3n}d_1) + f_n\xi_{3n}\cos(\xi_{3n}d_1)]a_3 \tag{12}$$

where

$$I_{3mn} = \int_{-\infty}^{\infty} \kappa_1 \cot(\kappa_1 d_1)a_2^2 F_m(\zeta a_2)a_3^2 F_n(-\zeta a_3)e^{-i\zeta(\alpha+\beta)}\mathrm{d}\zeta \tag{13}$$

$$I_{4mn} = \int_{-\infty}^{\infty} \kappa_1 \cot(\kappa_1 d_1)a_3^2 F_m(\zeta a_3)a_3^2 F_n(-\zeta a_3)\mathrm{d}\zeta \ . \tag{14}$$

The tangential E and H-field continuities between regions (II) and (IV) give

$$2\pi\xi_{2n}[b_n\sin(\xi_{2n}d_2) + c_n\cos(\xi_{2n}d_2)]a_2$$

$$= \sum_{m=1}^{\infty} [b_m \cos(\xi_{2m}d_2) - c_m \sin(\xi_{2m}d_2)]a_{2m}a_{2n}I_{5mn}$$

$$+ \sum_{m=1}^{\infty} [e_m \cos(\xi_{3m}d_2) - f_m \sin(\xi_{3m}d_2)]a_{3m}a_{2n}I_{6mn} \tag{15}$$

where

$$I_{5mn} = \int_{-\infty}^{\infty} \kappa_4 \cot \kappa_4 (d_3 - d_2) a_2^2 F_m(\zeta a_2) a_2^2 F_n(-\zeta a_2) \mathrm{d}\zeta \tag{16}$$

$$I_{6mn} = \int_{-\infty}^{\infty} \kappa_4 \cot \kappa_4 (d_3 - d_2) a_3^2 F_m(\zeta a_3) a_2^2 F_n(-\zeta a_2) e^{i\zeta(\alpha+\beta)} \mathrm{d}\zeta \ . \tag{17}$$

The tangential field continuities between regions (III) and (IV) give

$$\begin{aligned}
&2\pi\xi_{3n}[e_n \sin(\xi_{3n}d_2) + f_n \cos(\xi_{3n}d_2)]a_3 \\
&= \sum_{m=1}^{\infty} [b_m \cos(\xi_{2m}d_2) - c_m \cos(\xi_{2m}d_2)]a_{2m}a_{3n}I_{7mn} \\
&+ \sum_{m=1}^{\infty} [e_m \cos(\xi_{3m}d_2) - f_m \sin(\xi_{3m}d_2)]a_{3m}a_{3n}I_{8mn}
\end{aligned} \tag{18}$$

where

$$I_{7mn} = \int_{-\infty}^{\infty} \kappa_4 \cot \kappa_4 (d_3 - d_2) a_2^2 F_m(\zeta a_2) a_3^2 F_n(-\zeta a_3) e^{-i\zeta(\alpha+\beta)} \mathrm{d}\zeta \tag{19}$$

$$I_{8mn} = \int_{-\infty}^{\infty} \kappa_4 \cot \kappa_4 (d_3 - d_2) a_3^2 F_m(\zeta a_3) a_3^2 F_n(-\zeta a_3) \mathrm{d}\zeta \ . \tag{20}$$

The scattered field at $x = \pm\infty$ in region (I) is

$$E_y^I(\pm\infty, z) = \sum_{\nu} L_\nu^{\pm} \sin(k_{z\nu}z) e^{\pm i k_{x\nu} x} \tag{21}$$

where $1 \le \nu < k_1 d_1 / \pi$, $\nu = $ integer, $k_{z\nu} = \nu\pi/d_1$, $k_{x\nu} = \sqrt{k_1^2 - k_{z\nu}^2}$, and

$$\begin{aligned}
L_\nu^{\pm} &= \sum_{m=1}^{\infty} ia_{2m}[b_m \cos(\xi_{2m}d_1) - c_m \sin(\xi_{2m}d_1)] \\
&\quad \cdot \frac{a_2^2 k_{z\nu} e^{\pm i k_{x\nu} \alpha}}{d_1 k_{x\nu} \cos(k_{z\nu}d_1)} F_m(\mp k_{x\nu} a_2) \\
&\quad + \sum_{m=1}^{\infty} ia_{3m}[e_m \cos(\xi_{3m}d_1) - f_m \sin(\xi_{3m}d_1)] \\
&\quad \cdot \frac{a_3^2 k_{z\nu} e^{\mp i k_{x\nu} \beta}}{d_1 k_{x\nu} \cos(k_{z\nu}d_1)} F_m(\mp k_{x\nu} a_3) \ .
\end{aligned} \tag{22}$$

In region (IV) the scattered field at $x = \pm\infty$ is

$$E_y^{IV}(\pm\infty, z) = \sum_{\eta} K_\eta^{\pm} \sin(k_{z\eta}(z + d_3)) e^{\pm i k_{x\eta} x} \tag{23}$$

where $1 \le \eta < k_4(d_3 - d_2)/\pi$ ($\eta = $ integer), $k_{z\eta} = \eta\pi/(d_3 - d_2)$, $k_{x\eta} = \sqrt{k_4^2 - k_{z\eta}^2}$, and

$$K_\eta^\pm = - \sum_{m=1}^\infty ia_{2m}[b_m \cos(\xi_{2m}d_2) - c_m \sin(\xi_{2m}d_2)]$$

$$\cdot \frac{a_2^2 k_{z\eta} e^{\pm ik_{x\eta}\alpha} F_m(\mp k_{x\eta}a_2)}{(d_3 - d_2)\cos(k_{z\eta}(d_3 - d_2))k_{x\eta}}$$

$$- \sum_{m=1}^\infty ia_{3m}[e_m \cos(\xi_{3m}d_2) - f_m \sin(\xi_{3m}d_2)]$$

$$\cdot \frac{a_3^2 k_{z\eta} e^{\mp ik_{x\eta}\beta} F_m(\mp k_{x\eta}a_3)}{(d_3 - d_2)\cos(k_{z\eta}(d_3 - d_2))k_{x\eta}} . \tag{24}$$

The reflection and transmission coefficients are given as

$$\varrho = \frac{P_1}{P_i} = \frac{1}{k_{xs}} \sum_\nu k_{x\nu} |L_\nu^-|^2 \tag{25}$$

$$\tau_2 = \frac{P_i + P_2}{P_i} = |1 + L_s^+|^2 + \frac{1}{k_{xs}} \sum_{\nu \neq s} k_{x\nu} |L_\nu^+|^2 \tag{26}$$

$$\tau_3 = \frac{P_3}{P_i} = \frac{(d_3 - d_2)}{k_{xs}d_1} \sum_\eta k_{x\eta} |K_\eta^-|^2 \tag{27}$$

$$\tau_4 = \frac{P_4}{P_i} = \frac{(d_3 - d_2)}{k_{xs}d_1} \sum_\eta k_{x\eta} |K_\eta^+|^2 \tag{28}$$

where $1 \le \nu < k_1 d_1/\pi$ and $1 \le \eta < k_4(d_3 - d_2)/\pi$.

4.9 Parallel-Plate Multiple Slit Directional Coupler [29]

In Sect. 4.8 we examined TE wave coupling through a double slit between two parallel-plate waveguides. In this section we will consider TM (transverse magnetic to the z-axis) wave scattering from multiple slits in between two parallel-plate waveguides. A TM wave scattering analysis in this section is similar to the TE case in Sect. 4.8. A TM wave propagating along the z-direction is incident from port 1 onto multiple slits. In region (I) $(0 < x < d_1)$ the total H-field consists of the incident and scattered field components

$$H_y^i(x, z) = e^{ik_{zs}z} \cos k_{xs}(x - d_1) \tag{1}$$

$$H_y^I(x, z) = \frac{1}{2\pi} \int_{-\infty}^\infty \widetilde{H}_y^I(\zeta) \cos \kappa(x - d_1) e^{-i\zeta z} d\zeta \tag{2}$$

where $0 \le s < kd_1/\pi$, $k_{xs} = s\pi/d_1$, $k_{zs} = \sqrt{k^2 - k_{xs}^2}$, $\kappa = \sqrt{k^2 - \zeta^2}$, and k is the wavenumber. In region (II) $(-d_2 < x < 0$ and $T_l < z < T_l + a_l$: $l = 1, 2, \ldots, N)$ the total H-field is

$$H_y^{II}(x, z) = \sum_{m=0}^\infty [b_m^l \cos(\xi_{lm}x) + c_m^l \sin(\xi_{lm}x)] \cos a_{lm}(z - T_l) \tag{3}$$

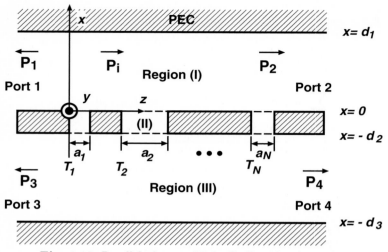

Fig. 4.10. Parallel-plate multiple slit directional coupler

where $a_{lm} = m\pi/a_l$ and $\xi_{lm} = \sqrt{k^2 - a_{lm}^2}$. In region (III) $(-d_3 < x < -d_2)$ the transmitted H-field is

$$H_y^{III}(x,z) = \frac{1}{2\pi} \int_{-\infty}^{\infty} \tilde{H}_y^{III}(\zeta) \cos \kappa(x + d_3) e^{-i\zeta z} \, d\zeta \ . \tag{4}$$

Between regions (I) and (II) the tangential field continuities give

$$\boxed{\begin{aligned} &(-1)^s \frac{ik_{zs}\left[1 - (-1)^n \exp(ik_{zs}a_r)\right]\exp(ik_{zs}T_r)}{k_{zs}^2 - a_{rn}^2} \\ &= \sum_{l=1}^{N} \sum_{m=0}^{\infty} \frac{c_m^l \xi_{lm}}{2\pi} I(d_1) - \frac{b_n^r \varepsilon_n a_r}{2} \end{aligned}} \tag{5}$$

where $\varepsilon_0 = 2$, $\varepsilon_l = 1$ $(l = 1, 2, 3, \dots)$, and

$$\begin{aligned} &I(d_1) \\ &= \int_{-\infty}^{\infty} \frac{\zeta^2 [1 - (-1)^n e^{-i\zeta a_r}][1 - (-1)^m e^{i\zeta a_l}] \exp[i\zeta(T_l - T_r)]}{\kappa \tan(\kappa d_1)(\zeta^2 - a_{lm}^2)(\zeta^2 - a_{rn}^2)} \, d\zeta \ . \end{aligned} \tag{6}$$

Utilizing the technique of contour integration, we obtain

$$I(d_1) = \frac{\pi\varepsilon_n a_l \delta_{nm}\delta_{rl}}{\sqrt{k^2 - a_{lm}^2}\,\tan(\sqrt{k^2 - a_{lm}^2}\,d_1)}$$
$$- \sum_{w=0}^{\infty} \mathrm{i}2\pi k_{zw}\frac{A_1}{\varepsilon_w d_1(k_{zw}^2 - a_{lm}^2)(k_{zw}^2 - a_{rn}^2)} \tag{7}$$

$$A_1 = \exp(\mathrm{i}k_{zw}|T_l - T_r|) + (-1)^{m+n}\exp(\mathrm{i}k_{zw}|a_l - a_r + T_l - T_r|)$$
$$-(-1)^m \exp(\mathrm{i}k_{zw}|T_l - T_r + a_l|)$$
$$-(-1)^n \exp(\mathrm{i}k_{zw}|T_l - T_r - a_r|)\;. \tag{8}$$

Similarly the tangential field continuities at $x = -d_2$ yield

$$\boxed{\begin{aligned}&\pi\left[b_n^r \cos(\xi_{rn}d_2) - c_n^r \sin(\xi_{rn}d_2)\right]\varepsilon_n a_r\\&+\sum_{l=1}^{N}\sum_{m=0}^{\infty}\left[b_m^l \xi_{lm}\sin(\xi_{lm}d_2) + c_m^l \xi_{lm}\cos(\xi_{lm}d_2)\right]I(d_3 - d_2) = 0\end{aligned}} \tag{9}$$

where

$$I(d_3 - d_2)$$
$$= \int_{-\infty}^{\infty} \frac{\zeta^2[1 - (-1)^n e^{-\mathrm{i}\zeta a_r}][1 - (-1)^m e^{\mathrm{i}\zeta a_l}]\exp[\mathrm{i}\zeta(T_l - T_r)]}{\kappa\tan(\kappa(d_3 - d_2))(\zeta^2 - a_{lm}^2)(\zeta^2 - a_{rn}^2)}\mathrm{d}\zeta\;. \tag{10}$$

The total transmitted and reflected fields at $z = \pm\infty$ in region (I) are

$$H_y^I(x, \pm\infty) = \sum_{\nu} L_\nu^\pm \cos(k_{x\nu}x)e^{\pm\mathrm{i}k_{z\nu}z} \tag{11}$$

where $0 \leq \nu < kd_1/\pi$, ν = integer, $k_{x\nu} = \nu\pi/d_1, k_{z\nu} = \sqrt{k^2 - k_{x\nu}^2}$, and

$$L_\nu^\pm$$
$$= \sum_{l=1}^{N}\sum_{m=0}^{\infty} c_m^l \xi_{lm}\frac{\mp[\exp(\mp\mathrm{i}k_{z\nu}T_l) - (-1)^m \exp(\mp\mathrm{i}k_{z\nu}(T_l + a_l))]}{\varepsilon_\nu d_1(k_{z\nu}^2 - a_{lm}^2)}\;. \tag{12}$$

In region (III) the scattered field at $z = \pm\infty$ is

$$H_y^{III}(x, \pm\infty) = \sum_{\eta} K_\eta^\pm \cos k_{x\eta}(x + d_2)e^{\pm\mathrm{i}k_{z\eta}z} \tag{13}$$

where $0 \leq \eta < k(d_3 - d_2)/\pi$ (η = integer), $k_{x\eta} = \eta\pi/(d_3 - d_2)$, $k_{z\eta} = \sqrt{k^2 - k_{x\eta}^2}$, and

$$K_\eta^\pm = \pm\sum_{l=1}^{N}\sum_{m=0}^{\infty}\frac{\exp(\mp\mathrm{i}k_{z\eta}T_l) - (-1)^m \exp[\mp\mathrm{i}k_{z\eta}(T_l + a_l)]}{\varepsilon_\eta(d_3 - d_2)(k_{z\eta}^2 - a_{lm}^2)}$$
$$\cdot\left[b_m^l \xi_{lm}\sin(\xi_{lm}d_2) + c_m^l \xi_{lm}\cos(\xi_{lm}d_2)\right]\;. \tag{14}$$

The reflection and transmission coefficients are

$$\varrho = \frac{P_1}{P_i} = \frac{1}{\varepsilon_s k_{zs}} \sum_\nu \varepsilon_\nu k_{z\nu} |L_\nu^-|^2 \tag{15}$$

$$\tau_2 = |1 + L_s^+|^2 + \sum_{\nu \neq s} \frac{\varepsilon_\nu k_{z\nu}}{\varepsilon_s k_{zs}} |L_\nu^+|^2 \tag{16}$$

$$\tau_3 = \frac{P_3}{P_i} = \frac{(d_3 - d_2)}{\varepsilon_s k_{zs} d_1} \sum_\eta \varepsilon_\eta k_{z\eta} |K_\eta^-|^2 \tag{17}$$

$$\tau_4 = \frac{P_4}{P_i} = \frac{(d_3 - d_2)}{\varepsilon_s k_{zs} d_1} \sum_\eta \varepsilon_\eta k_{z\eta} |K_\eta^+|^2 \tag{18}$$

where P_i is the incident power at port 1, $0 \leq \nu < k d_1 / \pi$, and $0 \leq \eta < k(d_3 - d_2)/\pi$.

It is possible to apply the present theory to a directional coupler consisting of wide-wall-slit rectangular waveguides with a dimension $a \times b$ with $a > b$. Our theory with a TM$_0$ incident wave ($s = 0$) is approximately applicable to a rectangular waveguide directional coupler with a TE$_{10}$ incident wave by replacing k in (5) and (9) with $\sqrt{k^2 - (\pi/b)^2}$.

References for Chapter 4

1. J. K. Park and H. J. Eom, "Fourier-transform analysis of inset dielectric guide with a conductor cover," *Microwave Opt. Technol. Lett.*, vol. 14, no. 6, pp. 324-327, 1997.
2. T. Rozzi and S. J. Hedge, "Rigorous analysis and network modeling of the inset dielectric guide," *IEEE Trans. Microwave Theory Tech.*, vol. 35, no. 9, pp. 823-833, Sept. 1987.
3. T. Rozzi and L. Ma, "Mode completeness, normalization, and Green's function of the inset dielectric guide," *IEEE Trans. Microwave Theory Tech.*, vol. 36, no. 3, pp. 542-551, March 1988.
4. B. T. Lee, J. W. Lee, H. J. Eom, and S. Y. Shin, "Fourier-transform analysis for rectangular groove guide," *IEEE Trans. Microwave Theory Tech.*, vol. 43, pp. 2162-2165, Sept. 1995.
5. T. Nakahara and N. Kurauchi, "Transmission modes in the grooved guide," *J. Inst. Electron. Commun. Eng. Jap.*, vol. 47, no. 7, pp. 43-51, July 1964.
6. A. A. Oliner and P. Lampariello, "The dominant mode properties of open groove guide: an improved solution," *IEEE Trans. Microwave Theory Tech.*, vol. 33, pp. 755-764, Sept. 1985.
7. Inderjit and M. Sachidananda, "Rigorous analysis of a groove guide," *IEE Proceedings-H*, vol. 139, no. 5, pp. 449-452, Oct. 1992.
8. H. J. Eom and Y. H. Cho, "Analysis of multiple groove guide," *Electron. Lett.*, vol. 35, no. 20, pp. 1749-1751, Sept. 1999.
9. D. J. Harris and K. W. Lee, "Theoretical and experimental characteristics of double-groove guide for 100GHz operation," *IEE Proc.*, vol. 128, pt. H. no. 1, pp. 6-10, Feb. 1981.
10. H. J. Eom, Y. C. Noh, and J. K. Park, "Scattering from multiple grooves in the inner conductor of coaxial line," *IEEE Trans. Microwave Theory Tech.*, vol. 48, no. 7, pp. 1151-1153, July 2000.

11. L. Young, "The practical realization of series capacitive coupling for microwave filters," *Microwave J.*, vol. 5, pp. 79-81, Dec. 1962.
12. H. E. Green, "The numerical solution of some important transmission-line problem," *IEEE Trans. Microwave Theory Tech.*, vol. 13, pp. 676-692, Sept. 1965.
13. H. J. Eom, Y. C. Noh, and J. K. Park, "Scattering analysis of a coaxial line terminated by a gap," *IEEE Microwave Guided Wave Lett.*, vol. 8, no. 6, pp. 218-219, June 1998.
14. H. N. Dawirs, "Equivalent circuit of a series gap in the center conductor of a coaxial transmission line," *IEEE Trans. Microwave Theory Tech.*, vol. 17, pp. 127-129, Feb. 1969.
15. S. Sen and P. K. Saha, "Equivalent circuit of a gap in the central conductor of a coaxial line," *IEEE Trans. Microwave Theory Tech.*, vol. 30, pp. 2026-2029, Nov. 1982.
16. H. J. Eom and K. W. Lee, "Scattering from a coaxial line with a cavity," *Microwave Opt. Technol. Lett.*, vol. 25, no. 4, pp. 285-287, May 2000.
17. A. G. Williamson, "Radial-line/coaxial-line junction: analysis and equivalent circuits," *Int. J. Electronics*, vol. 58, no. 1, pp. 91-104, 1985.
18. H. S. Lee and H. J. Eom, "Scattering from a cylindrical waveguide with rectangular corrugations," *IEEE Trans. Microwave Theory Tech.*, vol. 49, no. 2, Feb. 2001.
19. J. Esteban and J. M. Reboller, "Characterization of corrugated waveguides by modal analysis," *IEEE Trans. Microwave Theory Tech.*, vol. 39, no. 6, pp. 937-943, June 1991.
20. G. L. James, "Analysis and design of TE_{11} to HE_{11} corrugated cylindrical waveguide mode converters," *IEEE Trans. Microwave Theory Tech.*, vol. 29, no. 10, pp. 1059-1066, Oct. 1981.
21. L. C. Da Silva and M. G. Castello Branco, "A method of analysis of TE_{11} to HE_{11} mode converters," *IEEE Trans. Microwave Theory Tech.*, vol. 36, no. 3, pp. 480-488, Mar. 1988.
22. L. C. Da Silva and M. G. Castello Branco, "Analysis of the junction between smooth and corrugated waveguides in mode converters," *IEEE Trans. Microwave Theory Tech.*, vol. 38, no. 6, pp. 800-802, June 1990.
23. H. J. Eom and S. H. Min, "Coupling through a double slit between two parallel-plate waveguides," *Microwave Opt. Technol. Lett.*, vol. 24, no. 3, pp. 182-185, Feb. 2000.
24. A. J. Sangster and H. Wang, "A generalized analysis for a class of rectangular waveguide coupler employing narrow wall slots," *IEEE Trans. Microwave Theory Tech.*, vol. 44, no. 2, pp. 283-290, Feb. 1996.
25. V. M. Pandharipande and B. N. Das, "Coupling of waveguides through large aperture," *IEEE Trans. Microwave Theory Tech.*, vol. 26, no. 3, pp. 209-212, Mar. 1978.
26. A. M. Rajeek and A. Chakraborty, "Analysis of a wide compound slot-coupled parallel waveguide coupler and radiator," *IEEE Trans. Microwave Theory Tech.*, vol. 43, no. 4, pp. 802-809, Apr. 1995.
27. P. Alinikula and K.S. Kunz, "Analysis of waveguide aperture coupling using the finite-difference time-domain method," *IEEE Microwave Guided Wave Lett.*, vol. 1, no. 8, pp. 189-191, Aug. 1991.
28. H. Riblet, "Mathematical theory of directional coupler," *Proc. IRE*, vol. 35, pp. 1307-1313, Nov. 1947.
29. S. H. Min and H. J. Eom, "TM coupling by multiple slits between two parallel plates," *Microwave Opt. Technol. Lett.*, vol. 27, no. 3, pp. 195-197, Nov. 2000.

5. Junctions in Parallel-Plate/Rectangular Waveguide

5.1 T-Junction in a Parallel-Plate Waveguide

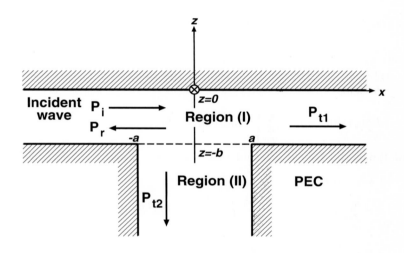

Fig. 5.1. T-junction in a parallel-plate waveguide

The T-junction in a rectangular waveguide is one of basic power-coupling components used in microwave circuit applications. The power-combining characteristics of the T-junction in a rectangular waveguide have been extensively studied in [1-3]. In the next two subsections we will consider scattering from the H-plane and E-plane T-junctions in a parallel-plate waveguide. A scattering study of the T-junction in a parallel-plate waveguide is directly applicable to the T-junction in a rectangular waveguide.

5.1.1 H-Plane T-Junction [4]

Assume that a TE wave $E_y^i(x,z)$ is incident on the parallel-plate waveguide T-junction in H-plane (wavenumber k). Scattering from the H-plane T-junction in a parallel-plate waveguide is equivalent to that from the H-plane T-junction in a rectangular waveguide with the lowest mode (TE$_{10}$ wave) excitation. The total E-field in region (I) ($-b < z < 0$) has the incident and scattered waves

$$E_y^i(x,z) = \exp(ik_{xs}x)\sin k_{zs}(z+b) \tag{1}$$

$$E_y^I(x,z) = \frac{1}{2\pi}\int_{-\infty}^{\infty} \widetilde{E}_y^I(\zeta)\sin(\kappa z)e^{-i\zeta x}\,d\zeta \tag{2}$$

where $k_{zs} = s\pi/b$, ($s = 1,2,3,\ldots$), $k_{xs} = \sqrt{k^2 - k_{zs}^2}$, and $\kappa = \sqrt{k^2 - \zeta^2}$. In region (II) ($-a < x < a$ and $z < -b$) the total transmitted field is

$$E_y^{II}(x,z) = \sum_{m=1}^{\infty} c_m \sin a_m(x+a)\exp(-i\xi_m z) \tag{3}$$

where $a_m = m\pi/(2a)$ and $\xi_m = \sqrt{k^2 - a_m^2}$.

Applying the Fourier transform to the tangential E-field continuity at $z = -b$ yields

$$\widetilde{E}_y^I(\zeta)\sin(-\kappa b) = \sum_{m=1}^{\infty} c_m a_m \exp(i\xi_m b)a^2 F_m(\zeta a) . \tag{4}$$

The tangential H-field continuity at ($-a < x < a$ and $z = -b$) is written as

$$k_{zs}\exp(ik_{xs}x)$$
$$-\sum_{m=1}^{\infty} c_m \frac{a_m \exp(i\xi_m b)}{2\pi}\int_{-\infty}^{\infty}\kappa\cot(\kappa b)a^2 F_m(\zeta a)e^{-i\zeta x}\,d\zeta$$
$$= -i\sum_{m=1}^{\infty} c_m\xi_m \exp(i\xi_m b)\sin a_m(x+a) . \tag{5}$$

Multiplying (5) by $\sin a_n(x+a)$ and integrating with respect to x from $-a$ to a, we get

$$\boxed{\begin{aligned}
&k_{zs}a_n a^2 F_n(k_{xs}a)\\
&= \sum_{m=1}^{\infty} c_m \frac{a_m a_n \exp(i\xi_m b)}{2\pi}a^2\Lambda_3(k) - ic_n\xi_n a \exp(i\xi_n b)
\end{aligned}} \tag{6}$$

where

$$\Lambda_3(k) = \int_{-\infty}^{\infty}\kappa\cot(\kappa b)a^2 F_m(\zeta a)F_n(-\zeta a)\,d\zeta . \tag{7}$$

For $m+n = $ odd, $\Lambda_3(k)=0$.
For $m+n = $ even,

$$\Lambda_3(k) = \frac{2\pi\xi_m}{aa_m^2 \tan(\xi_m b)} \delta_{nm}$$
$$- \sum_{l=1}^{\infty} \frac{\mathrm{i}4\pi(l\pi/b)^2 \left[1 - (-1)^m \exp\left(\mathrm{i}2\sqrt{k^2 - (l\pi/b)^2}a\right)\right]}{\sqrt{k^2 - (l\pi/b)^2}a^2 b \left[\xi_m^2 - (l\pi/b)^2\right]\left[\xi_n^2 - (l\pi/b)^2\right]} . \tag{8}$$

The total transmitted and reflected fields at $x = \pm\infty$ in region (I) are

$$E_y^I(\pm\infty, z) = \sum_v K_v^\pm \sin k_{zv}(z + b) \exp(\pm\mathrm{i}k_{xv}x) \tag{9}$$

where $1 \leq v < kb/\pi$, v: integer $(1,2,3,\ldots)$, $k_{zv} = v\pi/b$, $k_{xv} = \sqrt{k^2 - k_{zv}^2}$, and

$$K_v^\pm = \sum_{m=1}^{\infty} c_m \frac{\mathrm{i}a_m k_{zv} e^{\mathrm{i}\xi_m b}}{k_{xv}b} a^2 F_m(\mp k_{xv}a) . \tag{10}$$

The reflection (ϱ) and transmission (τ_1 and τ_2) coefficients are

$$\tau_1 = P_{t1}/P_i = |1 + K_s^+|^2 + \frac{1}{k_{xs}} \sum_{v \neq s} k_{xv}|K_v^+|^2 \tag{11}$$

$$\varrho = P_r/P_i = \frac{1}{k_{xs}} \sum_v k_{xv}|K_v^-|^2 \tag{12}$$

$$\tau_2 = P_{t2}/P_i = \frac{2a}{k_{xs}b} \sum_m \xi_m|c_m|^2 \tag{13}$$

where $1 \leq m < 2ak/\pi$, $1 \leq v < kb/\pi$, and m and v are integers.

5.1.2 E-Plane T-Junction [5]

Consider the E-plane T-junction in a parallel-plate waveguide when a TM wave $H_y^i(x, z)$ is incident on the junction. In region (I) $(-b < z < 0)$ the total H-field has the incident and scattered waves

$$H_y^i(x, z) = e^{\mathrm{i}k_{xs}x} \cos k_{zs}(z + b) \tag{14}$$

$$H_y^I(x, z) = \frac{1}{2\pi} \int_{-\infty}^{\infty} \widetilde{H}_y^I(\zeta) \cos(\kappa z) e^{-\mathrm{i}\zeta x} \, \mathrm{d}\zeta \tag{15}$$

where $k_{zs} = s\pi/b$, s: integer $(0, 1, 2, \ldots)$, $k_{xs} = \sqrt{k^2 - k_{zs}^2}$, and $\kappa = \sqrt{k^2 - \zeta^2}$. In region (II) $(-a < x < a$ and $z < -b)$ the transmitted field is

$$H_y^{II}(x, z) = \sum_{m=0}^{\infty} c_m \cos a_m(x + a) e^{-\mathrm{i}\xi_m z} \tag{16}$$

where $a_m = m\pi/(2a)$ and $\xi_m = \sqrt{k^2 - a_m^2}$.

The tangential E-field continuity at $z = -b$ yields

$$\widetilde{H}_y^I(\zeta)\kappa\sin(\kappa b) = \sum_{m=0}^{\infty} -c_m\xi_m e^{i\xi_m b}\zeta a^2 F_m(\zeta a) .\tag{17}$$

Multiplying the tangential H-field continuity along $(-a < x < a$ and $z = -b)$ by $\cos a_n(x + a)$ and integrating with respect to x from $-a$ to a, we obtain

$$k_{xs}a^2 F_n(k_{xs}a) = -\sum_{m=0}^{\infty} c_m\frac{\xi_m e^{i\xi_m b}}{2\pi}a^2\Omega_3(k) + ic_n\varepsilon_n a e^{i\xi_n b}\tag{18}$$

where

$$\Omega_3(k) = \int_{-\infty}^{\infty} \frac{\zeta^2}{\kappa\tan(\kappa b)}a^2 F_m(\zeta a)F_n(-\zeta a)\,\mathrm{d}\zeta .\tag{19}$$

For $m + n =$ odd, $\Omega_3(k)=0$.
For $m + n =$ even,

$$\Omega_3(k) = \frac{2\pi\varepsilon_m}{\xi_m a\tan(\xi_m b)}\delta_{nm}$$
$$+\sum_{l=0}^{\infty}\frac{-i4\pi\sqrt{k^2 - (l\pi/b)^2}\left[1 - (-1)^m\exp\left(i2\sqrt{k^2 - (l\pi/b)^2}a\right)\right]}{\varepsilon_l ba^2[\xi_m^2 - (l\pi/b)^2][\xi_n^2 - (l\pi/b)^2]} .\tag{20}$$

The total transmitted and reflected fields at $x = \pm\infty$ in region (I) are

$$H_y^I(\pm\infty, z) = \sum_v K_v^{\pm}\cos k_{zv}(z + b)\exp(\pm ik_{xv}x)\tag{21}$$

where $0 \le v < kb/\pi$, v: integer, $k_{zv} = v\pi/b$, $k_{xv} = \sqrt{k^2 - k_{zv}^2}$, and

$$K_v^{\pm} = \sum_{m=0}^{\infty}\frac{\mp ic_m\xi_m e^{i\xi_m b}}{\varepsilon_v b}a^2 F_m(\mp k_{xv}a) .\tag{22}$$

The transmission (τ_1 and τ_2) and reflection (ϱ) coefficients are

$$\tau_1 = P_{t1}/P_i = |1 + K_s^+|^2 + \frac{1}{\varepsilon_s k_{xs}}\sum_{v\ne s}\varepsilon_v k_{xv}|K_v^+|^2\tag{23}$$

$$\varrho = P_r/P_i = \frac{1}{\varepsilon_s k_{xs}}\sum_v\varepsilon_v k_{xv}|K_v^-|^2\tag{24}$$

$$\tau_2 = P_{t2}/P_i = \frac{2a}{\varepsilon_s b k_{xs}}\sum_m\varepsilon_m\xi_m|c_m|^2 .\tag{25}$$

where $0 \le m < 2ak/\pi$, $0 \le v < kb/\pi$, m and v: integers, and the incident power is $P_i = k_{xs}\varepsilon_s b/(4\omega\epsilon)$

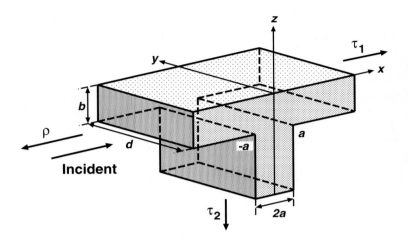

Fig. 5.2. E-plane T-junction in a rectangular waveguide

5.2 E-Plane T-Junction in a Rectangular Waveguide [6]

A study of electromagnetic wave scattering from the E-plane T-junction in a rectangular waveguide has been performed in [7]. In this section we will revisit the scattering problem of the E-plane T-junction in a rectangular waveguide. The TE_{10} wave is incident on the T-junction filled with a medium with wavenumber k $(= 2\pi/\lambda = \omega\sqrt{\mu\epsilon})$. The incident TE_{10} wave is given as

$$H_x^i(x,y) = \frac{k_y}{i\omega\mu}\cos(k_y y)e^{ik_x x} \tag{1}$$

where $k_y = \pi/d$ and $k_x = \sqrt{k^2 - k_y^2}$. The E-plane T-junction in a rectangular waveguide supports the propagation of a hybrid wave that is a combination of the TE and TM waves. In region (I) $(-b < z < 0$ and $0 < y < d)$ the scattered field consists of the TE (transverse electric to the x-direction) and TM (transverse magnetic to the x-direction) waves

$$H_x^I(x,y,z) = \frac{1}{2\pi}\int_{-\infty}^{\infty} a(\zeta)\cos(k_y y)\cos(\kappa_z z)e^{-i\zeta x}d\zeta \tag{2}$$

$$E_x^I(x,y,z) = \frac{1}{2\pi}\int_{-\infty}^{\infty} b(\zeta)\sin(k_y y)\sin(\kappa_z z)e^{-i\zeta x}d\zeta \tag{3}$$

where $\kappa_z = \sqrt{k^2 - k_y^2 - \zeta^2}$. In region (II) $(-a < x < a, z < -b,$ and $0 < y < d)$ assuming the longitudinal direction is z, we express the z-components as

$$H_z^{II}(x,y,z) = \sum_{n=0}^{\infty} c_n \cos k_n (x+a) \cos(k_y y)\mathrm{e}^{-\mathrm{i}\kappa_n z} \tag{4}$$

$$E_z^{II}(x,y,z) = \sum_{n=1}^{\infty} d_n \sin k_n (x+a) \sin(k_y y)\mathrm{e}^{-\mathrm{i}\kappa_n z} \tag{5}$$

where $k_n = n\pi/(2a)$ and $\kappa_n = \sqrt{k^2 - k_n^2 - k_y^2}$.

Applying the Fourier transform to the $E_x(x,y,-b)$ continuity

$$\int_{-\infty}^{\infty} E_x^{I}(x,y,-b)\mathrm{e}^{\mathrm{i}\zeta x}\mathrm{d}x = \int_{-a}^{a} E_x^{II}(x,y,-b)\mathrm{e}^{\mathrm{i}\zeta x}\mathrm{d}x \tag{6}$$

yields

$$b(\zeta) = \frac{a^2\zeta}{\sin(\kappa_z b)}$$
$$\cdot \left[\sum_{n=0}^{\infty} \frac{c_n \omega\mu k_y \mathrm{e}^{\mathrm{i}\kappa_n b} F_n(\zeta a)}{k^2 - \kappa_n^2} + \sum_{n=1}^{\infty} \frac{d_n k_n \kappa_n \mathrm{e}^{\mathrm{i}\kappa_n b} F_n(\zeta a)}{k^2 - \kappa_n^2} \right]. \tag{7}$$

Similarly applying the Fourier transform to the E_y continuity at $z = -b$ gives

$$a(\zeta) = \frac{a^2}{\kappa_z \sin(\kappa_z b)} \left[\sum_{n=0}^{\infty} \frac{c_n(k^2 k_n^2 - \zeta^2 k_n^2 - \zeta^2 k_y^2)\mathrm{e}^{\mathrm{i}\kappa_n b} F_n(\zeta a)}{k^2 - \kappa_n^2} \right.$$
$$\left. - \sum_{n=1}^{\infty} \frac{d_n k^2 k_n k_y \kappa_n \mathrm{e}^{\mathrm{i}\kappa_n b} F_n(\zeta a)}{\omega\mu(k^2 - \kappa_n^2)} \right]. \tag{8}$$

Substituting $a(\zeta)$ and $b(\zeta)$ into the H-field continuities

$$H_x^i(x,y) + H_x^I(x,y,-b) = H_x^{II}(x,y,-b) \tag{9}$$
$$H_y^i(x,y) + H_y^I(x,y,-b) = H_y^{II}(x,y,-b) \tag{10}$$

taking $\int_{-a}^{a}$ (9) $\sin k_t(x+a)\mathrm{d}x$ $(t = 1,2,\ldots)$, and $\int_{-a}^{a}$ (10) $\cos k_t(x+a)\mathrm{d}x$ $(t = 0,1,\ldots)$, we obtain

$$\sum_{n=0}^{\infty} c_n \frac{\omega\mu\mathrm{e}^{\mathrm{i}\kappa_n b}}{k^2 - \kappa_n^2}$$
$$\cdot \left(k^2 k_n^2 k_t I_1 - k_n^2 k_t I_2 - k_y^2 k_t I_2 - \frac{2\pi \mathrm{i} k_n \kappa_n}{a^3} \delta_{nt} \right)$$
$$+ \sum_{n=1}^{\infty} d_n \frac{k^2 \mathrm{e}^{\mathrm{i}\kappa_n b}}{k^2 - \kappa_n^2} \left(-k_y k_n k_t \kappa_n I_1 + \frac{2\pi \mathrm{i} k_y}{a^3} \delta_{nt} \right)$$
$$= 2\pi \mathrm{i} \frac{k_y k_t}{a^2} F_t(k_x a) \tag{11}$$

$$\sum_{n=0}^{\infty} c_n \frac{\omega\mu e^{i\kappa_n b}}{k^2 - \kappa_n^2}$$

$$\cdot \left(k^2 k_n^2 k_y I_3 - k_y k_n^2 I_4 - k_y^3 I_4 - k^2 k_y I_5 + \frac{2\pi i \varepsilon_n k_y \kappa_n}{a^3} \delta_{nt} \right)$$

$$- \sum_{n=1}^{\infty} d_n \frac{k^2 e^{i\kappa_n b}}{k^2 - \kappa_n^2} \left(k_y^2 k_n \kappa_n I_3 + k_n \kappa_n I_5 - \frac{2\pi i k_n}{a^3} \delta_{nt} \right)$$

$$= 2\pi i \frac{k_x^2}{a^2} F_t(k_x a) \tag{12}$$

where

$$I_1 = \int_{-\infty}^{\infty} \frac{\cot(\kappa_z b)}{\kappa_z} F_n(\zeta a) F_t(-\zeta a) \, d\zeta \tag{13}$$

$$I_2 = \int_{-\infty}^{\infty} \frac{\zeta^2 \cot(\kappa_z b)}{\kappa_z} F_n(\zeta a) F_t(-\zeta a) \, d\zeta \tag{14}$$

$$I_3 = \int_{-\infty}^{\infty} \frac{\zeta^2 \cot(\kappa_z b)}{\kappa_z (k^2 - \zeta^2)} F_n(\zeta a) F_t(-\zeta a) \, d\zeta \tag{15}$$

$$I_4 = \int_{-\infty}^{\infty} \frac{\zeta^4 \cot(\kappa_z b)}{\kappa_z (k^2 - \zeta^2)} F_n(\zeta a) F_t(-\zeta a) \, d\zeta \tag{16}$$

$$I_5 = \int_{-\infty}^{\infty} \frac{\zeta^2 \kappa_z \cot(\kappa_z b)}{(k^2 - \zeta^2)} F_n(\zeta a) F_t(-\zeta a) \, d\zeta \ . \tag{17}$$

It is possible to evaluate the integrals I_1 through I_5 using the residue calculus. The results are summarized below.
For $n + t =$ odd, $I_1 = I_2 = I_3 = I_4 = I_5 = 0$.
For $n + t =$ even,

$$I_1 = \frac{P_n}{k_n^2} \delta_{nt} - \frac{4\pi i}{a^4 b} \sum_{v=0}^{\infty} \frac{E(\zeta_v)}{\varepsilon_v \zeta_v}, \qquad t \neq 0 \quad \text{and} \quad n \neq 0 \tag{18}$$

$$I_2 = P_n \varepsilon_n \delta_{nt} - \frac{4\pi i}{a^4 b} \sum_{v=0}^{\infty} \frac{\zeta_v E(\zeta_v)}{\varepsilon_v} \tag{19}$$

$$I_3 = \frac{P_n \varepsilon_n}{(k^2 - k_n^2)} \delta_{nt} + \frac{2\pi i k \coth(k_y b)}{a^4 k_y} E(k)$$

$$- \frac{4\pi i}{a^4 b} \sum_{v=0}^{\infty} \frac{\zeta_v E(\zeta_v)}{\varepsilon_v (k^2 - \zeta_v^2)} \tag{20}$$

$$I_4 = \frac{2\pi i k^3 \coth(k_y b)}{a^4 k_y} E(k)$$

$$- \frac{4\pi i}{a^4 b} \sum_{v=0}^{\infty} \frac{\zeta_v^3 E(\zeta_v)}{\varepsilon_v (k^2 - \zeta_v^2)} + \begin{cases} 0, & n = t = 0 \\ \frac{P_n k_n^2}{(k^2 - k_n^2)} \delta_{nt}, & \text{otherwise} \end{cases} \tag{21}$$

$$I_5 = \frac{P_n \varepsilon_n \kappa_n^2}{(k^2 - k_n^2)} \delta_{nt} - \frac{2\pi i k k_y \coth(k_y b)}{a^4} E(k)$$

$$- \frac{4\pi i}{a^4 b} \sum_{v=1}^{\infty} \frac{\zeta_v (v\pi/b)^2 E(\zeta_v)}{(k^2 - \zeta_v^2)} \tag{22}$$

where $E(\zeta) = \dfrac{[1 - (-1)^n e^{i2\zeta a}]}{(\zeta^2 - k_n^2)(\zeta^2 - k_t^2)}$, $\zeta_v = \sqrt{k^2 - k_y^2 - (v\pi/b)^2}$, and

$P_n = \dfrac{2\pi \cot(\kappa_n b)}{a^3 \kappa_n}$.

The scattered fields for $x \leq -a$ and $x \geq a$ are

$$H_x^I(x, y, z) = \sum_l K_l^{\pm} \cos(k_y y) \cos(k_l z) e^{\pm i \zeta_l x} \tag{23}$$

$$E_x^I(x, y, z) = \sum_l J_l^{\pm} \sin(k_y y) \sin(k_l z) e^{\pm i \zeta_l x} \tag{24}$$

where

$$K_l^{\pm} = \sum_{n=1}^{\infty} \left(-c_n k_n + d_n \frac{k_y \kappa_n}{\omega \mu} \right) \frac{i a^2 k^2 k_n}{(-1)^l b \zeta_l \varepsilon_l (k^2 - \kappa_n^2)} F_n(\mp \zeta_l a) e^{i\kappa_n b}$$

$$+ \sum_{n=0}^{\infty} \frac{i a^2 c_n (k_n^2 + k_y^2) \zeta_l}{(-1)^l b \varepsilon_l (k^2 - \kappa_n^2)} F_n(\mp \zeta_l a) e^{i\kappa_n b} \tag{25}$$

$$J_l^{\pm} = \pm \sum_{n=0}^{\infty} \frac{c_n \omega \mu k_y i a^2 k_l}{(-1)^l b (k^2 - \kappa_n^2)} F_n(\mp \zeta_l a) e^{i\kappa_n b}$$

$$\pm \sum_{n=1}^{\infty} \frac{d_n k_n \kappa_n i a^2 k_l}{(-1)^l b (k^2 - \kappa_n^2)} F_n(\mp \zeta_l a) e^{i\kappa_n b} \tag{26}$$

$k_l = l\pi/b$, $0 \leq l$, and l: integer. The transmission (τ_1 and τ_2) and reflection (ϱ) coefficients are

$$\tau_1 = \left| 1 + \frac{i\omega \mu}{k_y} K_0^+ \right|^2 + \frac{\omega^2 \mu}{2k_x} \sum_{l=1} \frac{\zeta_l \left(\epsilon |J_l^+|^2 + \mu |K_l^+|^2 \right)}{k^2 - \zeta_l^2} \tag{27}$$

$$\tau_2 = \frac{2a(\omega \mu)^2}{b k_y^2} |c_0|^2 + \frac{a\omega^2 \mu}{b k_x} \sum_{n=1} \frac{\kappa_n (\mu |c_n|^2 + \epsilon |d_n|^2)}{k^2 - \kappa_n^2} \tag{28}$$

$$\varrho = \frac{(\omega \mu)^2}{k_y^2} |K_0^-|^2 + \frac{\omega^2 \mu}{2k_x} \sum_{l=1} \frac{\zeta_l \left(\epsilon |J_l^-|^2 + \mu |K_l^-|^2 \right)}{k^2 - \zeta_l^2} \tag{29}$$

where $0 \leq l < b/\pi \sqrt{k^2 - k_y^2}$, l: integer, $0 \leq n < 2a/\pi \sqrt{k^2 - k_y^2}$, and n: integer.

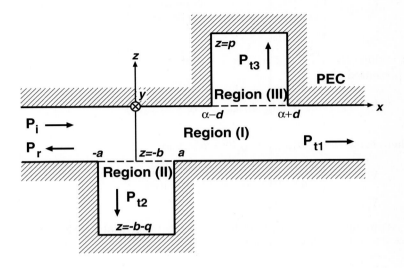

Fig. 5.3. H-plane double junction

5.3 H-Plane Double Junction [8]

In this section we will consider a problem of electromagnetic wave scattering from two junctions in the H-plane waveguide where two junctions are either short- or open-circuited. The two-junction scattering analysis given in this section is an extension of the single H-plane T-junction discussed in Subsect. 5.1.1. A TE wave $E_y^i(x, z)$, which is transverse electric (TE) to the x-axis, is incident on the junctions with wavenumber k $(=2\pi/\lambda=\omega\sqrt{\mu\epsilon})$. In region (I) $(-b < z < 0)$ the total E-field is assumed to have the incident and scattered components $E_y^i(x, z)$ and $E_y^I(x, z)$ as

$$E_y^i(x, z) = e^{ik_{xs}x} \sin k_{zs}(z + b) \tag{1}$$

$$E_y^I(x, z) = \frac{1}{2\pi} \int_{-\infty}^{\infty} \left[\widetilde{E}_y^{s+}(\zeta) e^{i\kappa z} + \widetilde{E}_y^{s-}(\zeta) e^{-i\kappa z} \right] e^{-i\zeta x} \, d\zeta \tag{2}$$

where $k_{zs} = s\pi/b$, s: integer, $k_{xs} = \sqrt{k^2 - k_{zs}^2}$, and $\kappa = \sqrt{k^2 - \zeta^2}$. In region (II) $(-a < x < a$ and $-q - b < z < -b)$ the transmitted field is

$$E_y^{II}(x, z) = \sum_{m=1}^{\infty} c_m \sin a_m(x + a) Y_q(z) \tag{3}$$

where $a_m = m\pi/(2a)$, $\xi_m = \sqrt{k^2 - a_m^2}$, and

$$Y_q(z) = \begin{cases} e^{-i\xi_m z}, & \text{open-circuited } (q \to \infty) \\ \sin \xi_m (z + b + q), & \text{short-circuited} \end{cases} \tag{4}$$

$$Y_q'(z) = dY_q(z)/dz . \tag{5}$$

In region (III) $(\alpha - d < x < \alpha + d$ and $0 < z < p)$ the transmitted field is represented as

$$E_y^{III}(x, z) = \sum_{m=1}^{\infty} d_m \sin b_m (x - \alpha + d) X_p(z) \tag{6}$$

where $b_m = m\pi/(2d)$, $\eta_m = \sqrt{k^2 - b_m^2}$, and

$$X_p(z) = \begin{cases} e^{i\eta_m z}, & \text{open-circuited } (p \to \infty) \\ \sin \eta_m (z - p), & \text{short-circuited} \end{cases} \tag{7}$$

$$X_p'(z) = dX_p(z)/dz . \tag{8}$$

Applying the Fourier transforms to the tangential E-field continuities at $z = -b$ and 0 yields respectively

$$\widetilde{E}_y^{s+}(\zeta)e^{-i\kappa b} + \widetilde{E}_y^{s-}(\zeta)e^{i\kappa b} = \sum_{m=1}^{\infty} c_m Y_q(-b) a_m a^2 F_m(\zeta a) \tag{9}$$

$$\widetilde{E}_y^{s+}(\zeta) + \widetilde{E}_y^{s-}(\zeta) = \sum_{m=1}^{\infty} d_m X_p(0) b_m e^{i\zeta \alpha} d^2 F_m(\zeta d) . \tag{10}$$

Substituting (9) and (10) into the tangential H-field continuities along $(-a < x < a$ and $z = -b)$ and along $(\alpha - d < x < \alpha + d$ and $z = 0)$ gives explicitly

$$k_{zs} e^{ik_{xs} x}$$

$$- \sum_{m=1}^{\infty} c_m Y_q(-b) \frac{a_m}{2\pi} \int_{-\infty}^{\infty} \kappa \cot(\kappa b) a^2 F_m(\zeta a) e^{-i\zeta x} \, d\zeta$$

$$+ \sum_{m=1}^{\infty} d_m X_p(0) \frac{b_m}{2\pi} \int_{-\infty}^{\infty} \frac{\kappa e^{i\zeta \alpha}}{\sin(\kappa b)} d^2 F_m(\zeta d) e^{-i\zeta x} \, d\zeta$$

$$= \sum_{m=1}^{\infty} c_m \sin a_m (x + a) Y_q'(-b) \tag{11}$$

$$k_{zs} e^{ik_{xs} x} \cos(k_{zs} b)$$

$$- \sum_{m=1}^{\infty} c_m Y_q(-b) \frac{a_m}{2\pi} \int_{-\infty}^{\infty} \frac{\kappa}{\sin(\kappa b)} a^2 F_m(\zeta a) e^{-i\zeta x} \, d\zeta$$

$$+ \sum_{m=1}^{\infty} d_m X_p(0) \frac{b_m}{2\pi} \int_{-\infty}^{\infty} \kappa \cot(\kappa b) e^{i\zeta \alpha} d^2 F_m(\zeta d) e^{-i\zeta x} \, d\zeta$$

$$= \sum_{m=1}^{\infty} d_m \sin b_m (x - \alpha + d) X_p'(0) . \tag{12}$$

Multiplying the H-field continuity at $z = -b$, (11), by $\sin a_n(x + a)$ and integrating with respect to x from $-a$ to a, we get

$$
k_{zs}a_n a^2 F_n(k_{xs}a) = \sum_{m=1}^{\infty} c_m Y_q(-b)\frac{a_m a_n}{2\pi} I_1
$$

$$
- \sum_{m=1}^{\infty} d_m X_p(0)\frac{b_m a_n}{2\pi} I_2 + c_n a Y_q'(-b)\delta_{nm} \tag{13}
$$

where

$$
I_1 = \int_{-\infty}^{\infty} \kappa \cot(\kappa b) a^2 F_m(\zeta a) a^2 F_n(-\zeta a)\,\mathrm{d}\zeta \tag{14}
$$

$$
I_2 = \int_{-\infty}^{\infty} \frac{\kappa e^{i\zeta\alpha}}{\sin(\kappa b)} d^2 F_m(\zeta d) a^2 F_n(-\zeta a)\,\mathrm{d}\zeta \ . \tag{15}
$$

Utilizing the technique of contour integration, we transform the integrals into fast-convergent series

$$
I_1 = \begin{cases} 0, & n + m = \text{ odd} \\ h_m \delta_{nm} + r_{nm}, & n + m = \text{ even} \end{cases} \tag{16}
$$

$$
I_2 = g_m \delta_{nm} \Delta + f_{nm} \tag{17}
$$

where

$$
\Delta = \begin{cases} 1, & d = a \text{ and } \alpha = 0 \\ 0, & \text{otherwise} \end{cases} \tag{18}
$$

$$
h_m = \frac{2\pi a \xi_m}{a_m^2 \tan(\xi_m b)} \tag{19}
$$

$$
g_m = \frac{B_m \eta_m}{b_m^2 \sin(\eta_m b)} \tag{20}
$$

$$
B_m = -\pi \big[|d - a| \exp(i\zeta|d - a|)
$$

$$
-(-1)^m |d + a| \exp(i\zeta|d + a|)\big]\big|_{\zeta=a_m} \tag{21}
$$

$$
r_{nm} = \sum_{l=1}^{\infty} 2\cos(l\pi)T_{\xi\xi}\left[1 - (-1)^m \exp\left(i2\sqrt{k^2 - (l\pi/b)^2}a\right)\right] \tag{22}
$$

$$
f_{nm} = \sum_{l=1}^{\infty} T_{\eta\xi}\big[(-1)^{m+n} \exp(i\zeta|d - a + \alpha|)
$$

$$
-(-1)^m \exp(i\zeta|d + a + \alpha|) - (-1)^n \exp(i\zeta|\alpha - a - d|)
$$

$$
+ \exp(i\zeta|\alpha + a - d|)\big]\big|_{\zeta=\sqrt{k^2-(l\pi/b)^2}} \tag{23}
$$

$$
T_{uv} = \frac{-i2\pi(l\pi/b)^2}{\sqrt{k^2 - (l\pi/b)^2}b[u_m^2 - (l\pi/b)^2][v_n^2 - (l\pi/b)^2]\cos(l\pi)} \tag{24}
$$

$$
u, v = \xi, \eta \ .
$$

Multiplying the H-field continuity at $z = 0$, (12), by $\sin b_n(x - \alpha + d)$ and integrating from $(\alpha - d)$ to $(\alpha + d)$ similarly, we obtain

$$
\boxed{
\begin{aligned}
&k_{zs}\cos(k_{zs}b)b_n \mathrm{e}^{\mathrm{i}k_{xs}\alpha}d^2 F_n(k_{xs}d) = \\
&\sum_{m=1}^{\infty} c_m Y_q(-b)\frac{a_m b_n}{2\pi}I_3 - \sum_{m=1}^{\infty} d_m X_p(0)\frac{b_m b_n}{2\pi}I_4 + d_n d X_p'(0)\delta_{nm}
\end{aligned}
}
\tag{25}
$$

where

$$
I_3 = \int_{-\infty}^{\infty} \frac{\kappa \mathrm{e}^{\mathrm{i}\zeta\alpha}}{\sin(\kappa b)}a^2 F_m(\zeta a)d^2 F_n(-\zeta d)\,\mathrm{d}\zeta
\tag{26}
$$

$$
I_4 = \int_{-\infty}^{\infty} \kappa \cot(\kappa b)d^2 F_m(\zeta d)d^2 F_n(-\zeta d)\,\mathrm{d}\zeta \ .
\tag{27}
$$

Note that

$$
I_3 = g_m \delta_{nm}\Delta + t_{nm}
\tag{28}
$$

$$
I_4 = \begin{cases} h_m \delta_{nm} + z_{nm}, & n + m = \text{even} \\ 0, & n + m = \text{odd} \end{cases}
\tag{29}
$$

where

$$
z_{nm} = \sum_{l=1}^{\infty} 2\cos(l\pi)T_{\eta\eta}\left[1 - (-1)^m \exp\left(\mathrm{i}2\sqrt{k^2 - (l\pi/b)^2}d\right)\right]
\tag{30}
$$

$$
\begin{aligned}
t_{nm} = \sum_{l=1}^{\infty} T_{\xi\eta}\big[&(-1)^{m+n}\exp(\mathrm{i}\zeta| - d + a - \alpha|) \\
&-(-1)^m \exp(\mathrm{i}\zeta|d + a - \alpha|) \\
&-(-1)^n \exp(\mathrm{i}\zeta| - \alpha - a - d|) \\
&+ \exp(\mathrm{i}\zeta| - \alpha - a + d|)\big]\Big|_{\zeta=\sqrt{k^2-(l\pi/b)^2}} \ .
\end{aligned}
\tag{31}
$$

The transmitted and reflected fields at $x = \pm\infty$ in region (I) are

$$
E_y^I(\pm\infty, z) = \sum_v L_v^{\pm} \sin k_{zv}(z + b)\mathrm{e}^{\pm \mathrm{i}k_{xv}x}
\tag{32}
$$

where $1 \le v < kb/\pi$, v: integer $(1, 2, 3, \dots)$, $k_{zv} = v\pi/b$, $k_{xv} = \sqrt{k^2 - k_{zv}^2}$, and

$$
\begin{aligned}
L_v^{\pm} = &\sum_{m=1}^{\infty} \mathrm{i}c_m Y_q(-b)\frac{a_m k_{zv}}{k_{xv}b}a^2 F_m(\mp k_{xv}a) \\
&- \sum_{m=1}^{\infty} \mathrm{i}d_m X_p(0)\frac{b_m k_{zv}\mathrm{e}^{\mp \mathrm{i}k_{xv}\alpha}}{k_{xv}b(-1)^v}d^2 F_m(\pm k_{xv}d) \ .
\end{aligned}
\tag{33}
$$

The transmission (τ_1, τ_2, and τ_3) and reflection (ϱ) coefficients are

$$\tau_1 = P_{t1}/P_i = |1 + L_s^+|^2 + \frac{1}{k_{xs}} \sum_{v \neq s} k_{xv} |L_v^+|^2 \tag{34}$$

$$\varrho = P_r/P_i = \frac{1}{k_{xs}} \sum_v k_{xv} |L_v^-|^2 \tag{35}$$

$$\tau_2 = P_{t2}/P_i = \frac{2a}{k_{xs}b} \sum_\gamma \xi_\gamma |c_\gamma|^2 \tag{36}$$

$$\tau_3 = P_{t3}/P_i = \frac{2d}{k_{xs}b} \sum_\sigma \eta_\sigma |d_\sigma|^2 \tag{37}$$

where $1 \leq \gamma < 2ak/\pi$, $1 \leq \sigma < 2dk/\pi$, $1 \leq v < kb/\pi$, and γ, σ, v are integers.

5.4 H-Plane Double Bend [9]

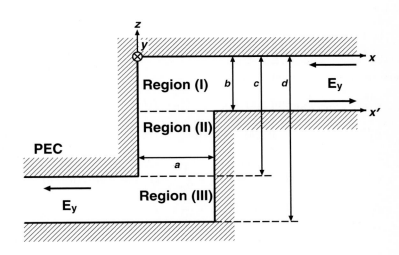

Fig. 5.4. H-plane double bend

Wave scattering from a rectangular waveguide double bend has been considered in [10] and applied to the problem of quantum waveguide structures. A TE wave $E_y^i(x, z)$ is incident on a right-angled double bend with the wavenumber $k = \omega\sqrt{\mu\epsilon}$. In region (I) ($0 < x$ and $-b < z < 0$) the field has the incident, reflected, and scattered components as

$$E_y^i(x, z) = e^{-ik_{zs}x} \sin k_{zs}(z + b) \tag{1}$$

$$E_y^r(x, z) = -e^{ik_{zs}x} \sin k_{zs}(z + b) \tag{2}$$

$$E_y^I(x, z) = \frac{2}{\pi} \int_0^\infty \widetilde{E}_y^I(\zeta) \sin \kappa z \sin \zeta x \, d\zeta \tag{3}$$

where $k_{zs} = s\pi/b$, s: integer, $k_{xs} = \sqrt{k^2 - k_{zs}^2}$, and $\kappa = \sqrt{k^2 - \zeta^2}$. In region (II) ($0 < x < a$ and $-c < z < -b$) the field is

$$E_y^{II}(x, z) = \sum_{m=1}^\infty \sin(a_m x)(\alpha_m \cos \xi_m z + \beta_m \sin \xi_m z) \tag{4}$$

where $a_m = m\pi/a$ and $\xi_m = \sqrt{k^2 - a_m^2}$. In region (III) ($x' < 0$, $x' = x - a$, and $-d < z < -c$) the field is

$$E_y^{III}(x, z) = \frac{2}{\pi} \int_0^\infty \widetilde{E}_y^{III}(\zeta) \sin \kappa(z + d) \sin \zeta x' \, d\zeta \ . \tag{5}$$

We apply the Fourier sine transform to the E-field continuities to represent $\widetilde{E}_y^I(\zeta)$ and $\widetilde{E}_y^{III}(\zeta)$ in terms of α_m and β_m. The field continuities at $z = -b$ and $-c$ yield

$$\begin{bmatrix} \Psi_{11} & \Psi_{12} \\ \Psi_{21} & \Psi_{22} \end{bmatrix} \begin{bmatrix} \alpha \\ \beta \end{bmatrix} = \begin{bmatrix} \Gamma \\ 0 \end{bmatrix} \tag{6}$$

where the matrix elements are

$$\psi_{11}^{nm} = \delta_{nm} \frac{a}{2} \xi_n \sin(\xi_n b) + \frac{1}{2\pi} \cos(\xi_m b) I(b) \tag{7}$$

$$\psi_{12}^{nm} = \delta_{nm} \frac{a}{2} \xi_n \cos(\xi_n b) - \frac{1}{2\pi} \sin(\xi_m b) I(b) \tag{8}$$

$$\psi_{21}^{nm} = \delta_{nm} \frac{a}{2} (-1)^n \xi_n \sin(\xi_n c) - \frac{1}{2\pi} (-1)^m \cos(\xi_m c) I(d - c) \tag{9}$$

$$\psi_{22}^{nm} = \delta_{nm} \frac{a}{2} (-1)^n \xi_n \cos(\xi_n c) + \frac{1}{2\pi} (-1)^m \cos(\xi_m c) I(d - c) \tag{10}$$

$$\gamma^n = \frac{k_{zs} a_n [(-1)^n e^{-ik_{xs}a} - (-1)^n e^{ik_{xs}a}]}{k_{xs}^2 - a_n^2} \tag{11}$$

$$I(u) = \sum_{l=1}^\infty \frac{-i2\pi(l\pi/u)^2(-1)^{m+n} a_m a_n \left[1 - \exp\left(i2\sqrt{k^2 - (l\pi/u)^2}a\right)\right]}{\sqrt{k^2 - (l\pi/u)^2}u[\xi_m^2 - (l\pi/u)^2][\xi_n^2 - (l\pi/u)^2]}$$
$$+ \delta_{nm} \pi a \xi_m \cot(\xi_m u) \ . \tag{12}$$

The reflected field for $x > 0$ and the transmitted field for $x' < 0$ are respectively

$$E_y^I(\infty, z) = \sum_p L_p^+ \sin(k_{zv}^+ z) e^{ik_{xv}^+ x} \tag{13}$$

$$E_y^{III}(-\infty, z) = \sum_q L_q^- \sin k_{zv}^-(z + d) e^{-ik_{xv}^- x'} \tag{14}$$

where $k_{zv}^+ = p\pi/b$, $k_{zv}^- = q\pi/(d-c)$, $k_{xv}^\pm = \sqrt{k^2 - (k_{zv}^\pm)^2}$, and

$$L_p^+ = \sum_{m=1}^\infty \frac{2k_{zv}^+(-1)^{m+p} a_m \sin(k_{xv}^+ a)(\alpha_m \cos \xi_m b - \beta_m \sin \xi_m b)}{k_{xv}^+ b[(k_{xv}^+)^2 - a_m^2]} \tag{15}$$

$$L_q^- = \sum_{m=1}^\infty \frac{2k_{zv}^-(-1)^q a_m \sin(k_{xv}^- a)(\alpha_m \cos \xi_m c - \beta_m \sin \xi_m c)}{k_{xv}^-(d-c)[(k_{xv}^-)^2 - a_m^2]} \;. \tag{16}$$

The transmission (τ) and reflection (ϱ) coefficients are

$$\tau = P_t/P_i = \sum_q \frac{k_{xv}^-(d-c)}{k_{xs}b} |L_q^-|^2 \tag{17}$$

$$\varrho = P_r/P_i = |1 - L_s^+|^2 + \sum_{p \neq s} \frac{k_{xv}^+}{k_{xs}} |L_p^+|^2 \tag{18}$$

where $1 \leq p < kb/\pi$, p: integer, $1 \leq q < k(d-c)/\pi$, and q: integer.

5.5 Acoustic Double Junction in a Rectangular Waveguide [11]

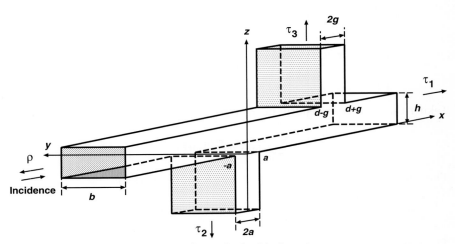

Fig. 5.5. Acoustic double junction

A study of acoustic wave scattering from a T-junction and a right-angled bend in a rectangular waveguide has been performed in [12-14]. In this section we will study acoustic wave scattering from two junctions in a hard surface rectangular waveguide. In region (I) ($0 < z < h$) an incident field (velocity potential) $\Phi^i(x, y, z)$ propagates along the x-direction and impinges on two junctions where the wavenumber is k ($= 2\pi/\lambda$, λ: wavelength). In region (I) the total field consists of the incident and scattered potentials

$$\Phi^i(x, y, z) = \cos(b_p y) \cos h_l(z - h) e^{i\beta_{pl}x} \tag{1}$$

$$\Phi^I(x, y, z) = \frac{1}{2\pi} \int_{-\infty}^{\infty} \left[\widetilde{A}(\zeta) \cos \kappa_p(z - h) + \widetilde{B}(\zeta) \cos(\kappa_p z) \right]$$
$$\cdot \cos(b_p y) e^{-i\zeta x} d\zeta \tag{2}$$

where $b_p = p\pi/b$, $h_l = l\pi/h$, $\beta_{pl} = \sqrt{k^2 - b_p^2 - h_l^2}$, and $\kappa_p = \sqrt{k^2 - \zeta^2 - b_p^2}$. In region (II) ($-a < x < a$ and $z < 0$) the transmitted field is

$$\Phi^{II}(x, y, z) = \sum_{m=0}^{\infty} C_m \cos a_m(x + a) \cos(b_p y) e^{-i\xi_{mp}z} \tag{3}$$

where $a_m = m\pi/(2a)$ and $\xi_{mp} = \sqrt{k^2 - a_m^2 - b_p^2}$. In region (III) ($d - g < x < d + g$ and $h < z$) the transmitted field is

$$\Phi^{III}(x, y, z) = \sum_{m=0}^{\infty} D_m \cos g_m(x - d + g) \cos(b_p y) e^{i\eta_{mp}z} \tag{4}$$

where $g_m = m\pi/(2g)$ and $\eta_{mp} = \sqrt{k^2 - g_m^2 - b_p^2}$.

The velocity continuity condition at $z = 0$ requires

$$\frac{\partial[\Phi^i(x, y, z) + \Phi^I(x, y, z)]}{\partial z} \bigg|_{z=0}$$
$$= \begin{cases} \dfrac{\partial \Phi^{II}(x, y, z)}{\partial z} \bigg|_{z=0}, & |x| < a, \ 0 < y < b \\ 0, & \text{otherwise} . \end{cases} \tag{5}$$

Taking the Fourier transform of (5) results in

$$\widetilde{A}(\zeta) = -\frac{a^2 \zeta}{\kappa_p \sin(\kappa_p h)} \sum_{m=0}^{\infty} \xi_{mp} C_m F_m(\zeta a) . \tag{6}$$

The velocity continuity condition at $z = h$ similarly yields

$$\widetilde{B}(\zeta) = \frac{-g^2 \zeta e^{i\zeta d}}{\kappa_p \sin(\kappa_p h)} \sum_{m=0}^{\infty} \eta_{mp} D_m F_m(\zeta g) e^{i\eta_{mp}h} . \tag{7}$$

The pressure continuity condition at $z = 0$

$$\Phi^i(x, y, 0) + \Phi^I(x, y, 0) = \Phi^{II}(x, y, 0), \qquad |x| < a, \ 0 < y < b \tag{8}$$

is written as

$$(-1)^l e^{i\beta_{pl}x} + \frac{1}{2\pi} \int_{-\infty}^{\infty} \left[\widetilde{A}(\zeta)\cos(\kappa_p h) + \widetilde{B}(\zeta) \right] e^{-i\zeta x}\, d\zeta$$

$$= \sum_{m=0}^{\infty} C_m \cos a_m (x+a) \,. \tag{9}$$

Similarly from the pressure continuity condition at $z = h$, we obtain

$$e^{i\beta_{pl}x} + \frac{1}{2\pi} \int_{-\infty}^{\infty} \left[\widetilde{A}(\zeta) + \widetilde{B}(\zeta)\cos(\kappa_p h) \right] e^{-i\zeta x}\, d\zeta$$

$$= \sum_{m=0}^{\infty} D_m \cos g_m (x-d+g) \exp(i\eta_{mp}h) \,. \tag{10}$$

We substitute $\widetilde{A}(\zeta)$ and $\widetilde{B}(\zeta)$ into (9), multiply (9) by $\cos a_n(x+a)$ ($n = 0, 1, 2, 3, \dots$), and perform integration from $-a$ to a to get

$$\boxed{\begin{aligned} i a \beta_{pl}(-1)^{l+1} F_n(\beta_{pl}a) &= \frac{ia}{2\pi} \sum_{m=0}^{\infty} \left[a^2 \xi_{mp} C_m I_{1nm}(a) \right. \\ &\left. + g^2 \eta_{mp} D_m e^{i\eta_{mp}h} I_{2nm} \right] + C_m \varepsilon_n \delta_{nm} \end{aligned}} \tag{11}$$

where

$$I_{1nm}(a) = \int_{-\infty}^{\infty} \frac{\zeta^2 \cot(\kappa_p h)}{\kappa_p} F_m(\zeta a) F_n(-\zeta a)\, d\zeta \tag{12}$$

$$I_{2nm} = \int_{-\infty}^{\infty} \frac{\zeta^2 e^{i\zeta d}}{\kappa_p \sin(\kappa_p h)} F_m(\zeta g) F_n(-\zeta a)\, d\zeta \tag{13}$$

$$\varepsilon_v = \begin{cases} 2, & v = 0 \\ 1, & v = 1, 2, 3, \dots \end{cases} \tag{14}$$

Note that $I_{1nm}(a)$ and I_{2nm} are transformed into rapidly-convergent series

$$I_{1nm}(a) = \begin{cases} 0, & n + m = \text{odd} \\ p_m(a)\delta_{nm} + q_{nm}(a), & n + m = \text{even} \end{cases} \tag{15}$$

$$I_{2nm} = X(\zeta) + \begin{cases} \left. 2\pi i P(\zeta) \right|_{\zeta=a_n} + \left. 2\pi i Q(\zeta) \right|_{\zeta=g_m}, & a_n \neq g_m \\[2mm] \left. \dfrac{2\pi i}{\varepsilon_n a^2 g^2} \dfrac{f'(\zeta)s(\zeta) - f(\zeta)s'(\zeta)}{s^2(\zeta)} \right|_{\zeta=a_n}, & a_n = g_m \end{cases} \tag{16}$$

where δ_{nm} is the Kronecker delta, $(\cdot)'$ denotes differentiation, and

$$p_m(a) = \frac{2\pi \varepsilon_m \cot(\xi_{mp}h)}{a^3 \xi_{mp}} \tag{17}$$

$$q_{nm}(a) = -\frac{4\pi i}{a^4 h} \sum_{v=0}^{\infty} \left. \frac{\zeta[1 - (-1)^m e^{i2\zeta a}]}{\varepsilon_v (\zeta^2 - a_m^2)(\zeta^2 - a_n^2)} \right|_{\zeta=\sqrt{k^2 - (v\pi/h)^2 - b_p^2}} \tag{18}$$

$$P(\zeta) = \frac{f_1(\zeta)}{a^2 g^2 \kappa_p \sin(h\kappa_p)(\zeta + a_n)(\zeta^2 - g_m^2)} \tag{19}$$

$$Q(\zeta) = \frac{f_1(\zeta)}{a^2 g^2 \kappa_p \sin(h\kappa_p)(\zeta + g_m)(\zeta^2 - a_n^2)} \tag{20}$$

$$f(\zeta) = \begin{cases} f_1(\zeta), & a_n = g_m \neq 0 \\ f_2(\zeta), & a_n = g_m = 0 \end{cases} \tag{21}$$

$$\begin{aligned} f_1(\zeta) = \zeta^2 [&(-1)^{m+n} \exp(i\zeta|d - a + g|) \\ &- (-1)^m \exp(i\zeta|d + a + g|) \\ &- (-1)^n \exp(i\zeta|d - a - g|) \\ &+ \exp(i\zeta|d + a - g|)] \end{aligned} \tag{22}$$

$$\begin{aligned} f_2(\zeta) = \big[&\exp(i\zeta|d - a + g|) - \exp(i\zeta|d + a + g|) \\ &- \exp(i\zeta|d - a - g|) + \exp(i\zeta|d + a - g|) \big] \end{aligned} \tag{23}$$

$$s(\zeta) = \begin{cases} \kappa_p \sin(h\kappa_p)(\zeta + a_n)^2, & a_n = g_m \neq 0 \\ \kappa_p \sin(h\kappa_p), & a_n = g_m = 0 \end{cases} \tag{24}$$

$$X(\zeta) = -\frac{2\pi i}{a^2 g^2 h} \sum_{v=0}^{\infty} \tag{25}$$

$$\frac{f_1(\zeta)}{\zeta \varepsilon_v (-1)^v [\xi_{np}^2 - (v\pi/h)^2][\eta_{mp}^2 - (v\pi/h)^2]} \Bigg|_{\zeta = \sqrt{k^2 - (v\pi/h)^2 - b_p^2}} \cdot$$

We similarly multiply (10) by $\cos g_n(x - d + g)$ $(n = 0, 1, 2, 3, \ldots)$ and perform integration to obtain

$$\begin{aligned} -ig\beta_{pl} e^{i\beta_{pl}d} F_n(\beta_{pl}g) = \frac{ig}{2\pi} \sum_{m=0}^{\infty} \Big[&a^2 \xi_{mp} C_m I_{2mn} \\ + g^2 \eta_{mp} D_m e^{i\eta_{mp}h} I_{1nm}(g) \Big] &+ e^{i\eta_{mp}h} D_m \varepsilon_n \delta_{nm} \cdot \end{aligned} \tag{26}$$

In low-frequency approximation, the lowest wave ($m = n = 0$) becomes dominant; thus (11) and (26) with $p = l = 0$ lead to

$$C_0 \approx \frac{\psi_4 \gamma_1 - \psi_2 \gamma_2}{\psi_1 \psi_4 - \psi_2 \psi_3} \tag{27}$$

$$D_0 \approx \frac{\psi_1 \gamma_2 - \psi_3 \gamma_1}{\psi_1 \psi_4 - \psi_2 \psi_3} \tag{28}$$

where

$$\psi_1 = 2\mathrm{i}\frac{\mathrm{e}^{-\mathrm{i}kh}}{\sin(kh)} + 2k\frac{a}{h}\sum_{v=0}^{\infty}\frac{F_0(-a\zeta_v)}{\varepsilon_v\zeta_v}\mathrm{e}^{\mathrm{i}\zeta_v a} \tag{29}$$

$$\psi_2 = \mathrm{e}^{\mathrm{i}kh}\left[-\frac{f_2'(0)}{2a\sin(kh)} + \frac{k}{ah}\sum_{v=0}^{\infty}\frac{f_2(\zeta_v)}{\varepsilon_v(-1)^v\zeta_v^3}\right] \tag{30}$$

$$\psi_3 = -\frac{f_2'(0)}{2g\sin(kh)} + \frac{k}{gh}\sum_{v=0}^{\infty}\frac{f_2(\zeta_v)}{\varepsilon_v(-1)^v\zeta_v^3} \tag{31}$$

$$\psi_4 = \mathrm{e}^{\mathrm{i}kh}\left[2\mathrm{i}\frac{\mathrm{e}^{-\mathrm{i}kh}}{\sin(kh)} + 2k\frac{g}{h}\sum_{v=0}^{\infty}\frac{F_0(-g\zeta_v)}{\varepsilon_v\zeta_v}\mathrm{e}^{\mathrm{i}\zeta_v g}\right] \tag{32}$$

$$\gamma_1 = -\mathrm{i}akF_0(ka) \tag{33}$$

$$\gamma_2 = -\mathrm{i}gk\mathrm{e}^{\mathrm{i}kd}F_0(kg) \tag{34}$$

$$\zeta_v = \sqrt{k^2 - (v\pi/h)^2}\ . \tag{35}$$

The scattered fields for $(x \le -a)$ and $(x \ge d+g)$ are given as

$$\Phi^I(x,y,z)$$
$$= \begin{cases} \displaystyle\sum_u K_u^+ \cos(b_p y)\cos h_u(z-h)\mathrm{e}^{\mathrm{i}\beta_{pu}x}, & x \ge d+g \\[2ex] \displaystyle\sum_u K_u^- \cos(b_p y)\cos h_u(z-h)\mathrm{e}^{-\mathrm{i}\beta_{pu}x}, & x \le -a \end{cases} \tag{36}$$

where

$$K_u^\pm = \mp\sum_{m=0}^{\infty}\frac{\mathrm{i}a^2\xi_{mp}C_m F_m(\mp\beta_{pu}a)}{h\varepsilon_u(-1)^u}$$
$$\mp\sum_{m=0}^{\infty}\frac{\mathrm{i}g^2\eta_{mp}D_m F_m(\mp\beta_{pu}g)\mathrm{e}^{\mp\mathrm{i}d\beta_{pu}}}{h\varepsilon_u}\mathrm{e}^{\mathrm{i}\eta_{mp}h} \tag{37}$$

$0 \le u$ and u: integer. The transmission (τ_1, τ_2, and τ_3) and reflection (ϱ) coefficients are

$$\tau_1 = |1 + K_l^+|^2 + \sum_{u \ne l}\frac{\varepsilon_u\beta_{pu}|K_u^+|^2}{\varepsilon_l\beta_{pl}} \tag{38}$$

$$\tau_2 = \frac{2a}{h\varepsilon_l\beta_{pl}}\sum_m |C_m|^2\varepsilon_m\xi_{mp} \tag{39}$$

$$\tau_3 = \frac{2g}{h\varepsilon_l\beta_{pl}}\sum_n |D_n|^2\varepsilon_n\eta_{np} \tag{40}$$

$$\varrho = \sum_u \frac{\varepsilon_u\beta_{pu}|K_u^-|^2}{\varepsilon_l\beta_{pl}} \tag{41}$$

where $0 \le u < (h/\pi)\sqrt{k^2 - b_p^2}$, u: integer, $0 \le m < (2a/\pi)\sqrt{k^2 - b_p^2}$, m: integer, $0 \le n < (2g/\pi)\sqrt{k^2 - b_p^2}$, and n: integer.

5.6 Acoustic Hybrid Junction in a Rectangular Waveguide [15]

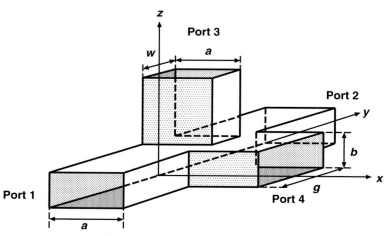

Fig. 5.6. Acoustic hybrid junction

The electromagnetic hybrid junction in a rectangular waveguide, also known as a magic T, is a microwave passive component (duplexer) to decouple powers between the E-plane and H-plane arms [16]. In this section we will consider a similar hybrid junction structure for acoustic wave scattering and investigate its transmission and reflection coefficients. Next two subsections present the scattering analyses for an acoustic hybrid junction in a rectangular waveguide consisting of hard and soft surfaces, respectively.

5.6.1 Hard-Surface Hybrid Junction

Consider a hybrid junction in a hard-surface rectangular waveguide where the wavenumber is k. In region (I) ($0 < x < a$ and $0 < z < b$) the incident fields (velocity potentials), $\Phi_1^i(x, y, z)$ and $\Phi_2^i(x, y, z)$, propagate along the $\pm y$ directions. Then the total field in region (I) consists of the incident and scattered components

$$\Phi_1^i(x, y, z) = \cos(A_1 x) \cos(B_1 z) e^{i\beta_1 y} \tag{1}$$

$$\Phi_2^i(x, y, z) = \cos(A_2 x) \cos(B_2 z) e^{-i\beta_2 y} \tag{2}$$

$$\Phi^I(x, y, z) = \frac{1}{2\pi} \int_{-\infty}^{\infty} \sum_{m=0}^{\infty} \left[\widetilde{A}_m(\zeta) \cos(\kappa_m x) \cos(b_m z) \right.$$

$$\left. + \widetilde{B}_m(\zeta) \cos(a_m x) \cos(\eta_m z) \right] e^{-i\zeta y} \mathrm{d}\zeta \tag{3}$$

where $A_1 = \alpha\pi/a$, $B_1 = \mu\pi/b$, $\beta_1 = \sqrt{k^2 - A_1^2 - B_1^2}$, $A_2 = \varrho\pi/a$, $B_2 = \nu\pi/b$,
$\beta_2 = \sqrt{k^2 - A_2^2 - B_2^2}$ $(\alpha, \mu, \varrho, \nu = 0, 1, 2, \ldots)$, $a_m = m\pi/a$, $b_m = m\pi/b$,
$\kappa_m = \sqrt{k^2 - \zeta^2 - b_m^2}$, and $\eta_m = \sqrt{k^2 - \zeta^2 - a_m^2}$. In region (II) $(0 < x <$
$a, -w/2 < y < w/2$, and $b < z$) the incident and transmitted fields through
port 3 are

$$\Phi_3^i(x, y, z) = \cos(A_3 x) \cos W_3(y + w/2) e^{-i\beta_3 z} \tag{4}$$

$$\Phi^{II}(x, y, z) = \sum_{p=0}^{\infty} \sum_{q=0}^{\infty} a_{pq} \cos(a_p x) \cos w_q(y + w/2) e^{i\xi_{pq} z} \tag{5}$$

where $A_3 = \phi\pi/a$, $W_3 = \tau\pi/w$, $\beta_3 = \sqrt{k^2 - A_3^2 - W_3^2}$ $(\phi, \tau = 0, 1, 2, \ldots)$,
$w_q = q\pi/w$, and $\xi_{pq} = \sqrt{k^2 - a_p^2 - w_q^2}$. In region (III) $(a < x, \ -g/2 < y <$
$g/2$, and $0 < z < b$) the incident and transmitted fields through port 4 are

$$\Phi_4^i(x, y, z) = \cos G_4(y + g/2) \cos(B_4 z) e^{-i\beta_4 x} \tag{6}$$

$$\Phi^{III}(x, y, z) = \sum_{p=0}^{\infty} \sum_{q=0}^{\infty} b_{pq} \cos g_p(y + g/2) \cos(b_q z) e^{i\gamma_{pq} x} \tag{7}$$

where $G_4 = \psi\pi/g$, $B_4 = \chi\pi/b$, $\beta_4 = \sqrt{k^2 - G_4^2 - B_4^2}$ $(\psi, \chi = 0, 1, 2, \ldots)$,
$g_p = p\pi/g$, and $\gamma_{pq} = \sqrt{k^2 - g_p^2 - b_q^2}$.

The velocity continuity condition at $z = b$ requires

$$\left. \frac{\partial[\Phi_1^i(x, y, z) + \Phi_2^i(x, y, z) + \Phi^I(x, y, z)]}{\partial z} \right|_{z=b}$$

$$= \begin{cases} \left. \dfrac{\partial[\Phi_3^i(x, y, z) + \Phi^{II}(x, y, z)]}{\partial z} \right|_{z=b}, & 0 < x < a, \ |y| < w/2 \\[2ex] 0, & \text{otherwise} . \end{cases} \tag{8}$$

Taking the Fourier transform of (8) yields

$$-\sum_{m=0}^{\infty} \widetilde{B}_m(\zeta) \eta_m \cos(a_m x) \sin(\eta_m b)$$

$$= -\beta_3 \zeta \cos(A_3 x) \Xi_\tau^w(\zeta) e^{-i\beta_3 b}$$

$$+ \sum_{p=0}^{\infty} \sum_{q=0}^{\infty} a_{pq} \xi_{pq} \zeta \cos(a_p x) \Xi_q^w(\zeta) e^{i\xi_{pq} b} \tag{9}$$

where

$$\Xi_q^w(\zeta) = \frac{(-1)^q e^{i\zeta w/2} - e^{-i\zeta w/2}}{\zeta^2 - w_q^2} . \tag{10}$$

We multiply (9) by $\cos(a_m x)$ and perform integration with respect to x from
0 to a to get

$$\widetilde{B}_m(\zeta) = \frac{\zeta\beta_3}{\eta_\phi \sin(\eta_\phi b)}\Xi_\tau^w(\zeta)\mathrm{e}^{-\mathrm{i}\beta_3 b}$$

$$-\sum_{q=0}^{\infty} a_{mq}\frac{\zeta\xi_{mq}}{\eta_m \sin(\eta_m b)}\Xi_q^w(\zeta)\mathrm{e}^{\mathrm{i}\xi_{mq} b} \ . \tag{11}$$

Similarly we apply the Fourier transform to the velocity continuity condition at $x = a$, multiply by $\cos(b_m z)$, and perform integration from 0 to b to get

$$\widetilde{A}_m(\zeta) = \frac{\zeta\beta_4}{\kappa_\chi \sin(\kappa_\chi a)}\Xi_\psi^g(\zeta)\mathrm{e}^{-\mathrm{i}\beta_4 a}$$

$$-\sum_{p=0}^{\infty} b_{pm}\frac{\zeta\gamma_{pm}}{\kappa_m \sin(\kappa_m a)}\Xi_p^g(\zeta)\mathrm{e}^{\mathrm{i}\gamma_{pm} a} \ . \tag{12}$$

Note that the pressure continuity condition at $z = b$, $0 < x < a$, and $|y| < w/2$ requires

$$\Phi_1^i(x,y,b) + \Phi_2^i(x,y,b) + \Phi^I(x,y,b) = \Phi_3^i(x,y,b) + \Phi^{II}(x,y,b) \ . \tag{13}$$

Re-expressing (13) explicitly yields

$$\cos(A_1 x)(-1)^\mu \mathrm{e}^{\mathrm{i}\beta_1 y} + \cos(A_2 x)(-1)^\nu \mathrm{e}^{-\mathrm{i}\beta_2 y}$$

$$+\frac{1}{2\pi}\int_{-\infty}^{\infty}\sum_{m=0}^{\infty}[\widetilde{A}_m(\zeta)\cos(\kappa_m x)(-1)^m$$

$$+\widetilde{B}_m(\zeta)\cos(a_m x)\cos(\eta_m b)]\mathrm{e}^{-\mathrm{i}\zeta y}\mathrm{d}\zeta$$

$$= \cos(A_3 x)\cos W_3\left(y + \frac{w}{2}\right)\mathrm{e}^{-\mathrm{i}\beta_3 b}$$

$$+\sum_{p=0}^{\infty}\sum_{q=0}^{\infty} a_{pq}\cos(a_p x)\cos w_q\left(y + \frac{w}{2}\right)\mathrm{e}^{\mathrm{i}\xi_{pq} b} \ . \tag{14}$$

We substitute (11) and (12) into (14), and perform

$$\int_{-\frac{w}{2}}^{\frac{w}{2}}\int_0^a (14)\cos(a_s x)\cos w_t\left(y + \frac{w}{2}\right)\mathrm{d}x\mathrm{d}y, \ (s = 0, 1, \ldots \text{ and } t = 0, 1, \ldots), \text{ to}$$

obtain

$$\begin{aligned} &-\mathrm{i}(-1)^\mu \frac{a}{2}\beta_1\varepsilon_\alpha\delta_{as}\Xi_t^w(\beta_1) + \mathrm{i}(-1)^\nu \frac{a}{2}\beta_2\varepsilon_\varrho\delta_{\varrho s}\Xi_t^w(-\beta_2)\\ &+\frac{a}{2}\delta_{\phi s}\mathrm{e}^{-\mathrm{i}\beta_3 b}\left(\mathrm{i}\varepsilon_s\beta_3 I_{1\phi\tau t}^{baw} - \frac{w}{2}\varepsilon_\phi\varepsilon_\tau\delta_{\tau t}\right)\\ &+\mathrm{i}(-1)^{\chi+s}\beta_4\mathrm{e}^{-\mathrm{i}\beta_4 a}I_{2s\chi\psi t}^{gw}\\ &= \sum_{p=0}^{\infty}\sum_{q=0}^{\infty}\left[a_{pq}\frac{a}{2}\varepsilon_s\delta_{ps}\mathrm{e}^{\mathrm{i}\xi_{pq} b}\left(\mathrm{i}\xi_{pq}I_{1pqt}^{baw} + \frac{w}{2}\varepsilon_t\delta_{qt}\right)\right.\\ &\left.+b_{pq}\mathrm{i}(-1)^{q+s}\gamma_{pq}\mathrm{e}^{\mathrm{i}\gamma_{pq} a}I_{2sqpt}^{gw}\right] \end{aligned} \tag{15}$$

where δ_{ms} is the Kronecker delta and

$$\varepsilon_s = \begin{cases} 2, & s = 0 \\ 1, & s = 1, 2, 3, \dots \end{cases} \tag{16}$$

The pressure continuity condition at $x = a$, $|y| < g/2$, and $0 < z < b$ requires

$$\Phi_1^i(a, y, z) + \Phi_2^i(a, y, z) + \Phi^I(a, y, z) = \Phi_4^i(a, y, z) + \Phi^{III}(a, y, z) . \tag{17}$$

We substitute (11) and (12) into (17), and perform

$$\int_{-\frac{g}{2}}^{\frac{g}{2}} \int_0^b (17) \cos(b_h z) \cos g_n \left(y + \frac{g}{2} \right) \mathrm{d}z \mathrm{d}y, \ (h = 0, 1, \dots \text{ and } n = 0, 1, \dots) \text{ to}$$

get

$$
\begin{aligned}
&- \mathrm{i}(-1)^\alpha \frac{b}{2} \beta_1 \varepsilon_\mu \delta_{\mu h} \Xi_n^g(\beta_1) + \mathrm{i}(-1)^\varrho \frac{b}{2} \beta_2 \varepsilon_\nu \delta_{\nu h} \Xi_n^g(-\beta_2) \\
&+ \mathrm{i}(-1)^{\phi+h} \beta_3 \mathrm{e}^{-\mathrm{i}\beta_3 b} I_{2\phi h \tau n}^{wg} \\
&+ \frac{b}{2} \delta_{\chi h} \mathrm{e}^{-\mathrm{i}\beta_4 a} \left(\mathrm{i}\varepsilon_h \beta_4 I_{1\chi\psi n}^{abg} - \frac{g}{2} \varepsilon_\chi \varepsilon_\psi \delta_{\psi n} \right) \\
&= \sum_{p=0}^\infty \sum_{q=0}^\infty \left[a_{pq} \mathrm{i}(-1)^{p+h} \xi_{pq} \mathrm{e}^{\mathrm{i}\xi_{pq} b} I_{2phqn}^{wg} \right. \\
&\left. + b_{pq} \frac{b}{2} \varepsilon_h \delta_{qh} \mathrm{e}^{\mathrm{i}\gamma_{pq} a} \left(\mathrm{i}\gamma_{pq} I_{1qpn}^{abg} + \frac{g}{2} \varepsilon_n \delta_{pn} \right) \right]
\end{aligned} \tag{18}
$$

where

$$I_{1pqt}^{baw} = \frac{1}{2\pi} \int_{-\infty}^\infty \frac{\zeta^2 \cot \left(b\sqrt{k^2 - \zeta^2 - a_p^2} \right)}{\sqrt{k^2 - \zeta^2 - a_p^2}} \Xi_q^w(\zeta) \Xi_t^w(-\zeta) \, \mathrm{d}\zeta \tag{19}$$

$$I_{2sqpt}^{gw} = \frac{1}{2\pi} \int_{-\infty}^\infty \frac{\zeta^2}{\beta_{sq}^2 - \zeta^2} \Xi_p^g(\zeta) \Xi_t^w(-\zeta) \, \mathrm{d}\zeta . \tag{20}$$

In Subsect. 5.6.3 Appendix, we transform I_{1pqt}^{baw} and I_{2sqpt}^{gw} into rapidly-convergent series.

The scattered fields for $(y \leq -g/2)$ and $(y \geq g/2)$ are obtained with a residue calculus as

$$\Phi^I(x, y \geq g/2, z) = \sum_{u=0}^\infty \sum_{m=0}^\infty K_{um}^+ \cos(a_u x) \cos(b_m z) \mathrm{e}^{\mathrm{i}\beta_{um} y} \tag{21}$$

$$\Phi^I(x, y \leq -g/2, z) = \sum_{u=0}^\infty \sum_{m=0}^\infty K_{um}^- \cos(a_u x) \cos(b_m z) \mathrm{e}^{-\mathrm{i}\beta_{um} y} \tag{22}$$

where

$$
\begin{aligned}
K_{um}^\pm = \mp \sum_{q=0}^\infty &\left(a_{uq} \mathrm{e}^{\mathrm{i}\xi_{uq} b} - \mathrm{e}^{-\mathrm{i}\beta_3 b} \delta_{u\phi} \delta_{q\tau} \right) \frac{\mathrm{i}\xi_{uq} \Xi_q^w(\mp \beta_{um})}{b \varepsilon_m (-1)^m} \\
\mp \sum_{p=0}^\infty &\left(b_{pm} \mathrm{e}^{\mathrm{i}\gamma_{pm} a} - \mathrm{e}^{-\mathrm{i}\beta_4 a} \delta_{p\psi} \delta_{m\chi} \right) \frac{\mathrm{i}\gamma_{pm} \Xi_p^g(\mp \beta_{um})}{a \varepsilon_u (-1)^u}
\end{aligned} \tag{23}
$$

$0 \leq u$, $0 \leq m$, and u, m : integers.

5.6.2 Soft-Surface Hybrid Junction

Consider acoustic scattering from a soft-surface hybrid junction. In region (I) the total field consists of the incident and scattered components

$$\Phi_1^i(x,y,z) = \sin(A_1 x)\sin(B_1 z)e^{i\beta_1 y} \tag{24}$$

$$\Phi_2^i(x,y,z) = \sin(A_2 x)\sin(B_2 z)e^{-i\beta_2 y} \tag{25}$$

$$\Phi^I(x,y,z) = \frac{1}{2\pi}\int_{-\infty}^{\infty}\sum_{m=1}^{\infty}\left[\widetilde{A}_m(\zeta)\sin(\kappa_m x)\sin(b_m z)\right.$$

$$\left. +\widetilde{B}_m(\zeta)\sin(a_m x)\sin(\eta_m z)\right]e^{-i\zeta y}\mathrm{d}\zeta \tag{26}$$

where $\alpha,\mu,\varrho,\nu = 1,2,3,\ldots$. In regions (II) and (III) the incident and transmitted fields take the forms of

$$\Phi_3^i(x,y,z) = \sin(A_3 x)\sin W_3(y+w/2)e^{-i\beta_3 z} \tag{27}$$

$$\Phi^{II}(x,y,z) = \sum_{p=1}^{\infty}\sum_{q=1}^{\infty}a_{pq}\sin(a_p x)\sin w_q(y+w/2)e^{i\xi_{pq}z} \tag{28}$$

$$\Phi_4^i(x,y,z) = \sin G_4(y+g/2)\sin(B_4 z)e^{-i\beta_4 x} \tag{29}$$

$$\Phi^{III}(x,y,z) = \sum_{p=1}^{\infty}\sum_{q=1}^{\infty}b_{pq}\sin g_p(y+g/2)\sin(b_q z)e^{i\gamma_{pq}x} \tag{30}$$

where $\phi,\tau,\psi,\chi = 1,2,3,\ldots$.

From the pressure continuity conditions at $z=b$ and $x=a$, we get

$$\widetilde{B}_m(\zeta) = \frac{W_3}{\sin(\eta_\phi b)}\Xi_\tau^w(\zeta)e^{-i\beta_3 b} + \sum_{q=1}^{\infty}a_{mq}\frac{w_q}{\sin(\eta_m b)}\Xi_q^w(\zeta)e^{i\xi_{mq}b} \tag{31}$$

$$\widetilde{A}_m(\zeta) = \frac{G_4}{\sin(\kappa_\chi a)}\Xi_\psi^g(\zeta)e^{-i\beta_4 a} + \sum_{p=1}^{\infty}b_{pm}\frac{g_p}{\sin(\kappa_m a)}\Xi_p^g(\zeta)e^{i\gamma_{pm}a}\;. \tag{32}$$

The velocity continuity conditions at $z=b$ and $x=a$, respectively, give

$$\boxed{\begin{aligned} &B_1(-1)^\mu\frac{a}{2}w_t\delta_{\alpha s}\Xi_t^w(\beta_1) + B_2(-1)^\nu\frac{a}{2}w_t\delta_{\varrho s}\Xi_t^w(-\beta_2) \\ &+\frac{a}{2}\delta_{\phi s}e^{-i\beta_3 b}\left(W_3 w_t I_{3\phi\tau t}^{baw} + i\frac{w}{2}\beta_3\delta_{\tau t}\right) \\ &+(-1)^{\chi+s}a_s B_4 G_4 w_t e^{-i\beta_4 a}I_{4s\chi\psi t}^{gw} \\ &= \sum_{p=1}^{\infty}\sum_{q=1}^{\infty}\left[a_{pq}\frac{a}{2}\delta_{ps}e^{i\xi_{pq}b}\left(-w_q w_t I_{3pqt}^{baw} + i\frac{w}{2}\xi_{pq}\delta_{qt}\right)\right. \\ &\left. -b_{pq}(-1)^{q+s}a_s b_q g_p w_t e^{i\gamma_{pq}a}I_{4sqpt}^{gw}\right] \end{aligned}} \tag{33}$$

$$A_1(-1)^\alpha \frac{b}{2} g_n \delta_{\mu h} \Xi_n^g(\beta_1) + A_2(-1)^\varrho \frac{b}{2} g_n \delta_{\nu h} \Xi_n^g(-\beta_2)$$

$$+(-1)^{\phi+h} A_3 b_h g_n W_3 e^{-i\beta_3 b} I_{4\phi h\tau n}^{wg}$$

$$+\frac{b}{2} \delta_{\chi h} e^{-i\beta_4 a} \left(G_4 g_n I_{3\chi\psi n}^{abg} + i\frac{g}{2} \beta_4 \delta_{\psi n} \right)$$

$$= \sum_{p=1}^{\infty} \sum_{q=1}^{\infty} \left[-a_{pq}(-1)^{p+h} a_{pb} b_h g_n w_q e^{i\xi_{pq} b} I_{4phqn}^{wq} \right.$$

$$\left. + b_{pq} \frac{b}{2} \delta_{qh} e^{i\gamma_{pq} a} \left(-g_p g_n I_{3qpn}^{abg} + i\frac{g}{2} \gamma_{pq} \delta_{pn} \right) \right] \tag{34}$$

where $q \neq 0$, $t \neq 0$, $h \neq 0$, $n \neq 0$, and

$$I_{3pqt}^{baw} = \frac{1}{2\pi} \int_{-\infty}^{\infty} \frac{\sqrt{k^2 - \zeta^2 - a_p^2}}{\tan\left(b\sqrt{k^2 - \zeta^2 - a_p^2}\right)} \Xi_q^w(\zeta) \Xi_t^w(-\zeta) \, d\zeta \tag{35}$$

$$I_{4sqpt}^{gw} = \frac{1}{2\pi} \int_{-\infty}^{\infty} \frac{1}{\beta_{sq}^2 - \zeta^2} \Xi_p^g(\zeta) \Xi_t^w(-\zeta) \, d\zeta . \tag{36}$$

The rapidly-convergent series expressions for I_{3pqt}^{baw} and I_{4sqpt}^{gw} are available in Subsect. 5.6.3 Appendix.

The scattered fields for $(y \leq -g/2)$ and $(y \geq g/2)$ are

$$\Phi^I(x, y \geq g/2, z) = \sum_{u=1}^{\infty} \sum_{m=1}^{\infty} K_{um}^+ \sin(a_u x) \sin(b_m z) e^{i\beta_{um} y} \tag{37}$$

$$\Phi^I(x, y \leq -g/2, z) = \sum_{u=1}^{\infty} \sum_{m=1}^{\infty} K_{um}^- \sin(a_u x) \sin(b_m z) e^{-i\beta_{um} y} \tag{38}$$

where

$$K_{um}^\pm = -\sum_{q=1}^{\infty} \left(a_{uq} e^{i\xi_{uq} b} + e^{-i\beta_3 b} \delta_{u\phi} \delta_{q\tau} \right) \frac{i w_q b_m \Xi_q^w(\mp\beta_{um})}{b\beta_{um}(-1)^m}$$

$$-\sum_{p=1}^{\infty} \left(b_{pm} e^{i\gamma_{pm} a} + e^{-i\beta_4 a} \delta_{p\psi} \delta_{m\chi} \right) \frac{i g_p a_u \Xi_p^g(\mp\beta_{um})}{a\beta_{um}(-1)^u} \tag{39}$$

$0 < u$, $0 < m$, and u, m: integers.

5.6.3 Appendix

Consider

$$I_{1pqt}^{baw} = \frac{1}{2\pi} \int_{-\infty}^{\infty} \frac{\zeta^2 \cot\left(b\sqrt{k^2 - \zeta^2 - a_p^2}\right)}{\sqrt{k^2 - \zeta^2 - a_p^2}} \Xi_q^w(\zeta) \Xi_t^w(-\zeta) \, d\zeta \tag{40}$$

$$I_{2sqpt}^{gw} = \frac{1}{2\pi} \int_{-\infty}^{\infty} \frac{\zeta^2}{\beta_{sq}^2 - \zeta^2} \Xi_p^g(\zeta) \Xi_t^w(-\zeta) \, d\zeta . \tag{41}$$

The integral I_{1pqt}^{baw} contains simple poles at $\zeta = \pm w_q$ and an infinite number of poles corresponding to $\sin\left(b\sqrt{k^2 - \zeta^2 - a_p^2}\right) = 0$. The integral I_{2sqpt}^{gw} contains simple poles at $\zeta = \pm\beta_{sq}$ and at $\zeta = \pm g_p, \pm w_t$ (double poles when $g_p = w_t$). Using the residue calculus, we obtain the series representations as

$$I_{1pqt}^{baw} = \frac{w\varepsilon_q \cot(b\sqrt{k^2 - w_q^2 - a_p^2})}{2\sqrt{k^2 - w_q^2 - a_p^2}}\delta_{qt}$$

$$-\frac{i}{b}\sum_{v=0}^{\infty}\left\{\zeta[(-1)^{q+t} - (-1)^q e^{i\zeta|w|} - (-1)^t e^{i\zeta|w|} + 1]\right\}$$

$$\cdot\left\{\varepsilon_v(\zeta^2 - w_q^2)(\zeta^2 - w_t^2)\right\}^{-1}\Big|_{\zeta=\sqrt{k^2 - a_p^2 - b_v^2}} \tag{42}$$

$$I_{2sqpt}^{gw} = X_1(\zeta) + \begin{cases} iP_1(\zeta)|_{\zeta=g_p} + iQ_1(\zeta)|_{\zeta=w_t}, & g_p \neq w_t \\ \dfrac{i}{\varepsilon_p}\dfrac{f_1'(\zeta)s(\zeta) - f_1(\zeta)s'(\zeta)}{s^2(\zeta)}\Big|_{\zeta=g_p}, & g_p = w_t \end{cases} \tag{43}$$

Similarly we get

$$I_{3pqt}^{baw} = \frac{w\sqrt{k^2 - w_q^2 - a_p^2}}{2w_q^2 \tan\left(b\sqrt{k^2 - w_q^2 - a_p^2}\right)}\delta_{qt}$$

$$-\frac{i}{b}\sum_{v=1}^{\infty}(k^2 - \zeta^2 - a_p^2)$$

$$\cdot[(-1)^{q+t} - (-1)^q e^{i\zeta w} - (-1)^t e^{i\zeta w} + 1]$$

$$\cdot[\zeta(\zeta^2 - w_q^2)(\zeta^2 - w_t^2)]^{-1}\Big|_{\zeta=\sqrt{k^2 - a_p^2 - b_v^2}} \tag{44}$$

$$I_{4sqpt}^{gw} = X_2(\zeta) + \begin{cases} iP_2(\zeta)|_{\zeta=g_p} + iQ_2(\zeta)|_{\zeta=w_t}, & g_p \neq w_t \\ i\dfrac{f_2'(\zeta)s(\zeta) - f_2(\zeta)s'(\zeta)}{s^2(\zeta)}\Big|_{\zeta=g_p}, & g_p = w_t \end{cases} \tag{45}$$

where $(\cdot)'$ denotes differentiation and

$$X_j(\zeta) = \frac{-if_j(\zeta)}{(\beta_{sq} + \zeta)(\zeta^2 - g_p^2)(\zeta^2 - w_t^2)}\Big|_{\zeta=\beta_{sq}}, \qquad j = 1, 2 \tag{46}$$

$$P_j(\zeta) = \frac{f_j(\zeta)}{(\beta_{sq}^2 - \zeta^2)(\zeta + g_p)(\zeta^2 - w_t^2)} \tag{47}$$

$$Q_j(\zeta) = \frac{f_j(\zeta)}{(\beta_{sq}^2 - \zeta^2)(\zeta + w_t)(\zeta^2 - g_p^2)} \tag{48}$$

$$s(\zeta) = \begin{cases} (\beta_{sq}^2 - \zeta^2)(\zeta + g_p)^2, & g_p = w_t \neq 0 \\[2mm] \beta_{sq}^2 - \zeta^2, & g_p = w_t = 0 \end{cases} \tag{49}$$

$$f_1(\zeta) = \begin{cases} \zeta^2 \Big[(-1)^{p+t} \exp(i\zeta|g - w|/2) - (-1)^p \exp(i\zeta|g + w|/2) \\[1mm] \quad -(-1)^t \exp(i\zeta| - g - w|/2) + \exp(i\zeta| - g + w|/2) \Big], \\[2mm] \qquad\qquad\qquad g_p = w_t \neq 0 \\[3mm] \exp(i\zeta|g - w|/2) - \exp(i\zeta|g + w|/2) \\[2mm] \quad - \exp(i\zeta| - g - w|/2) + \exp(i\zeta| - g + w|/2), \\[1mm] \qquad\qquad\qquad g_p = w_t = 0 \end{cases} \tag{50}$$

$$f_2(\zeta) = \Big[(-1)^{p+t} \exp(i\zeta|g - w|/2) - (-1)^p \exp(i\zeta|g + w|/2) \\ \quad -(-1)^t \exp(i\zeta| - g - w|/2) + \exp(i\zeta| - g + w|/2) \Big]. \tag{51}$$

References for Chapter 5

1. N. Marcuvitz, *Waveguide Handbook*, Radiation Laboratory Series, vol. 10, New York: McGraw-Hill, 1951.
2. X. P. Liang, K. A. Zaki, and A. E. Atia, "A rigorous three plane mode-matching technique for characterizing waveguide T-junctions and its application in multiplexer design," *IEEE Trans. Microwave Theory Tech.*, vol. 39, no. 12, pp. 2138-2147, Dec. 1991.
3. M. Koshiba and M. Suzuki, "Application of the boundary-element method to waveguide discontinuities," *IEEE Trans. Microwave Theory Tech.*, vol. 34, no. 2, pp. 301-307, Feb. 1986.
4. K. H. Park and H. J. Eom, "An analytic series solution for H-plane waveguide T-junction," *IEEE Microwave Guided Wave Lett.*, vol. 3, no. 4, pp. 104-106, April 1993.
5. K. H. Park, H. J. Eom, and Y. Yamaguchi "An analytic series solution for E-plane T-junction in parallel-plate waveguide," *IEEE Trans. Microwave Theory Tech.*, vol. 42, no. 2, pp. 356-358, Feb. 1994.
6. H. J. Eom, K. H. Park, J. Y. Kwon and Y. Yamaguchi "Fourier transform analysis for E-plane T-junction in a rectangular waveguide," *Microwave Opt. Technol. Lett.*, vol. 26, no. 1, pp. 34-37, July 2000.
7. F. Arndt, I. Ahrens, U. Papziner, U. Wiechmann, and R. Wilkeit, "Optimized E-plane T-junction series power dividers," *IEEE Trans. Microwave Theory Tech.*, vol. 35, no. 11, pp. 1052-1059, Nov. 1987.
8. J. W. Lee and H. J. Eom, "TE-mode scattering from two junctions in H-plane waveguide," *IEEE Trans. Microwave Theory Tech.*, vol. 42, no. 4, pp. 601-606, April 1994.
9. J. W. Lee, H. J. Eom and K. Uchida, "Right-angle H-plane waveguide double bend," *IEICE Trans. Commun.*, vol. 77, no. 12, 1994.

10. A. Weisshaar, J. Lary, S. M. Goodnick, and V. K. Tripathi, "Analysis and modeling of quantum waveguide structures and devices," *J. Appl. Phys.*, vol. 70, no. 1, pp. 355-366, July 1991.
11. J. Y. Kwon, H. H. Park, and H. J. Eom, "Acoustic scattering from two junctions in a rectangular waveguide," *J. Acoust. Soc. Am.*, vol. 103, pp. 1209-1212, 1998.
12. J. W. Miles, "The diffraction of sound due to right-angled joints in rectangular tubes," *J. Acoust. Soc. Am.*, vol. 19, no. 4, pp. 572-579, July 1947.
13. J. C. Bruggeman, "The propagation of low-frequency sound in a two-dimensional duct system with T joints and right angle bends: Theory and experiment," *J. Acoust. Soc. Am.*, vol. 82, no. 3, pp. 1045-1051, Sept. 1987.
14. C. Thompson, "Linear inviscid wave propagation in a waveguide having a single boundary discontinuity: Part II: Application," *J. Acoust. Soc. Am.*, vol. 75, no. 2, pp. 356-362, Feb. 1984.
15. J. Y. Kwon and H. J. Eom, "Acoustic hybrid junction in a rectangular waveguide," *J. Acoust. Soc. Am.*, vol. 107, no. 4, pp. 1868-1873, April 2000.
16. T. Sieverding and F. Arndt, "Modal analysis of the magic tee," *IEEE Microwave Guided Wave Lett.*, vol. 3, no. 5, pp. 150-152, May 1993.

6. Rectangular Apertures in a Plane

6.1 Static Potential Through a Rectangular Aperture in a Plane

A low-frequency field penetration into an aperture is often modeled in terms of the electrostatic or magnetostatic potential that is governed by Laplace's equation. A static potential distribution through a thin rectangular aperture in a conducting plane was studied in [1,2] using approximate methods. In the next two subsections, we will consider electrostatic and magnetostatic potential distributions through a thick rectangular aperture and derive their polarizabilities.

6.1.1 Electrostatic Distribution [3]

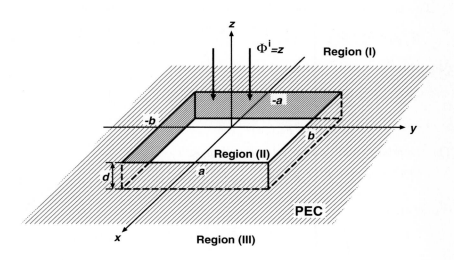

Fig. 6.1. A rectangular aperture in a conducting plane

In this section we will investigate an electrostatic potential distribution through a rectangular aperture in a thick conducting plane. In region (I) $(z > 0)$ an incident potential $\Phi^i(x, y, z)$ impinges on a rectangular aperture in a perfectly-conducting thick plane at zero potential. In region (I) the electrostatic potential has the incident (primary) and scattered (perturbed) potentials

$$\Phi^i(x, y, z) = z \tag{1}$$

$$\Phi^s(x, y, z) = \frac{1}{(2\pi)^2} \int_{-\infty}^{\infty} \int_{-\infty}^{\infty} \widetilde{\Phi}^s(\zeta, \eta)$$
$$\cdot \exp\left(-i\zeta x - i\eta y - \sqrt{\zeta^2 + \eta^2} z\right) d\zeta d\eta . \tag{2}$$

The electrostatic potential in region (II) $(-d < z < 0, |x| < a, \text{ and } |y| < b)$ is

$$\Phi^d(x, y, z) = \sum_{m=1}^{\infty} \sum_{n=1}^{\infty} [c_{mn} \sinh k_{mn}(z + d) + d_{mn} \cosh k_{mn}(z + d)]$$
$$\cdot \sin a_m(x + a) \sin b_n(y + b) \tag{3}$$

where $a_m = m\pi/(2a)$, $b_n = n\pi/(2b)$, and $k_{mn} = \sqrt{a_m^2 + b_n^2}$. The transmitted electrostatic potential in region (III) $(z < -d)$ is

$$\Phi^t(x, y, z) = \frac{1}{(2\pi)^2} \int_{-\infty}^{\infty} \int_{-\infty}^{\infty} \widetilde{\Phi}^t(\zeta, \eta)$$
$$\cdot \exp\left[-i\zeta x - i\eta y + \sqrt{\zeta^2 + \eta^2}(z + d)\right] d\zeta d\eta . \tag{4}$$

The boundary condition on the field continuity at $z = 0$ requires

$$\Phi^i(x, y, 0) + \Phi^s(x, y, 0) = \begin{cases} \Phi^d(x, y, 0), & |x| < a, |y| < b \\ 0, & \text{otherwise} . \end{cases} \tag{5}$$

Applying the Fourier transform to (5) yields

$$\widetilde{\Phi}^s(\zeta, \eta) = \sum_{m=1}^{\infty} \sum_{n=1}^{\infty} [c_{mn} \sinh(k_{mn}d) + d_{mn} \cosh(k_{mn}d)]$$
$$\cdot a_m b_n (ab)^2 F_m(\zeta a) F_n(\eta b) . \tag{6}$$

The boundary condition for $|x| < a$ and $|y| < b$

$$\left. \frac{\partial[\Phi^i(x, y, z) + \Phi^s(x, y, z)]}{\partial z} \right|_{z=0} = \left. \frac{\partial \Phi^d(x, y, z)}{\partial z} \right|_{z=0} \tag{7}$$

is rewritten as

$$1 - \frac{1}{(2\pi)^2} \int_{-\infty}^{\infty} \int_{-\infty}^{\infty} \sqrt{\zeta^2 + \eta^2} \widetilde{\Phi}^s(\zeta, \eta) \exp(-i\zeta x - i\eta y) d\zeta d\eta$$
$$= \sum_{m=1}^{\infty} \sum_{n=1}^{\infty} k_{mn} [c_{mn} \cosh(k_{mn}d) + d_{mn} \sinh(k_{mn}d)]$$
$$\cdot \sin a_m(x + a) \sin b_n(y + b) . \tag{8}$$

We multiply (8) by $\sin a_p(x+a) \sin b_q(y+b)$ $(p, q = 1, 2, 3, \ldots)$ and integrate over the aperture to obtain

$$
\boxed{
\begin{aligned}
&\gamma - \frac{a_p b_q}{(2\pi)^2}(ab)^3 \sum_{m=1}^{\infty} \sum_{n=1}^{\infty} \\
&a_m b_n [c_{mn} \sinh(k_{mn}d) + d_{mn} \cosh(k_{mn}d)]I \\
&= k_{pq}[c_{pq} \cosh(k_{pq}d) + d_{pq} \sinh(k_{pq}d)]
\end{aligned}
}
\tag{9}
$$

where

$$
\gamma = \frac{4\,[1-(-1)^p]\,[1-(-1)^q]}{pq\pi^2}
\tag{10}
$$

$$
I = \int_{-\infty}^{\infty} \int_{-\infty}^{\infty} \sqrt{\zeta^2 + \eta^2}\, F_m(\zeta a) F_p(-\zeta a) F_n(\eta b) F_q(-\eta b)\, \mathrm{d}\zeta \mathrm{d}\eta \ .
\tag{11}
$$

Also the boundary conditions at $z = -d$ yield

$$
\boxed{
\frac{a_p b_q}{(2\pi)^2}(ab)^3 \sum_{m=1}^{\infty} \sum_{n=1}^{\infty} a_m b_n d_{mn} I = k_{pq} c_{pq} \ .
}
\tag{12}
$$

The electric polarizability is shown to be

$$
\begin{aligned}
\chi(z) &\equiv 2 \int_{-b}^{b} \int_{-a}^{a} \Phi^d(x, y, z)\, \mathrm{d}x \mathrm{d}y \\
&= 2ab \sum_{m=1}^{\infty} \sum_{n=1}^{\infty} \gamma_{mn} \\
&\quad \cdot [c_{mn} \sinh k_{mn}(z+d) + d_{mn} \cosh k_{mn}(z+d)] \ .
\end{aligned}
\tag{13}
$$

6.1.2 Magnetostatic Distribution

A problem of magnetostatic potential distribution through various apertures has been studied extensively [4-6]. An incident magnetostatic potential $\Phi^i(x, y, z)$ impinges on a rectangular aperture (thickness: d and area: $2a$ by $2b$) in a thick conducting plane. Regions (II) $(-d < z < 0$, $|x| < a$, and $|y| < b$) and (III) $(z < -d)$ denote an aperture interior and a half-space, respectively. In region (I) the total magnetostatic potential is a sum of the incident and scattered potentials

$$
\Phi^i(x, y, z) = x \cos \theta_i + y \sin \theta_i
\tag{14}
$$

$$
\begin{aligned}
\Phi^s(x, y, z) &= \frac{1}{(2\pi)^2} \int_{-\infty}^{\infty} \int_{-\infty}^{\infty} \widetilde{\Phi}^s(\zeta, \eta) \\
&\quad \cdot \exp\left(-i\zeta x - i\eta y - \sqrt{\zeta^2 + \eta^2}\, z\right) \mathrm{d}\zeta \mathrm{d}\eta \ .
\end{aligned}
\tag{15}
$$

In region (II) the magnetostatic potential is

$$\Phi^d(x,y,z) = \sum_{m=0}^{\infty} \sum_{n=0}^{\infty} [c_{mn} \sinh k_{mn}(z+d) + d_{mn} \cosh k_{mn}(z+d)]$$
$$\cdot \cos a_m(x+a) \cos b_n(y+b) \tag{16}$$

where $a_m = m\pi/(2a)$, $b_n = n\pi/(2b)$, and $k_{mn} = \sqrt{a_m^2 + b_n^2}$. In region (III) the total transmitted magnetostatic potential is

$$\Phi^t(x,y,z) = \frac{1}{(2\pi)^2} \int_{-\infty}^{\infty} \int_{-\infty}^{\infty} \tilde{\Phi}^t(\zeta,\eta)$$
$$\cdot \exp\left[-i\zeta x - i\eta y + \sqrt{\zeta^2 + \eta^2}\,(z+d)\right] d\zeta d\eta \,. \tag{17}$$

The boundary condition on the continuity of normal derivative of the potential at $z = 0$ is

$$\frac{\partial}{\partial z}\left[\Phi^i(x,y,z) + \Phi^s(x,y,z)\right]_{z=0}$$
$$= \begin{cases} \dfrac{\partial}{\partial z}\left[\Phi^d(x,y,z)\right]_{z=0}, & |x| < a, \ |y| < b \\ \\ 0, & \text{otherwise}\,. \end{cases} \tag{18}$$

Applying the Fourier transform to (18) yields

$$\tilde{\Phi}^s(\zeta,\eta) = \sum_{m=0}^{\infty} \sum_{n=0}^{\infty} k_{mn}[c_{mn} \cosh(k_{mn}d) + d_{mn} \sinh(k_{mn}d)]$$
$$\cdot (ab)^2 \frac{\zeta\eta}{\sqrt{\zeta^2 + \eta^2}} F_m(\zeta a) F_n(\eta b) \,. \tag{19}$$

The boundary condition on the continuity of the potential across the aperture is

$$\Phi^i(x,y,0) + \Phi^s(x,y,0) = \Phi^d(x,y,0), \qquad |x| < a, \ |y| < b\,. \tag{20}$$

We multiply (20) by $\cos a_p(x+a) \cos b_q(y+b)$ $(p,q = 0,1,2,3,\dots)$, and perform integration to get

$$\boxed{\begin{aligned} &\gamma - \frac{(ab)^4}{(2\pi)^2} \sum_{m=0}^{\infty} \sum_{n=0}^{\infty} k_{mn}\left[c_{mn} \cosh(k_{mn}d) + d_{mn} \sinh(k_{mn}d)\right] I \\ &= \epsilon\left[c_{pq} \sinh(k_{pq}d) + d_{pq} \cosh(k_{pq}d)\right] \end{aligned}} \tag{21}$$

where

$$\gamma = \begin{cases} 2b \cos\theta_i \dfrac{[(-1)^p - 1]}{a_p^2}, & (p \neq 0, \ q = 0) \\ \\ 2a \sin\theta_i \dfrac{[(-1)^q - 1]}{b_q^2}, & (p = 0, \ q \neq 0) \end{cases} \tag{22}$$

$$I = \int_{-\infty}^{\infty} \int_{-\infty}^{\infty} \frac{(\zeta\eta)^2}{\sqrt{\zeta^2 + \eta^2}} F_m(\zeta a) F_n(\eta b) F_p(-\zeta a) F_q(-\eta b) \, \mathrm{d}\zeta \mathrm{d}\eta \qquad (23)$$

$$\epsilon = \begin{cases} 4ab, & (p = 0, \ q = 0) \\ ab, & (p \neq 0, \ q \neq 0) \\ 2ab, & (p \neq 0, \ q = 0) \ \ \text{or} \ \ (p = 0, \ q \neq 0) \ . \end{cases} \qquad (24)$$

Also the boundary conditions at $z = -d$ give

$$\boxed{\frac{(ab)^4}{(2\pi)^2} \sum_{m=0}^{\infty} \sum_{n=0}^{\infty} k_{mn} c_{mn} I = \epsilon d_{pq} \ .} \qquad (25)$$

The longitudinal magnetic polarizability ψ_x and the transverse magnetic polarizability ψ_y are given by

$$\psi_x \equiv -\int_{-b}^{b} \int_{-a}^{a} x \frac{\partial}{\partial z} \left[\Phi^d(x, y, z) \right] \, \mathrm{d}x\mathrm{d}y$$

$$= 2b \sum_{m=1}^{\infty} \frac{1 - (-1)^m}{a_m}$$

$$\cdot \left[c_{m0} \cosh a_m(z + d) + d_{m0} \sinh a_m(z + d) \right] \qquad (26)$$

$$\psi_y \equiv -\int_{-b}^{b} \int_{-a}^{a} y \frac{\partial}{\partial z} \left[\Phi^d(x, y, z) \right] \, \mathrm{d}x\mathrm{d}y$$

$$= 2a \sum_{n=1}^{\infty} \frac{1 - (-1)^n}{b_n} \left[c_{0n} \cosh b_n(z + d) + d_{0n} \sinh b_n(z + d) \right] \ . \qquad (27)$$

6.2 Acoustic Scattering from a Rectangular Aperture in a Hard Plane [7]

A study of acoustic wave scattering from a rectangular aperture in a thick hard plane was performed in [8] using an approximate technique based on the radiation impedance. In this section we will revisit the problem of acoustic wave scattering from a rectangular aperture in a thick hard plane. In region (I) $(z > 0)$ a field (velocity potential) $\Phi^i(x, y, z)$ is incident on a rectangular aperture with wavenumber $k(= 2\pi/\lambda$ and λ: wavelength). In region (I) the total field has the incident, reflected, and scattered waves

$$\Phi^i(x, y, z) = \exp(\mathrm{i}k_x x + \mathrm{i}k_y y - \mathrm{i}k_z z) \qquad (1)$$

$$\Phi^r(x, y, z) = \exp(\mathrm{i}k_x x + \mathrm{i}k_y y + \mathrm{i}k_z z) \qquad (2)$$

$$\Phi^s(x, y, z) = \frac{1}{(2\pi)^2} \int_{-\infty}^{\infty} \int_{-\infty}^{\infty} \tilde{\Phi}^s(\zeta, \eta)$$

$$\cdot \exp(-\mathrm{i}\zeta x - \mathrm{i}\eta y + \mathrm{i}\kappa z) \, \mathrm{d}\zeta \mathrm{d}\eta \qquad (3)$$

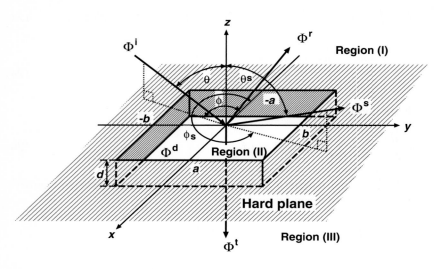

Fig. 6.2. A rectangular aperture in a hard plane

where $k_x = k\cos\phi\sin\theta$, $k_y = k\sin\phi\sin\theta$, $k_z = k\cos\theta$, and $\kappa = \sqrt{k^2 - \zeta^2 - \eta^2}$. In region (II) ($-d < z < 0$, $|x| \le a$, and $|y| \le b$) the field is

$$\Phi^d(x,y,z) = \sum_{m=0}^{\infty}\sum_{n=0}^{\infty}[c_{mn}\cos\xi_{mn}(z+d) + d_{mn}\sin\xi_{mn}(z+d)]$$
$$\cdot \cos a_m(x+a)\cos b_n(y+b) \tag{4}$$

where $a_m = m\pi/(2a)$, $b_n = n\pi/(2b)$, and $\xi_{mn} = \sqrt{k^2 - a_m^2 - b_n^2}$. In region (III) ($z < -d$) the transmitted field is

$$\Phi^t(x,y,z) = \frac{1}{(2\pi)^2}\int_{-\infty}^{\infty}\int_{-\infty}^{\infty}\widetilde{\Phi}^t(\zeta,\eta)$$
$$\cdot \exp[-i\zeta x - i\eta y - i\kappa(z+d)]\,d\zeta d\eta . \tag{5}$$

The velocity continuity condition at $z=0$ requires

$$\frac{\partial[\Phi^i(x,y,z) + \Phi^r(x,y,z) + \Phi^s(x,y,z)]}{\partial z}\Bigg|_{z=0}$$
$$= \begin{cases} \dfrac{\partial\Phi^d(x,y,z)}{\partial z}\Bigg|_{z=0}, & |x| < a,\ |y| < b \\ 0, & \text{otherwise} . \end{cases} \tag{6}$$

Applying the Fourier transform to (6) yields

$$\widetilde{\Phi}^s(\zeta,\eta) = -i(ab)^2\frac{\zeta\eta}{\kappa}\sum_{m=0}^{\infty}\sum_{n=0}^{\infty}$$
$$\xi_{mn}[c_{mn}\sin(\xi_{mn}d) - d_{mn}\cos(\xi_{mn}d)]F_m(\zeta a)F_n(\eta b) . \tag{7}$$

The pressure continuity condition at $z = 0$ for $|x| < a$ and $|y| < b$ is

$$\Phi^i(x, y, 0) + \Phi^r(x, y, 0) + \Phi^s(x, y, 0) = \Phi^d(x, y, 0) . \tag{8}$$

We substitute $\widetilde{\Phi}^s(\zeta, \eta)$ into (8), multiply (8) by $\cos a_p(x + a) \cos b_q(y + b)$ $(p, q = 0, 1, 2, 3, \ldots)$, and perform integration to obtain

$$
\begin{aligned}
&\gamma + \frac{\mathrm{i}(ab)^3}{(2\pi)^2} \sum_{m=0}^{\infty} \sum_{n=0}^{\infty} \xi_{mn}[c_{mn} \sin(\xi_{mn}d) - d_{mn} \cos(\xi_{mn}d)]I \\
&= \varepsilon_p \varepsilon_q [c_{pq} \cos(\xi_{pq}d) + d_{pq} \sin(\xi_{pq}d)]
\end{aligned}
\tag{9}
$$

where

$$\gamma = -2abk_x k_y F_p(k_x a) F_q(k_y b) \tag{10}$$

$$I = \int_{-\infty}^{\infty} \int_{-\infty}^{\infty} \frac{(\zeta\eta)^2}{\kappa} F_m(\zeta a) F_p(-\zeta a) F_n(\eta b) F_q(-\eta b) \, \mathrm{d}\zeta \mathrm{d}\eta . \tag{11}$$

The boundary conditions at $z = -d$ similarly yield

$$\frac{\mathrm{i}(ab)^3}{(2\pi)^2} \sum_{m=0}^{\infty} \sum_{n=0}^{\infty} \xi_{mn} d_{mn} I = \varepsilon_p \varepsilon_q c_{pq} . \tag{12}$$

The reflection coefficient ϱ and the transmission coefficient τ are

$$
\begin{aligned}
\varrho = \frac{1}{4k_z} Im \Bigg\{ &\sum_{p=0}^{\infty} \sum_{q=0}^{\infty} \varepsilon_p \varepsilon_q \xi_{pq}^* \big[|c_{pq}|^2 \sin^*(\xi_{pq}d) \cos(\xi_{pq}d) \\
&+ c_{pq}^* d_{pq} |\sin(\xi_{pq}d)|^2 - c_{pq} d_{pq}^* |\cos(\xi_{pq}d)|^2 \\
&- |d_{pq}|^2 \sin(\xi_{pq}d) \cos^*(\xi_{pq}d) \big] \Bigg\}
\end{aligned}
\tag{13}
$$

$$\tau = \frac{1}{4k_z} Im \left\{ \sum_{p=0}^{\infty} \sum_{q=0}^{\infty} \varepsilon_p \varepsilon_q \xi_{pq}^* c_{pq} d_{pq}^* \right\} . \tag{14}$$

6.3 Electrostatic Potential Through Rectangular Apertures in a Plane [9]

An electrostatic potential penetration into a single rectangular aperture in a conducting plane was considered in Sect. 6.1. In this section we will study a potential distribution through multiple rectangular apertures by extending the theoretical analysis given in Sect. 6.1. An incident electrostatic potential $\Phi^i(x, y, z)$ impinges on rectangular apertures in a perfectly-conducting thick plane at zero potential. In region (I) $(z > 0)$ the potential consists of the incident and scattered components as

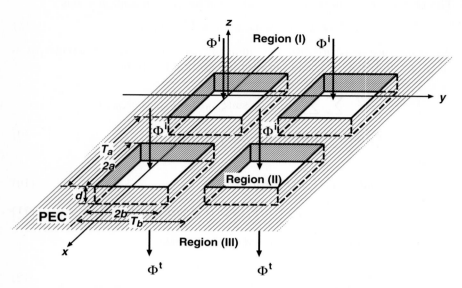

Fig. 6.3. Multiple rectangular apertures in a conducting plane

$$\Phi^i(x, y, z) = z \tag{1}$$

$$\Phi^s(x, y, z) = \frac{1}{4\pi^2} \int_{-\infty}^{\infty} \int_{-\infty}^{\infty} \widetilde{\Phi}^s(\zeta, \eta)$$
$$\cdot \exp\left(-\mathrm{i}\zeta x - \mathrm{i}\eta y - \sqrt{\zeta^2 + \eta^2}\, z\right) \mathrm{d}\zeta \mathrm{d}\eta . \tag{2}$$

In region (II) $(-d < z < 0,\ |x - lT_a| < a,\ |y - uT_b| < b\colon l = 0, \dots, M_l,$ $u = 0, \dots, M_u,$ and M_l and M_u are integers) the total potential is

$$\Phi^d(x, y, z) = \sum_{m=1}^{\infty} \sum_{n=1}^{\infty} [c_{mn}^{lu} \sinh k_{mn}(z + d) + d_{mn}^{lu} \cosh k_{mn}(z + d)]$$
$$\cdot \sin a_m(x - lT_a + a) \sin b_n(y - uT_b + b) \tag{3}$$

where $a_m = m\pi/(2a)$, $b_n = n\pi/(2b)$, and $k_{mn} = \sqrt{a_m^2 + b_n^2}$. The transmitted potential in region (III) $(z < -d)$ is

$$\Phi^t(x, y, z) = \frac{1}{4\pi^2} \int_{-\infty}^{\infty} \int_{-\infty}^{\infty} \widetilde{\Phi}^t(\zeta, \eta)$$
$$\cdot \exp\left[-\mathrm{i}\zeta x - \mathrm{i}\eta y + \sqrt{\zeta^2 + \eta^2}(z + d)\right] \mathrm{d}\zeta \mathrm{d}\eta . \tag{4}$$

From the boundary conditions on the field continuities at $z = 0$, we obtain

$$
\gamma_{pq} - \frac{(ab)^3}{4\pi^2} a_p b_q \sum_{l=0}^{M_l} \sum_{u=0}^{M_u} \sum_{m=1}^{\infty} \sum_{n=1}^{\infty} a_m b_n
$$
$$
\cdot \left[c_{mn}^{lu} \sinh(k_{mn}d) + d_{mn}^{lu} \cosh(k_{mn}d) \right] I
$$
$$
= k_{pq} \left[c_{pq}^{rv} \cosh(k_{pq}d) + d_{pq}^{rv} \sinh(k_{pq}d) \right] \tag{5}
$$

where $p = 1, 2, \ldots$, $q = 1, 2, \ldots$, $r = 0, 1, \ldots, M_l$, $v = 0, 1, \ldots, M_u$, and

$$
\gamma_{pq} = \frac{4[1 - (-1)^p][1 - (-1)^q]}{pq\pi^2} \tag{6}
$$

$$
I = \int_{-\infty}^{\infty} \int_{-\infty}^{\infty} \sqrt{\zeta^2 + \eta^2} F_m(\zeta a) F_p(-\zeta a) F_n(\eta b) F_q(-\eta b)
$$
$$
\cdot \exp[i\zeta(l - r)T_a + i\eta(u - v)T_b] \, d\zeta d\eta \ . \tag{7}
$$

The boundary conditions at $z = -d$ similarly yield

$$
\frac{(ab)^3}{4\pi^2} a_p b_q \sum_{l=0}^{M_l} \sum_{u=0}^{M_u} \sum_{m=1}^{\infty} \sum_{n=1}^{\infty} a_m b_n d_{mn}^{lu} I = k_{pq} c_{pq}^{rv} \ . \tag{8}
$$

The total electric polarizability is

$$
\chi(z) \equiv 2 \sum_{l=0}^{M_l} \sum_{u=0}^{M_u} \int_{uT_b-b}^{uT_b+b} \int_{lT_a-a}^{lT_a+a} \Phi^d(x, y, z) \, dx dy
$$
$$
= 2ab \sum_{l=0}^{M_l} \sum_{u=0}^{M_u} \sum_{m=1}^{\infty} \sum_{n=1}^{\infty}
$$
$$
\gamma_{mn} \left[c_{mn}^{lu} \sinh k_{mn}(z + d) + d_{mn}^{lu} \cosh k_{mn}(z + d) \right] \ . \tag{9}
$$

6.4 Magnetostatic Potential Through Rectangular Apertures in a Plane [10]

The magnetic polarizability is a useful concept for approximately estimating the shielding effectiveness of an enclosure with apertures in low-frequency regime. In this section we will derive the magnetic polarizability of multiple rectangular apertures in a thick conducting plane. A magnetostatic potential analysis in this section is similar to the electrostatic case considered in Sect. 6.3. Consider an incident magnetostatic potential $\Phi^i(x, y, z)$ impinging on multiple rectangular apertures in a thick conducting plane. In region (I) ($z > 0$) the total magnetostatic potential is a sum of the incident and scattered potentials. The incident potential is

$$
\Phi^i(x, y, z) = x \cos \theta_i + y \sin \theta_i - A^{l,u} \tag{1}
$$

for $|x - lT_a| < a$, $|y - uT_b| < b$, $l = 0, \ldots, M_l$, and $u = 0, \ldots, M_u$. In the multiply-connected structure as considered in this section, the magnetostatic

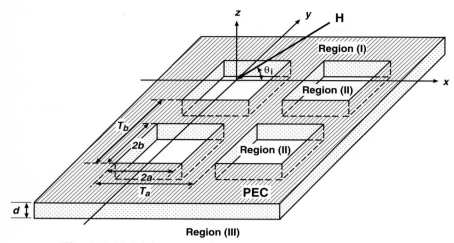

Fig. 6.4. Multiple rectangular apertures in a conducting plane

potential is a multi-valued function. In order to make the magnetostatic potential a single-valued function, a constant $A^{l,u}$ needs to be determined later in matching the boundary conditions. The scattered potential is

$$\Phi^s(x,y,z) = \frac{1}{4\pi^2} \int_{-\infty}^{\infty} \int_{-\infty}^{\infty} \widetilde{\Phi}^s(\zeta,\eta)$$
$$\cdot \exp\left(-\mathrm{i}\zeta x - \mathrm{i}\eta y - \sqrt{\zeta^2 + \eta^2}\,z\right) \, \mathrm{d}\zeta \mathrm{d}\eta \ . \tag{2}$$

In each aperture of region (II) the potential take the form of

$$\Phi^d(x,y,z) = \sum_{m=0}^{\infty} \sum_{n=0}^{\infty} \left[c_{mn}^{lu} \sinh k_{mn}(z+d) + d_{mn}^{lu} \cosh k_{mn}(z+d) \right]$$
$$\cdot \cos a_m(x + a - lT_a) \cos b_n(y + b - uT_b) \tag{3}$$

where $a_m = m\pi/(2a)$, $b_n = n\pi/(2b)$, and $k_{mn} = \sqrt{a_m^2 + b_n^2}$. Note that T_a and T_b denote the periods of apertures in the x- and y-directions, respectively. In region (III) ($z < -d$) the transmitted potential is

$$\Phi^t(x,y,z) = \frac{1}{4\pi^2} \int_{-\infty}^{\infty} \int_{-\infty}^{\infty} \widetilde{\Phi}^t(\zeta,\eta)$$
$$\cdot \exp\left[-\mathrm{i}\zeta x - \mathrm{i}\eta y + \sqrt{\zeta^2 + \eta^2}(z+d)\right] \, \mathrm{d}\zeta \mathrm{d}\eta \ . \tag{4}$$

The boundary conditions at $z = 0$ yield

$$\gamma_{pq} - \frac{(ab)^4}{4\pi^2} \sum_{l=0}^{M_l} \sum_{u=0}^{M_u} \sum_{m=0}^{\infty} \sum_{n=0}^{\infty}$$

$$k_{mn} \left[c_{mn}^{lu} \cosh(k_{mn}d) + d_{mn}^{lu} \sinh(k_{mn}d) \right] I$$
$$= \epsilon \left[c_{pq}^{rv} \sinh(k_{pq}d) + d_{pq}^{rv} \cosh(k_{pq}d) \right] ab \tag{5}$$

where $p = 0, 1, 2, \ldots,\ q = 0, 1, 2, \ldots,\ r = 0, 1, \ldots, M_l,\ v = 0, 1, \ldots, M_u,$ and

$$\gamma_{pq} = \begin{cases} 4ab(rT_a \cos\theta_i + vT_b \sin\theta_i - A^{l,u}), & (p = 0,\ q = 0) \\ 2b \cos\theta_i[(-1)^p - 1]/a_p^2, & (p \neq 0,\ q = 0) \\ 2a \sin\theta_i[(-1)^q - 1]/b_q^2, & (p = 0,\ q \neq 0) \\ 0, & (p \neq 0,\ q \neq 0) \end{cases} \tag{6}$$

$$I = \int_{-\infty}^{\infty} \int_{-\infty}^{\infty} \frac{(\zeta\eta)^2}{\sqrt{\zeta^2 + \eta^2}} F_m(\zeta a) F_n(\eta b) F_p(-\zeta a) F_q(-\eta b)$$
$$\cdot \exp[i\zeta(l - r)T_a + i\eta(u - v)T_b]\, d\zeta d\eta . \tag{7}$$

Note that ϵ is 4 for $(p = 0,\ q = 0)$, 1 for $(p \neq 0,\ q \neq 0)$, and 2 for others. Similarly the boundary conditions at $z = -d$ give

$$\frac{(ab)^4}{4\pi^2} \sum_{l=0}^{M_l} \sum_{u=0}^{M_u} \sum_{m=0}^{\infty} \sum_{n=0}^{\infty} k_{mn} c_{mn}^{lu} I = \epsilon d_{pq}^{rv} ab . \tag{8}$$

We solve (5) and (8) for the modal coefficients c_{mn}^{lu} and d_{mn}^{lu} when $(p \neq 0)$ or $(q \neq 0)$, and then subsequently determine the constant terms $d_{0,0}^{l,u}$ and $A^{l,u}$.

The total longitudinal (ψ_x) and transverse (ψ_y) magnetic polarizabilities are

$$\psi_x \equiv \sum_{l=0}^{M_l} \sum_{u=0}^{M_u} \int_{uT_b-b}^{uT_b+b} \int_{lT_a-a}^{lT_a+a} -x \frac{\partial}{\partial z} \left[\Phi^d(x, y, z) \right]\, dx dy$$

$$= 2b \sum_{l=0}^{M_l} \sum_{u=0}^{M_u} \sum_{m=1}^{\infty} \frac{1 - (-1)^m}{a_m}$$
$$\cdot \left[c_{m0}^{lu} \cosh a_m(z + d) + d_{m0}^{lu} \sinh a_m(z + d) \right] \tag{9}$$

$$\psi_y = 2a \sum_{l=0}^{M_l} \sum_{u=0}^{M_u} \sum_{n=1}^{\infty} \frac{1 - (-1)^n}{b_n}$$
$$\cdot \left[c_{0n}^{lu} \cosh b_n(z + d) + d_{0n}^{lu} \sinh b_n(z + d) \right] . \tag{10}$$

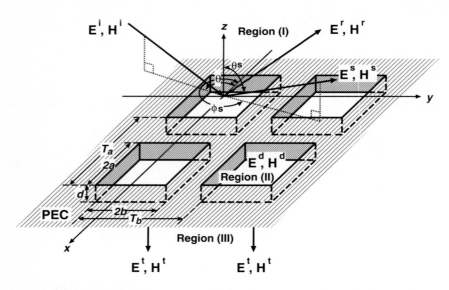

Fig. 6.5. Multiple rectangular apertures in a conducting plane

6.5 EM Scattering from Rectangular Apertures in a Conducting Plane [11]

A study of electromagnetic penetration into an aperture in a conducting plane is important in EMI/EMC-related problems [12-17]. In this section we will analyze electromagnetic wave scattering from multiple rectangular apertures in a perfectly-conducting thick plane. Consider multiple rectangular apertures with area $2a \times 2b$ and thickness d where T_a and T_b are the periods in the x- and y-directions, respectively. Assume that an electromagnetic wave is obliquely incident on multiple rectangular apertures. The incident and reflected fields in region (I) take the forms of

$$\begin{aligned} \boldsymbol{E}^i = Z_0 \{ & (u \sin \phi + v \cos \phi \cos \theta) \hat{x} - (u \cos \phi - v \sin \phi \cos \theta) \hat{y} \\ & + v \sin \theta \hat{z} \} \exp(\mathrm{i} k_x x + \mathrm{i} k_y y - \mathrm{i} k_z z) \end{aligned} \quad (1)$$

$$\begin{aligned} \boldsymbol{E}^r = -Z_0 \{ & (u \sin \phi + v \cos \phi \cos \theta) \hat{x} - (u \cos \phi - v \sin \phi \cos \theta) \hat{y} \\ & - v \sin \theta \hat{z} \} \exp(\mathrm{i} k_x x + \mathrm{i} k_y y + \mathrm{i} k_z z) \end{aligned} \quad (2)$$

where $\hat{x}$, $\hat{y}$, and $\hat{z}$ are the unit vectors, $k_x = k_0 \cos \phi \sin \theta$, $k_y = k_0 \sin \phi \sin \theta$, $k_z = k_0 \cos \theta$, $k_0 = \omega \sqrt{\mu_0 \epsilon_0} = 2\pi/\lambda$ is the wavenumber, and $Z_0 = \sqrt{\mu_0/\epsilon_0}$ is the free-space intrinsic impedance. A polarization state of the incident field is determined by the choice of u and v. It is convenient to represent the fields in terms of the electric and magnetic vector potentials. The relations between the electric and magnetic vector potentials and the fields are given in

Appendix A.1. The scattered $\bar{E}^s$- and $\bar{H}^s$-fields in region (I) are obtained from the z-components of the scattered electric and magnetic vector potentials, $F_z^s(x, y, z)$ and $A_z^s(x, y, z)$

$$F_z^s(x, y, z) = \frac{1}{4\pi^2} \int_{-\infty}^{\infty} \int_{-\infty}^{\infty} \widetilde{F}_z^s(\zeta, \eta)$$
$$\cdot \exp(-i\zeta x - i\eta y + i\kappa z) d\zeta d\eta \tag{3}$$

$$A_z^s(x, y, z) = \frac{1}{4\pi^2} \int_{-\infty}^{\infty} \int_{-\infty}^{\infty} \widetilde{A}_z^s(\zeta, \eta)$$
$$\cdot \exp(-i\zeta x - i\eta y + i\kappa z) d\zeta d\eta \tag{4}$$

where $\kappa = \sqrt{k_0^2 - \zeta^2 - \eta^2}$. In region (II) ($l$th and kth aperture in the x and y directions) the vector potentials $F_z^d(x, y, z)$ and $A_z^d(x, y, z)$ within the rectangular aperture are

$$F_z^d(x, y, z) = \sum_{m=0}^{\infty} \sum_{n=0}^{\infty} \left[c_{mn}^{lk} \cos \xi_{mn}(z + d) + d_{mn}^{lk} \sin \xi_{mn}(z + d) \right]$$
$$\cdot \cos a_m(x - lT_a + a) \cos b_n(y - kT_b + b) \tag{5}$$

$$A_z^d(x, y, z) = \sum_{m=1}^{\infty} \sum_{n=1}^{\infty} \left[\bar{c}_{mn}^{lk} \cos \xi_{mn}(z + d) + \bar{d}_{mn}^{lk} \sin \xi_{mn}(z + d) \right]$$
$$\cdot \sin a_m(x - lT_a + a) \sin b_n(y - kT_b + b) \tag{6}$$

where $(m, n) \neq (0, 0)$, $\xi_{mn} = \sqrt{k_1^2 - a_m^2 - b_n^2}$, $k_1 = \omega\sqrt{\mu_1 \epsilon_1}$, $a_m = m\pi/(2a)$, and $b_n = n\pi/(2b)$. In region (III) the transmitted vector potentials $F_z^t(x, y, z)$ and $A_z^t(x, y, z)$ are

$$F_z^t(x, y, z) = \frac{1}{4\pi^2} \int_{-\infty}^{\infty} \int_{-\infty}^{\infty} \widetilde{F}_z^t(\zeta, \eta)$$
$$\cdot \exp[-i\zeta x - i\eta y - i\kappa(z + d)] d\zeta d\eta \tag{7}$$

$$A_z^t(x, y, z) = \frac{1}{4\pi^2} \int_{-\infty}^{\infty} \int_{-\infty}^{\infty} \widetilde{A}_z^t(\zeta, \eta)$$
$$\cdot \exp[-i\zeta x - i\eta y - i\kappa(z + d)] d\zeta d\eta . \tag{8}$$

The $E_{x,y}(x, y, 0)$ field continuity is given by

$$E_{x,y}^i(x, y, 0) + E_{x,y}^r(x, y, 0) + E_{x,y}^s(x, y, 0)$$
$$= \begin{cases} E_{x,y}^d(x, y, 0), & |x - lT_a| < a \text{ and } |y - kT_b| < b \\ 0, & \text{otherwise} . \end{cases} \tag{9}$$

Applying the Fourier transform to (9) and solving for $\widetilde{F}_z^s(\zeta, \eta)$ and $\widetilde{A}_z^s(\zeta, \eta)$, we obtain

$$\widetilde{F}_z^s(\zeta, \eta) = -\frac{(ab)^2 \epsilon_0}{\epsilon_1} \sum_{l=0}^{M_l} \sum_{k=0}^{M_k} \sum_{m=0}^{\infty} \sum_{n=0}^{\infty} \frac{\zeta \eta (a_m^2 + b_n^2)}{\zeta^2 + \eta^2}$$
$$\cdot \left[c_{mn}^{lk} \cos(\xi_{mn}d) + d_{mn}^{lk} \sin(\xi_{mn}d) \right]$$
$$\cdot F_m(\zeta a) F_n(\eta b) \exp(i\zeta lT_a + i\eta kT_b) \tag{10}$$

$$\widetilde{A}_z^s(\zeta,\eta) = \frac{(ab)^2\omega\mu_0\epsilon_0}{\epsilon_1}\sum_{l=0}^{M_l}\sum_{k=0}^{M_k}\sum_{m=0}^{\infty}\sum_{n=0}^{\infty}\frac{(a_m^2\eta^2 - b_n^2\zeta^2)}{(\zeta^2+\eta^2)\kappa}$$

$$\cdot\left[c_{mn}^{lk}\cos(\xi_{mn}d) + d_{mn}^{lk}\sin(\xi_{mn}d)\right]$$

$$\cdot F_m(\zeta a)F_n(\eta b)\exp(\mathrm{i}\zeta lT_a + \mathrm{i}\eta kT_b)$$

$$+\frac{\mathrm{i}(ab)^2\mu_0\epsilon_0}{\mu_1\epsilon_1}\sum_{l=0}^{M_l}\sum_{k=0}^{M_k}\sum_{m=1}^{\infty}\sum_{n=1}^{\infty}\frac{a_m b_n\xi_{mn}}{\kappa}$$

$$\cdot\left[\bar{c}_{mn}^{lk}\sin(\xi_{mn}d) - \bar{d}_{mn}^{lk}\cos(\xi_{mn}d)\right]$$

$$\cdot F_m(\zeta a)F_n(\eta b)\exp(\mathrm{i}\zeta lT_a + \mathrm{i}\eta kT_b)\ . \tag{11}$$

The boundary condition on the $H_{x,y}(x,y,0)$ field continuity requires

$$H_{x,y}^i(x,y,0) + H_{x,y}^r(x,y,0) + H_{x,y}^s(x,y,0)$$

$$= H_{x,y}^d(x,y,0),\qquad |x - lT_a| < a \text{ and } |y - kT_b| < b\ . \tag{12}$$

We substitute $\widetilde{A}_z^s(\zeta,\eta)$ and $\widetilde{F}_z^s(\zeta,\eta)$ into (12), multiply (12) by $\cos a_p(x - rT_a + a)\sin b_q(y - wT_b + b)$ or $\sin a_p(x - rT_a + a)\cos b_q(y - wT_b + b)$, and perform integration over the area of aperture to obtain

$$\gamma + \frac{(ab)^3 b_q}{4\pi^2}\sum_{l=0}^{M_l}\sum_{k=0}^{M_k}\sum_{m=0}^{\infty}\sum_{n=0}^{\infty}$$

$$\left[c_{mn}^{lk}\cos(\xi_{mn}d) + d_{mn}^{lk}\sin(\xi_{mn}d)\right]$$

$$\cdot\left[\frac{\omega\epsilon_0}{\epsilon_1}b_n^2 I_2 - \frac{(a_m^2 + b_n^2)}{\omega\mu_0\epsilon_1}I_1\right]$$

$$-\frac{\mathrm{i}(ab)^3\epsilon_0 b_q}{4\pi^2\mu_1\epsilon_1}\sum_{l=0}^{M_l}\sum_{k=0}^{M_k}\sum_{m=1}^{\infty}\sum_{n=1}^{\infty}$$

$$\left[\bar{c}_{mn}^{lk}\sin(\xi_{mn}d) - \bar{d}_{mn}^{lk}\cos(\xi_{mn}d)\right]a_m b_n\xi_{mn}I_2$$

$$= \frac{\mathrm{i}\varepsilon_p b_q}{\omega\mu_1\epsilon_1}\left[c_{pq}^{rw}\sin(\xi_{pq}d) - d_{pq}^{rw}\cos(\xi_{pq}d)\right]\xi_{pq}$$

$$-\frac{\varepsilon_p a_p}{\mu_1}\left[\bar{c}_{pq}^{rw}\cos(\xi_{pq}d) + \bar{d}_{pq}^{rw}\sin(\xi_{pq}d)\right],$$

$$p = 0, 1, \ldots \text{ and } q = 1, 2, \ldots \tag{13}$$

$$\bar{\gamma} + \frac{(ab)^3 a_p}{4\pi^2} \sum_{l=0}^{M_l} \sum_{k=0}^{M_k} \sum_{m=0}^{\infty} \sum_{n=0}^{\infty}$$

$$\left[c_{mn}^{lk} \cos(\xi_{mn}d) + d_{mn}^{lk} \sin(\xi_{mn}d) \right] \left[\frac{\omega\epsilon_0}{\epsilon_1} a_m^2 I_3 - \frac{(a_m^2 + b_n^2)}{\omega\mu_0\epsilon_1} I_1 \right]$$

$$+ \frac{i(ab)^3 \epsilon_0 a_p}{4\pi^2 \mu_1 \epsilon_1} \sum_{l=0}^{M_l} \sum_{k=0}^{M_k} \sum_{m=1}^{\infty} \sum_{n=1}^{\infty}$$

$$\left[\bar{c}_{mn}^{lk} \sin(\xi_{mn}d) - \bar{d}_{mn}^{lk} \cos(\xi_{mn}d) \right] a_m b_n \xi_{mn} I_3$$

$$= \frac{i\varepsilon_q a_p}{\omega\mu_1\epsilon_1} \left[c_{pq}^{rw} \sin(\xi_{pq}d) - d_{pq}^{rw} \cos(\xi_{pq}d) \right] \xi_{pq}$$

$$+ \frac{\varepsilon_q b_q}{\mu_1} \left[\bar{c}_{pq}^{rw} \cos(\xi_{pq}d) + \bar{d}_{pq}^{rw} \sin(\xi_{pq}d) \right],$$

$$p = 1, 2, \ldots \quad \text{and} \quad q = 0, 1, \ldots \tag{14}$$

where

$$\gamma = 2iab(u \sin\phi \cos\theta + v \cos\phi)k_x b_q$$
$$\cdot F_p(k_x a) F_q(k_y b) \exp(ik_x rT_a + ik_y wT_b) \tag{15}$$

$$\bar{\gamma} = 2iab(u \cos\phi \cos\theta - v \sin\phi)k_y a_p$$
$$\cdot F_p(k_x a) F_q(k_y b) \exp(ik_x rT_a + ik_y wT_b) \tag{16}$$

$$I_1 = \int_{-\infty}^{\infty} \int_{-\infty}^{\infty} \frac{(\zeta\eta)^2}{\kappa} L(\zeta, \eta)$$
$$\cdot \exp[i\zeta(l-r)T_a + i\eta(k-w)T_b] \, d\zeta d\eta \tag{17}$$

$$I_2 = \int_{-\infty}^{\infty} \int_{-\infty}^{\infty} \frac{\zeta^2}{\kappa} L(\zeta, \eta)$$
$$\cdot \exp[i\zeta(l-r)T_a + i\eta(k-w)T_b] \, d\zeta d\eta \tag{18}$$

$$I_3 = \int_{-\infty}^{\infty} \int_{-\infty}^{\infty} \frac{\eta^2}{\kappa} L(\zeta, \eta)$$
$$\cdot \exp[i\zeta(l-r)T_a + i\eta(k-w)T_b] \, d\zeta d\eta \tag{19}$$

with $L(\zeta, \eta) = F_m(\zeta a) F_p(-\zeta a) F_n(\eta b) F_q(-\eta b)$. In high-frequency limit for a single aperture case, $I_1, I_2,$ and I_3 have dominant contributions from the poles of $L(\zeta, \eta)$ at $m = p$ and $n = q$ when $l = r$ and $k = w$, leading to $I_1 \rightarrow \frac{4\pi^2}{(ab)^3} \frac{1}{\xi_{pq}} \varepsilon_p \varepsilon_q$, $I_2 \rightarrow \frac{4\pi^2}{(ab)^3} \frac{1}{b_q^2 \xi_{pq}} \varepsilon_p$, and $I_3 \rightarrow \frac{4\pi^2}{(ab)^3} \frac{1}{a_p^2 \xi_{pq}} \varepsilon_q$. The boundary conditions on the $E_{x,y}$ and $H_{x,y}$ continuities at $z = -d$ similarly yield

$$\frac{(ab)^3 b_q}{4\pi^2} \sum_{l=0}^{M_l} \sum_{k=0}^{M_k} \sum_{m=0}^{\infty} \sum_{n=0}^{\infty} c_{mn}^{lk} \left[\frac{\omega\epsilon_0}{\epsilon_1} b_n^2 I_2 - \frac{(a_m^2 + b_n^2)}{\omega\mu_0\epsilon_1} I_1 \right]$$

$$+ \frac{\mathrm{i}(ab)^3 \epsilon_0 b_q}{4\pi^2 \mu_1\epsilon_1} \sum_{l=0}^{M_l} \sum_{k=0}^{M_k} \sum_{m=1}^{\infty} \sum_{n=1}^{\infty} \bar{d}_{mn}^{lk} a_m b_n \xi_{mn} I_2$$

$$= \frac{\mathrm{i}\varepsilon_p b_q}{\omega\mu_1\epsilon_1} d_{pq}^{rw} \xi_{pq} + \frac{\varepsilon_p a_p}{\mu_1} \bar{c}_{pq}^{rw},$$

$$p = 0, 1, \ldots \quad \text{and} \quad q = 1, 2, \ldots \tag{20}$$

$$\frac{(ab)^3 a_p}{4\pi^2} \sum_{l=0}^{M_l} \sum_{k=0}^{M_k} \sum_{m=0}^{\infty} \sum_{n=0}^{\infty} c_{mn}^{lk} \left[\frac{\omega\epsilon_0}{\epsilon_1} a_m^2 I_3 - \frac{(a_m^2 + b_n^2)}{\omega\mu_0\epsilon_1} I_1 \right]$$

$$- \frac{\mathrm{i}(ab)^3 \epsilon_0 a_p}{4\pi^2 \mu_1\epsilon_1} \sum_{l=0}^{M_l} \sum_{k=0}^{M_k} \sum_{m=1}^{\infty} \sum_{n=1}^{\infty} \bar{d}_{mn}^{lk} a_m b_n \xi_{mn} I_3$$

$$= \frac{\mathrm{i}\varepsilon_q a_p}{\omega\mu_1\epsilon_1} d_{pq}^{rw} \xi_{pq} - \frac{\varepsilon_q b_q}{\mu_1} \bar{c}_{pq}^{rw},$$

$$p = 1, 2, \ldots \quad \text{and} \quad q = 0, 1, \ldots \tag{21}$$

When region (III) is filled with a perfect conducting medium (PEC), the field representations for $F_z^t(x, y, z)$ and $A_z^t(x, y, z)$ are unnecessary. This implies that we need to match the boundary conditions at $z = 0$ only, thus leading to (13) and (14) with $c_{mn}^{lk} = \bar{d}_{mn}^{lk} = 0$.

The transmission coefficient τ of an aperture, a ratio of the total transmitted power to that incident on an aperture, is shown to be

$$\tau = \frac{1}{4\omega\mu_1\epsilon_1 Z_0 \cos\theta} Im \left\{ \sum_{m=0}^{\infty} \sum_{n=0}^{\infty} \frac{1}{\epsilon_1} (\varepsilon_n a_m^2 + \varepsilon_m b_n^2) \xi_{mn}^* c_{mn}^{lk} d_{mn}^{lk*} \right.$$

$$\left. - \sum_{m=1}^{\infty} \sum_{n=1}^{\infty} \frac{1}{\mu_1} (a_m^2 + b_n^2) \xi_{mn} \bar{c}_{mn}^{lk*} \bar{d}_{mn}^{lk} \right\}. \tag{22}$$

6.6 EM Scattering from Rectangular Apertures in a Rectangular Cavity [18]

A subject of electromagnetic wave penetration into a cavity with apertures is a canonical problem in electromagnetic compatibility and interference. Various approaches [19-23] have been used to estimate the shielding effectiveness of a rectangular enclosure with apertures. The present section provides an analysis for electromagnetic wave penetration into a three-dimensional cavity with multiple rectangular apertures in a conducting plane. An electromagnetic wave is normally incident on the cavity-backed multiple rectangular

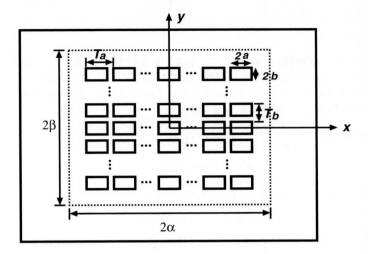

(a) Front view

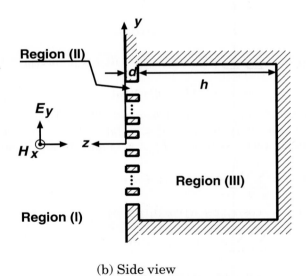

(b) Side view

Fig. 6.6. Multiple rectangular apertures in a rectangular cavity

apertures with periods T_a and T_b in the x- and y-directions, respectively. The total field in region (I) ($z > 0$) consists of the incident, reflected, and scattered waves

$$E_y^i = Z_0 e^{-ik_0 z} \tag{1}$$

$$E_y^r = -Z_0 e^{ik_0 z} \tag{2}$$

where the wavenumber is $k_0 = \omega\sqrt{\mu_0 \epsilon_0} = 2\pi/\lambda$ and $Z_0 = \sqrt{\mu_0/\epsilon_0}$ is the free-space intrinsic impedance. The scattered E^s-and H^s-fields in region (I) are given in terms of the z-components of the electric and magnetic vector potentials, $F_z^s(x,y,z)$ and $A_z^s(x,y,z)$

$$F_z^s(x,y,z) = \frac{1}{4\pi^2} \int_{-\infty}^{\infty} \int_{-\infty}^{\infty} \tilde{F}_z^s(\zeta,\eta)$$
$$\cdot \exp(-i\zeta x - i\eta y + i\kappa z)\mathrm{d}\zeta\mathrm{d}\eta \tag{3}$$

$$A_z^s(x,y,z) = \frac{1}{4\pi^2} \int_{-\infty}^{\infty} \int_{-\infty}^{\infty} \tilde{A}_z^s(\zeta,\eta)$$
$$\cdot \exp(-i\zeta x - i\eta y + i\kappa z)\mathrm{d}\zeta\mathrm{d}\eta \tag{4}$$

where $\kappa = \sqrt{k_0^2 - \zeta^2 - \eta^2}$. In region (II) ($l$th and kth aperture in the x and y directions) the vector potentials $F_z^d(x,y,z)$ and $A_z^d(x,y,z)$ are

$$F_z^d(x,y,z) = \sum_{m=0}^{\infty}\sum_{n=0}^{\infty} \left[c_{mn}^{lk}\cos\xi_{mn}(z+d) + d_{mn}^{lk}\sin\xi_{mn}(z+d)\right]$$
$$\cdot \cos a_m(x - lT_a + a)\cos b_n(y - kT_b + b) \tag{5}$$

$$A_z^d(x,y,z) = \sum_{m=1}^{\infty}\sum_{n=1}^{\infty} \left[\bar{c}_{mn}^{lk}\cos\xi_{mn}(z+d) + \bar{d}_{mn}^{lk}\sin\xi_{mn}(z+d)\right]$$
$$\cdot \sin a_m(x - lT_a + a)\sin b_n(y - kT_b + b) \tag{6}$$

where $(m,n) \neq (0,0)$, $a_m = m\pi/(2a)$, $b_n = n\pi/(2b)$, and $\xi_{mn} = \sqrt{k_0^2 - a_m^2 - b_n^2}$. In region (III) of a cavity interior ($|x| < \alpha$, $|y| < \beta$, and $-d - h < z < -d$) the corresponding vector potentials $F_z^c(x,y,z)$ and $A_z^c(x,y,z)$ are

$$F_z^c(x,y,z) = \sum_{g=0}^{\infty}\sum_{j=0}^{\infty} e_{gj}\sin\gamma_{gj}(z+d+h)$$
$$\cdot \cos\alpha_g(x+\alpha)\cos\beta_j(y+\beta) \tag{7}$$

$$A_z^c(x,y,z) = \sum_{g=1}^{\infty}\sum_{j=1}^{\infty} \bar{e}_{gj}\cos\gamma_{gj}(z+d+h)$$
$$\cdot \sin\alpha_g(x+\alpha)\sin\beta_j(y+\beta) \tag{8}$$

where $(g,j) \neq (0,0)$, $\alpha_g = g\pi/(2\alpha)$, $\beta_j = j\pi/(2\beta)$, and $\gamma_{gj} = \sqrt{k_0^2 - \alpha_g^2 - \beta_j^2}$.

The boundary conditions on the $E_{x,y}$ and $H_{x,y}$ continuities at $z = 0$ require respectively

$$E_{x,y}^i(x,y,0) + E_{x,y}^r(x,y,0) + E_{x,y}^s(x,y,0)$$
$$= \begin{cases} E_{x,y}^d(x,y,0), & |x - lT_a| < a \text{ and } |y - kT_b| < b \\ 0, & \text{otherwise} \end{cases} \tag{9}$$

and

$$H_{x,y}^i(x,y,0) + H_{x,y}^r(x,y,0) + H_{x,y}^s(x,y,0) = H_{x,y}^d(x,y,0), \tag{10}$$
$$|x - lT_a| < a \text{ and } |y - kT_b| < b \, .$$

Applying the Fourier transform to (9) and solving for $\widetilde{F}_z^s(\zeta,\eta)$ and $\widetilde{A}_z^s(\zeta,\eta)$, we get

$$\widetilde{F}_z^s(\zeta,\eta) = -(ab)^2 \sum_{l=0}^{M_l} \sum_{k=0}^{M_k} \sum_{m=0}^{\infty} \sum_{n=0}^{\infty}$$
$$\cdot \frac{\zeta\eta(a_m^2 + b_n^2)}{\zeta^2 + \eta^2} \left[c_{mn}^{lk} \cos(\xi_{mn}d) + d_{mn}^{lk} \sin(\xi_{mn}d) \right]$$
$$\cdot F_m(\zeta a)F_n(\eta b)\exp(i\zeta lT_a + i\eta kT_b) \tag{11}$$

$$\widetilde{A}_z^s(\zeta,\eta) = (ab)^2 \omega\mu_0 \sum_{l=0}^{M_l} \sum_{k=0}^{M_k} \sum_{m=0}^{\infty} \sum_{n=0}^{\infty}$$
$$\cdot \frac{(a_m^2\eta^2 - b_n^2\zeta^2)}{(\zeta^2 + \eta^2)\kappa} \left[c_{mn}^{lk} \cos(\xi_{mn}d) + d_{mn}^{lk} \sin(\xi_{mn}d) \right]$$
$$\cdot F_m(\zeta a)F_n(\eta b)\exp(i\zeta lT_a + i\eta kT_b)$$
$$+ i(ab)^2 \sum_{l=0}^{M_l} \sum_{k=0}^{M_k} \sum_{m=1}^{\infty} \sum_{n=1}^{\infty} \frac{a_m b_n \xi_{mn}}{\kappa}$$
$$\cdot \left[\bar{c}_{mn}^{lk} \sin(\xi_{mn}d) - \bar{d}_{mn}^{lk} \cos(\xi_{mn}d) \right]$$
$$\cdot F_m(\zeta a)F_n(\eta b)\exp(i\zeta lT_a + i\eta kT_b) \, . \tag{12}$$

We substitute (11) and (12) into (10), multiply (10) by $\cos a_p(x - rT_a + a)\sin b_q(y - wT_b + b)$ or $\sin a_p(x - rT_a + a)\cos b_q(y - wT_b + b)$, and perform integration over the area of apertures to get

$$\frac{(ab)^3 b_q}{4\pi^2} \sum_{l=0}^{M_l} \sum_{k=0}^{M_k} \sum_{m=0}^{\infty} \sum_{n=0}^{\infty}$$
$$\cdot \left[c_{mn}^{lk} \cos(\xi_{mn}d) + d_{mn}^{lk} \sin(\xi_{mn}d) \right] \left[\omega b_n^2 I_2 - \frac{(a_m^2 + b_n^2)}{\omega\mu_0\epsilon_0} I_1 \right]$$
$$- \frac{i(ab)^3 b_q}{4\pi^2 \mu_0} \sum_{l=0}^{M_l} \sum_{k=0}^{M_k} \sum_{m=1}^{\infty} \sum_{n=1}^{\infty}$$
$$\cdot \left[\bar{c}_{mn}^{lk} \sin(\xi_{mn}d) - \bar{d}_{mn}^{lk} \cos(\xi_{mn}d) \right] a_m b_n \xi_{mn} I_2$$
$$= \frac{i\varepsilon_p b_q}{\omega\mu_0\epsilon_0} \left[c_{pq}^{rw} \sin(\xi_{pq}d) - d_{pq}^{rw} \cos(\xi_{pq}d) \right] \xi_{pq}$$
$$- \frac{\varepsilon_p a_p}{\mu_0} \left[\bar{c}_{pq}^{rw} \cos(\xi_{pq}d) + \bar{d}_{pq}^{rw} \sin(\xi_{pq}d) \right] ,$$
$$p = 0, 1, \ldots \text{ and } q = 1, 2, \ldots \tag{13}$$

$$\frac{(ab)^3 a_p}{4\pi^2} \sum_{l=0}^{M_l} \sum_{k=0}^{M_k} \sum_{m=0}^{\infty} \sum_{n=0}^{\infty}$$

$$\cdot \left[c_{mn}^{lk} \cos(\xi_{mn}d) + d_{mn}^{lk} \sin(\xi_{mn}d) \right] \left[\omega a_m^2 I_3 - \frac{(a_m^2 + b_n^2)}{\omega\mu_0\epsilon_0} I_1 \right]$$

$$+\frac{\mathrm{i}(ab)^3 a_p}{4\pi^2 \mu_0} \sum_{l=0}^{M_l} \sum_{k=0}^{M_k} \sum_{m=1}^{\infty} \sum_{n=1}^{\infty}$$

$$\cdot \left[\bar{c}_{mn}^{lk} \sin(\xi_{mn}d) - \bar{d}_{mn}^{lk} \cos(\xi_{mn}d) \right] a_m b_n \xi_{mn} I_3 - s_{pq}^{rw}$$

$$= \frac{\mathrm{i}\varepsilon_q a_p}{\omega\mu_0\epsilon_0} \left[c_{pq}^{rw} \sin(\xi_{pq}d) - d_{pq}^{rw} \cos(\xi_{pq}d) \right] \xi_{pq}$$

$$+\frac{\varepsilon_q b_q}{\mu_0} \left[\bar{c}_{pq}^{rw} \cos(\xi_{pq}d) + \bar{d}_{pq}^{rw} \sin(\xi_{pq}d) \right],$$

$$p = 1, 2, \ldots \text{ and } q = 0, 1, \ldots \qquad (14)$$

where δ_{mp} is the Kronecker delta, $\varepsilon_0 = 2$, $\varepsilon_1 = \varepsilon_2 = \cdots = 1$, and

$$s_{pq}^{rw} = \begin{cases} 8[1 - (-1)^p]/(p\pi), & q = 0 \\ 0, & \text{otherwise} . \end{cases} \qquad (15)$$

Note that $I_1, I_2,$ and I_3 are given by (17), (18), and (19) in Sect. 6.5. The boundary conditions on the $E_{x,y}$ and $H_{x,y}$ continuities at $z = -d$ for $|x| < \alpha$ and $|y| < \beta$ require

$$E_{x,y}^c(x, y, -d)$$
$$= \begin{cases} E_{x,y}^d(x, y, -d), & |x - lT_a| < a \text{ and } |y - kT_b| < b \\ 0, & \text{otherwise} \end{cases} \qquad (16)$$

and

$$H_{x,y}^c(x, y, -d) = H_{x,y}^d(x, y, -d), \qquad |x - lT_a| < a, \ |y - kT_b| < b . \quad (17)$$

Performing $\displaystyle\int_{-\beta}^{\beta} \int_{-\alpha}^{\alpha} \{E_x \text{ continuity in (16)}\} \cos\alpha_{g'}(x + \alpha) \sin\beta_{j'}(y + \beta) \mathrm{d}x \mathrm{d}y$

and $\displaystyle\int_{-\beta}^{\beta} \int_{-\alpha}^{\alpha} \{E_y \text{ continuity in (16)}\} \sin\alpha_{g'}(x + \alpha) \cos\beta_{j'}(y + \beta) \mathrm{d}x \mathrm{d}y$, and solving for e_{gj} and $\bar{e}_{gj}$, we get

$$e_{gj} = \frac{1}{\alpha\beta} \sum_{l=0}^{M_l} \sum_{k=0}^{M_k} \sum_{m=0}^{\infty} \sum_{n=0}^{\infty} \frac{\left(b_n\beta_j \Upsilon_{mn}^{lk} + a_m\alpha_g \bar{\Upsilon}_{mn}^{lk} \right)}{\sin(\gamma_{gj}h)(\varepsilon_j\alpha_g^2 + \varepsilon_g\beta_j^2)} c_{mn}^{lk}$$

$$+\frac{\mathrm{i}}{\omega\mu_0\alpha\beta} \sum_{l=0}^{M_l} \sum_{k=0}^{M_k} \sum_{m=1}^{\infty} \sum_{n=1}^{\infty}$$

$$\frac{\left(a_m\beta_j \Upsilon_{mn}^{lk} - b_n\alpha_g \bar{\Upsilon}_{mn}^{lk} \right)}{\sin(\gamma_{gj}h)(\varepsilon_j\alpha_g^2 + \varepsilon_g\beta_j^2)} \xi_{mn} \bar{d}_{mn}^{lk}, \qquad (18)$$

$$g = 0, 1, 2, \ldots, \ j = 0, 1, 2, \ldots, \ (g, j) \neq (0, 0)$$

$$\bar{e}_{gj} = \frac{i\omega\mu_0}{\alpha\beta} \sum_{l=0}^{M_l} \sum_{k=0}^{M_k} \sum_{m=0}^{\infty} \sum_{n=0}^{\infty} \frac{(b_n\alpha_g\Upsilon_{mn}^{lk} - a_m\beta_j\bar{\Upsilon}_{mn}^{lk})}{\gamma_{gj}\sin(\gamma_{gj}h)(\alpha_g^2 + \beta_j^2)} c_{mn}^{lk}$$

$$-\frac{1}{\alpha\beta} \sum_{l=0}^{M_l} \sum_{k=0}^{M_k} \sum_{m=1}^{\infty} \sum_{n=1}^{\infty} \frac{(a_m\alpha_g\Upsilon_{mn}^{lk} + b_n\beta_j\bar{\Upsilon}_{mn}^{lk})}{\gamma_{gj}\sin(\gamma_{gj}h)(\alpha_g^2 + \beta_j^2)} \xi_{mn}\bar{d}_{mn}^{lk}, \tag{19}$$

$$g = 1, 2, 3, \ldots, \; j = 1, 2, 3, \ldots$$

where

$$\Upsilon_{mn}^{lk} = \begin{cases} \dfrac{2ab_n}{(\beta_j^2 - b_n^2)} \big\{ \sin[\beta_j(b - \beta - kT_b)] \\ + (-1)^n \sin[\beta_j(b + \beta + kT_b)] \big\}, \\ \qquad\qquad g = m = 0 \\[6pt] \dfrac{\alpha_g b_n}{(\alpha_g^2 - a_m^2)(\beta_j^2 - b_n^2)} \big\{ \sin[\alpha_g(a - \alpha - lT_a)] \\ + (-1)^m \sin[\alpha_g(a + \alpha + lT_a)] \big\} \\ \cdot \big\{ \sin[\beta_j(b - \beta - kT_b)] + (-1)^n \sin[\beta_j(b + \beta + kT_b)] \big\}, \\ \qquad\qquad \text{otherwise} \end{cases} \tag{20}$$

$$\bar{\Upsilon}_{mn}^{lk} = \begin{cases} \dfrac{2ba_m}{(\alpha_g^2 - a_m^2)} \big\{ \sin[\alpha_g(a - \alpha - lT_a)] \\ + (-1)^m \sin[\alpha_g(a + \alpha + lT_a)] \big\}, \\ \qquad\qquad j = n = 0 \\[6pt] \dfrac{a_m\beta_j}{(\alpha_g^2 - a_m^2)(\beta_j^2 - b_n^2)} \big\{ \sin[\alpha_g(a - \alpha - lT_a)] \\ + (-1)^m \sin[\alpha_g(a + \alpha + lT_a)] \big\} \\ \cdot \big\{ \sin[\beta_j(b - \beta - kT_b)] + (-1)^n \sin[\beta_j(b + \beta + kT_b)] \big\}, \\ \qquad\qquad \text{otherwise} . \end{cases} \tag{21}$$

We substitute (18) and (19) into (17), multiply by $\cos a_p(x - rT_a + a)\sin b_q(y - wT_b + b)$ or $\sin a_p(x - rT_a + a)\cos b_q(y - wT_b + b)$, and perform integration over the area of apertures to get

$$\frac{i}{\alpha\beta ab} \sum_{l=0}^{M_l} \sum_{k=0}^{M_k} \left(\sum_{m=0}^{\infty} \sum_{n=0}^{\infty} c_{mn}^{lk}Q + \frac{i}{\omega\mu_0} \sum_{m=1}^{\infty} \sum_{n=1}^{\infty} \xi_{mn}\bar{d}_{mn}^{lk}\bar{Q} \right)$$

$$= \frac{i\varepsilon_p}{\omega\mu_0\epsilon_0} b_q\xi_{pq}d_{pq}^{rw} + \frac{1}{\mu_0}a_p\bar{c}_{pq}^{rw},$$

$$p, g = 0, 1, 2, \ldots \text{ and } q, j = 1, 2, 3, \ldots \tag{22}$$

$$\frac{i}{\alpha\beta ab}\sum_{l=0}^{M_l}\sum_{k=0}^{M_k}\left(\sum_{m=0}^{\infty}\sum_{n=0}^{\infty}c_{mn}^{lk}P+\frac{i}{\omega\mu_0}\sum_{m=1}^{\infty}\sum_{n=1}^{\infty}\xi_{mn}\bar{d}_{mn}^{lk}\bar{P}\right)$$

$$=\frac{i\varepsilon_q}{\omega\mu_0\epsilon_0}a_p\xi_{pq}d_{pq}^{rw}-\frac{1}{\mu_0}b_q\bar{c}_{pq}^{rw},$$

$$p,g=1,2,3,\ldots \text{ and } q,j=0,1,2,\ldots \qquad (23)$$

where

$$Q=\frac{1}{\omega\mu_0\epsilon_0}\sum_{g=0}^{\infty}\sum_{j=1}^{\infty}\beta_j\gamma_{gj}\Upsilon_{pq}^{rw}\frac{(b_n\beta_j\Upsilon_{mn}^{lk}+a_m\alpha_g\bar{\Upsilon}_{mn}^{lk})}{\tan(\gamma_{gj}h)(\alpha_g^2+\varepsilon_g\beta_j^2)}$$

$$+\omega\sum_{g=1}^{\infty}\sum_{j=1}^{\infty}\alpha_g\Upsilon_{pq}^{rw}\frac{(b_n\alpha_g\Upsilon_{mn}^{lk}-a_m\beta_j\bar{\Upsilon}_{mn}^{lk})}{\gamma_{gj}\tan(\gamma_{gj}h)(\alpha_g^2+\beta_j^2)} \qquad (24)$$

$$\bar{Q}=\frac{1}{\omega\mu_0\epsilon_0}\sum_{g=0}^{\infty}\sum_{j=1}^{\infty}\beta_j\gamma_{gj}\Upsilon_{pq}^{rw}\frac{(a_m\beta_j\Upsilon_{mn}^{lk}-b_n\alpha_g\bar{\Upsilon}_{mn}^{lk})}{\tan(\gamma_{gj}h)(\alpha_g^2+\varepsilon_g\beta_j^2)}$$

$$+\omega\sum_{g=1}^{\infty}\sum_{j=1}^{\infty}\alpha_g\Upsilon_{pq}^{rw}\frac{(a_m\alpha_g\Upsilon_{mn}^{lk}+b_n\beta_j\bar{\Upsilon}_{mn}^{lk})}{\gamma_{gj}\tan(\gamma_{gj}h)(\alpha_g^2+\beta_j^2)} \qquad (25)$$

$$P=\frac{1}{\omega\mu_0\epsilon_0}\sum_{g=1}^{\infty}\sum_{j=0}^{\infty}\alpha_g\gamma_{gj}\bar{\Upsilon}_{pq}^{rw}\frac{(b_n\beta_j\Upsilon_{mn}^{lk}+a_m\alpha_g\bar{\Upsilon}_{mn}^{lk})}{\tan(\gamma_{gj}h)(\varepsilon_j\alpha_g^2+\beta_j^2)}$$

$$-\omega\sum_{g=1}^{\infty}\sum_{j=1}^{\infty}\beta_j\bar{\Upsilon}_{pq}^{rw}\frac{(b_n\alpha_g\Upsilon_{mn}^{lk}-a_m\beta_j\bar{\Upsilon}_{mn}^{lk})}{\gamma_{gj}\tan(\gamma_{gj}h)(\alpha_g^2+\beta_j^2)} \qquad (26)$$

$$\bar{P}=\frac{1}{\omega\mu_0\epsilon_0}\sum_{g=1}^{\infty}\sum_{j=0}^{\infty}\alpha_g\gamma_{gj}\bar{\Upsilon}_{pq}^{rw}\frac{(a_m\beta_j\Upsilon_{mn}^{lk}-b_n\alpha_g\bar{\Upsilon}_{mn}^{lk})}{\tan(\gamma_{gj}h)(\varepsilon_j\alpha_g^2+\beta_j^2)}$$

$$-\omega\sum_{g=1}^{\infty}\sum_{j=1}^{\infty}\beta_j\bar{\Upsilon}_{pq}^{rw}\frac{(a_m\alpha_g\Upsilon_{mn}^{lk}+b_n\beta_j\bar{\Upsilon}_{mn}^{lk})}{\gamma_{gj}\tan(\gamma_{gj}h)(\alpha_g^2+\beta_j^2)}. \qquad (27)$$

References for Chapter 6

1. N. A. McDonald, "Polynomial approximations for the electric polarizabilities of some small apertures," *IEEE Trans. Microwave Theory Tech.*, vol. 33, no. 11, pp. 1146-1149, November 1985.
2. S. B. Cohn, "The electric polarizability of aperture of arbitrary shape," *Proc. I.R.E.*, vol. 40, pp. 1069-1071, 1952.
3. H. H. Park and H. J. Eom, "Electrostatic potential distribution through a rectangular aperture in a thick conducting plane," *IEEE Trans. Microwave Theory Tech.*, vol. 44, no. 10, pp. 1745-1747, Oct. 1996.
4. S. B. Cohn, "Determination of aperture parameters by electrolytic-tank measurements," *Proc. I.R.E.*, vol. 39, pp. 1416-1421, 1951.

5. R. L. Gluckstern and R. K. Cooper, "Electric polarizability and magnetic susceptibility of small holes in a thin screen," *IEEE Trans. Microwave Theory Tech.*, vol. 38, no. 2, pp. 186-192, Feb. 1990.
6. L. K. Warne and K. C. Chen, "Relation between equivalent antenna radius and transverse line dipole moments of a narrow slot aperture having depth," *IEEE Trans. Electromagn. Compat.*, vol. 30, no. 3, pp. 364-370, Aug. 1988.
7. H. H. Park and H. J. Eom, "Acoustic scattering from a rectangular aperture in a thick hard screen," *J. Acoust. Soc. Am.*, vol. 101, no. 1, pp. 595-598, Jan. 1997.
8. A. Sauter, Jr. and W. W. Soroka, "Sound transmission through rectangular slots of finite depth between reverberant rooms," *J. Acoust. Soc. Am.*, vol. 47, pp. 5-11, 1970.
9. H. H. Park and H. J. Eom, "Electrostatic potential distribution through thick multiple rectangular apertures," *Electron. Lett.*, vol. 34, no. 15, pp. 1500-1501, 23rd, July 1998.
10. J. G. Lee and H. J. Eom, "Magnetic polarisability of thick multiple rectangular apertures," *Electron. Lett.*, vol. 35, no. 21, pp. 1850-1851, Oct. 1999.
11. H. H. Park and H. J. Eom, "Electromagnetic scattering from multiple rectangular apertures in a thick conducting screen," *IEEE Trans. Antennas Propagat.*, vol. 47, no. 6, pp. 1056-1060, June 1999.
12. A. El-Hajj and K. Kabalan, "Characteristic modes of a rectangular aperture in a perfectly conducting plane," *IEEE Trans. Antennas Propagat.*, vol. 42, no. 10, pp. 1447-1450, Oct. 1994.
13. K. Barkeshli and J. L. Volakis, "Electromagnetic scattering from an aperture formed by a rectangular cavity recessed in a ground plane," *J. Electromagn. Waves Appl.*, vol. 5, no. 7, pp. 715-735, 1991.
14. J. M. Jin and J. L. Volakis, "Electromagnetic scattering by the transmission through a three-dimensional slot in a thick conducting plane," *IEEE Trans. Antennas Propagat.*, vol. 39, no. 4, pp. 1544-1550, Apr. 1991.
15. C. C. Chen, "Transmission of microwave through perforated flat plates of finite thickness," *IEEE Trans. Microwave Theory Tech.*, vol. 21, no. 1, pp. 1-6, Jan. 1973.
16. T. K. Sarkar, M. F. Costa, C-L. I, and R. F. Harrington, "Electromagnetic transmission through mesh covered apertures and arrays of apertures in a conducting screen," *IEEE Trans. Antennas Propagat.*, vol. 32, no. 9, pp. 908-913, Sept. 1984.
17. T. Andersson, "Moment-method calculations on apertures using basis singular functions," *IEEE Trans. Antennas Propagat.*, vol. 41, no. 12, pp. 1709-1716, Dec. 1993.
18. H. H. Park and H. J. Eom, "Electromagnetic penetration into a rectangular cavity with multiple rectangular apertures in a conducting plane," *IEEE Trans. Electromagn. Compat.*, vol. 42, no. 3, pp. 303-307, Aug. 2000.
19. T. Wang, R. F. Harrington, and J. R. Mautz, "Electromagnetic scattering from and transmission through arbitrary apertures in conducting bodies," *IEEE Trans. Antennas Propagat.*, vol. 38, no. 11, pp. 1805-1814, Nov. 1990.
20. K. P. Ma, M. Li, J. L. Drewniak, T. H. Hubing, and T. P. Van Doren, "Comparison of FDTD algorithms for subcellular modeling of slots in shielding enclosures," *IEEE Trans. Electromagn. Compat.*, vol. 39, no. 2, pp. 147-155, May 1997.
21. H. Kogure, H. Nakano, K. Koshiji, and E. Shu, "Analysis of electromagnetic field inside equipment housing with an aperture," *IEICE Trans. Commun.*, vol. E80-B, no. 11, pp. 1620-1623, Nov. 1997.

22. B. W. Kim, Y. C. Chung, and T. K. Kang, "Analysis of electromagnetic penetration through apertures of shielded enclosure using finite element method," *14th Annual Review of Progress in Applied Computational Electromagnetics*, Monterey, CA, vol. II, pp. 795-798, Mar. 1998.
23. M. P. Robinson, T. M. Benson, C. Christopoulos, J. F. Dawson, M. D. Ganley, A. C. Marvin, S. J. Porter, and D. W. P. Thomas, "Analytical formulation for the shielding effectiveness of enclosures with apertures," *IEEE Trans. Electromagn. Compat.*, vol. 40, no. 3, pp. 240-248, Aug. 1998.

7. Circular Apertures in a Plane

7.1 Static Potential Through a Circular Aperture in a Plane

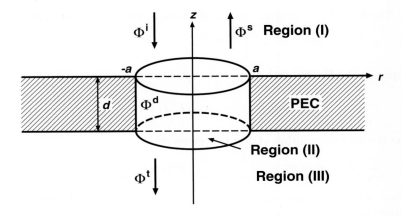

Fig. 7.1. A circular aperture in a conducting plane

7.1.1 Electrostatic Distribution [1]

A potential penetration into a circular aperture in a thick conducting plane has been studied in [2-3] due to its application in low-frequency field leakage problems. In this section we will study an electrostatic potential penetration into a circular aperture in a thick perfectly-conducting plane. In region (I) $(z > 0)$ an incident potential $\Phi^i(r, z)$ impinges on a circular aperture (radius:

a and depth: d) in a thick perfectly-conducting plane at zero potential. Regions (II) ($-d < z < 0$ and $r < a$) and (III) ($z < -d$) represent a circular aperture and a half-space, respectively. The total electrostatic potential in region (I) has the incident and scattered potential components as

$$\Phi^i(r, z) = z \tag{1}$$

$$\Phi^s(r, z) = \int_0^\infty \widetilde{\Phi}^s(\zeta) J_0(\zeta r) e^{-\zeta z} \zeta d\zeta . \tag{2}$$

The total electrostatic potential in region (II) is

$$\Phi^d(r, z) = \sum_{n=1}^\infty [b_n \sinh k_n(z + d) + c_n \cosh k_n(z + d)] J_0(k_n r) \tag{3}$$

where k_n is determined by the characteristic equation $J_0(k_n a) = 0$. The total transmitted electrostatic potential in region (III) is

$$\Phi^t(r, z) = \int_0^\infty \widetilde{\Phi}^t(\zeta) J_0(\zeta r) e^{\zeta(z+d)} \zeta d\zeta . \tag{4}$$

The boundary condition on the field continuity at $z = 0$ requires

$$\Phi^i(r, 0) + \Phi^s(r, 0) = \begin{cases} \Phi^d(r, 0), & r < a \\ 0, & r > a . \end{cases} \tag{5}$$

Applying the Hankel transform to (5) yields

$$\widetilde{\Phi}^s(\zeta) = \sum_{n=1}^\infty (b_n \sinh k_n d + c_n \cosh k_n d) \left[\frac{-a k_n J_0(\zeta a) J_1(k_n a)}{\zeta^2 - k_n^2} \right] . \tag{6}$$

The boundary condition at $z = 0$ is given as

$$\frac{\partial}{\partial z} \left[\Phi^i(r, z) + \Phi^s(r, z) \right]_{z=0} = \frac{\partial}{\partial z} \left[\Phi^d(r, z) \right]_{z=0}, \quad r < a . \tag{7}$$

Substituting $\widetilde{\Phi}^s(\zeta)$ into (7), multiplying (7) by $J_0(k_p r) r$, and integrating with respect to r from 0 to a, we get

$$\boxed{ \begin{aligned} & \frac{a}{k_p} - \sum_{n=1}^\infty (b_n \sinh k_n d + c_n \cosh k_n d) a^2 k_n k_p J_1(k_n a) I \\ & = \frac{a^2}{2} k_p J_1(k_p a)(b_p \cosh k_p d + c_p \sinh k_p d) \end{aligned} } \tag{8}$$

where

$$I = \int_0^\infty \frac{J_0^2(\zeta a) \zeta^2}{(\zeta^2 - k_n^2)(\zeta^2 - k_p^2)} d\zeta . \tag{9}$$

The additional boundary conditions at $z = -d$ give

$$\boxed{ \sum_{n=1}^\infty c_n k_n J_1(k_n a) I = \frac{1}{2} b_p J_1(k_p a) . } \tag{10}$$

The electric polarizability is shown to be

$$\chi(z) \equiv 2\pi \int_0^a \Phi^d(r, z) r \mathrm{d}r$$

$$= 2\pi a \sum_{n=1}^{\infty} [b_n \sinh k_n(z + d) + c_n \cosh k_n(z + d)] J_1(k_n a)/k_n . \quad (11)$$

7.1.2 Magnetostatic Distribution [4]

A behavior of magnetostatic potential penetration into a circular aperture was studied in [2,3,5]. Consider an incident magnetostatic potential $\Phi^i(x, y, z)$ impinging on a circular aperture (radius: a and depth: d) in a thick conducting plane. Regions (II) ($-d < z < 0$ and $r < a$) and (III) ($z < -d$) are a circular aperture and a half-space, respectively. In region (I) ($z > 0$) the total magnetostatic potential consists of the incident and scattered potentials

$$\Phi^i(r, \phi, z) = x = r \cos \phi \quad (12)$$

$$\Phi^s(r, \phi, z) = \cos \phi \int_0^{\infty} \widetilde{\Phi}^s(\zeta) J_1(\zeta r) e^{-\zeta z} \zeta \mathrm{d}\zeta . \quad (13)$$

The total magnetostatic potential in region (II) is

$$\Phi^d(r, \phi, z)$$
$$= \cos \phi \sum_{n=1}^{\infty} [b_n \sinh k_n(z + d) + c_n \cosh k_n(z + d)] J_1(k_n r) \quad (14)$$

where the constant k_n is determined by $\dfrac{\partial}{\partial r} [J_1(k_n r)]_{r=a} = 0$. The total transmitted magnetostatic potential in region (III) is

$$\Phi^t(r, \phi, z) = \cos \phi \int_0^{\infty} \widetilde{\Phi}^t(\zeta) J_1(\zeta r) e^{\zeta(z+d)} \zeta \mathrm{d}\zeta . \quad (15)$$

The continuity of normal derivative of the potential at $z = 0$ requires

$$\frac{\partial}{\partial z} [\Phi^i(r, \phi, z) + \Phi^s(r, \phi, z)]_{z=0}$$

$$= \begin{cases} \dfrac{\partial}{\partial z} [\Phi^d(r, \phi, z)]_{z=0}, & r < a \\ \\ 0, & r > a . \end{cases} \quad (16)$$

Applying the Hankel transform to (16) yields

$$\widetilde{\Phi}^s(\zeta) = -\frac{1}{\zeta} \sum_{n=1}^{\infty} k_n(b_n \cosh k_n d + c_n \sinh k_n d)$$

$$\cdot \left[\frac{J_1(k_n a)[J_1(\zeta a) - a\zeta J_0(\zeta a)]}{\zeta^2 - k_n^2} \right] . \quad (17)$$

The continuity of the potential across the aperture is

$$\Phi^i(r,\phi,0) + \Phi^s(r,\phi,0) = \Phi^d(r,\phi,0), \quad r < a . \tag{18}$$

Substituting $\widetilde{\Phi}^s(\zeta)$ into (18), multiplying (18) by $J_1(k_p r)r$ $(p = 1, 2, 3, \ldots)$, and integrating with respect to r from 0 to a, we get

$$\frac{a^2 J_2(k_p a)}{k_p}$$

$$- \sum_{n=1}^{\infty} k_n (b_n \cosh k_n d + c_n \sinh k_n d) J_1(k_n a) J_1(k_p a) I$$

$$= \frac{a^2}{2} \left[1 - \frac{1}{(k_p a)^2} \right] [J_1(k_p a)]^2 (b_p \sinh k_p d + c_p \cosh k_p d) \tag{19}$$

where

$$I = \int_0^{\infty} \frac{[J_1(\zeta a) - a\zeta J_0(\zeta a)]^2}{(\zeta^2 - k_n^2)(\zeta^2 - k_p^2)} d\zeta . \tag{20}$$

Similarly the boundary conditions at $z = -d$ yield

$$\sum_{n=1}^{\infty} b_n k_n J_1(k_n a) I = \frac{a^2}{2} c_p \left[1 - \frac{1}{(k_p a)^2} \right] J_1(k_p a) . \tag{21}$$

The magnetic polarizability of a circular aperture is

$$\psi(z) \equiv \int_0^{2\pi} \int_0^a r \cos\phi \frac{\partial}{\partial z} [\Phi^d(r,\phi,z)] r dr d\phi$$

$$= \pi a^2 \sum_{n=1}^{\infty} [b_n \cosh k_n(z+d) + c_n \sinh k_n(z+d)] J_2(k_n a) . \tag{22}$$

7.2 Acoustic Scattering from a Circular Aperture in a Hard Plane [6]

Acoustic and scalar wave scattering from a circular aperture in a hard plane has been studied in [7-9]. In this section we will consider acoustic scattering from a circular aperture in a thick hard plane. In region (I) $(z > 0)$ a field (velocity potential) $\Phi^i(r,\phi,z)$ is incident on a circular aperture (radius: a and depth: d) in a thick hard plane. Regions (II) $(-d < z < 0$ and $r < a)$ and (III) $(z < -d)$ denote a circular aperture and a half-space, respectively, where the wavenumber is k $(= 2\pi/\lambda$ and λ: wavelength). The total field in region (I) consists of the incident, reflected, and scattered waves

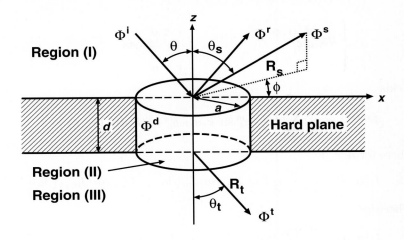

Fig. 7.2. A circular aperture in a hard plane

$$\Phi^i(r,\phi,z) = \exp(\mathrm{i}kx\sin\theta - \mathrm{i}kz\cos\theta)$$

$$= \exp(-\mathrm{i}kz\cos\theta) \sum_{m=-\infty}^{\infty} \mathrm{i}^m J_m(kr\sin\theta)\mathrm{e}^{\mathrm{i}m\phi} \tag{1}$$

$$\Phi^r(r,\phi,z) = \exp(\mathrm{i}kx\sin\theta + \mathrm{i}kz\cos\theta)$$

$$= \exp(\mathrm{i}kz\cos\theta) \sum_{m=-\infty}^{\infty} \mathrm{i}^m J_m(kr\sin\theta)\mathrm{e}^{\mathrm{i}m\phi} \tag{2}$$

$$\Phi^s(r,\phi,z) = \sum_{m=-\infty}^{\infty} \mathrm{e}^{\mathrm{i}m\phi} \int_0^\infty \widetilde{\Phi}^{sm}(\zeta) J_m(\zeta r)\mathrm{e}^{\mathrm{i}\kappa z}\zeta\,\mathrm{d}\zeta \tag{3}$$

where $\kappa = \sqrt{k^2 - \zeta^2}$. The total field in region (II) is

$$\Phi^d(r,\phi,z)$$

$$= \sum_{m=-\infty}^{\infty} \mathrm{e}^{\mathrm{i}m\phi} \sum_{n=1}^{\infty} [b_n^m \sin k_z(z+d) + c_n^m \cos k_z(z+d)] J_m(k_n r) \tag{4}$$

where $k_z = \sqrt{k^2 - k_n^2}$, $J_m'(k_n a) = 0$, and the prime denotes differentiation with respect to the argument. The total transmitted field in region (III) is

$$\Phi^t(r,\phi,z) = \sum_{m=-\infty}^{\infty} \mathrm{e}^{\mathrm{i}m\phi} \int_0^\infty \widetilde{\Phi}^{tm}(\zeta) J_m(\zeta r) \exp\left[-\mathrm{i}\kappa(z+d)\right]\zeta\,\mathrm{d}\zeta \;. \tag{5}$$

Applying the Hankel transform to

$$\frac{\partial}{\partial z} \left[\Phi^i(r,\phi,z) + \Phi^r(r,\phi,z) + \Phi^s(r,\phi,z) \right]_{z=0}$$

$$= \begin{cases} \dfrac{\partial}{\partial z} \left[\Phi^d(r,\phi,z) \right]_{z=0}, & r < a \\[2mm] 0, & r > a \end{cases} \tag{6}$$

yields

$$\widetilde{\Phi}^{sm}(\zeta) = \frac{-\mathrm{i}}{\kappa} \sum_{n=1}^{\infty} k_z(-c_n^m \sin k_z d + b_n^m \cos k_z d) I_1(k_n, \zeta) \tag{7}$$

where for $\alpha \neq \beta$

$$I_1(\alpha, \beta) = \frac{a}{\alpha^2 - \beta^2} \left[\alpha J_{m+1}(\alpha a) J_m(\beta a) - \beta J_m(\alpha a) J_{m+1}(\beta a) \right] \tag{8}$$

otherwise

$$I_1(\alpha, \alpha) = \frac{a^2}{2} \left[J_m^2(\alpha a) - J_{m-1}(\alpha a) J_{m+1}(\alpha a) \right] . \tag{9}$$

The boundary condition for $r < a$ at $z = 0$ requires

$$\Phi^i(r,\phi,0) + \Phi^r(r,\phi,0) + \Phi^s(r,\phi,0) = \Phi^d(r,\phi,0) . \tag{10}$$

Substituting $\widetilde{\Phi}^{sm}(\zeta)$ into (10), multiplying (10) by $J_m(k_p r)r$, and integrating with respect to r from 0 to a, we get

$$\boxed{\begin{aligned} 2\mathrm{i}^m I_1(k\sin\theta, k_p) &- \sum_{n=1}^{\infty} \mathrm{i}k_z \left(-c_n^m \sin k_z d + b_n^m \cos k_z d \right) I_2 \\ &= (c_p^m \cos k_z d + b_p^m \sin k_z d) I_1(k_p, k_p) \end{aligned}} \tag{11}$$

where

$$I_2 = \int_0^\infty \kappa^{-1} I_1(k_n, \zeta) I_1(k_p, \zeta) \zeta \mathrm{d}\zeta . \tag{12}$$

Applying the Hankel transform to

$$\frac{\partial}{\partial z} \left[\Phi^t(r,\phi,z) \right]_{z=-d} = \begin{cases} \dfrac{\partial}{\partial z} \left[\Phi^d(r,\phi,z) \right]_{z=-d}, & r < a \\[2mm] 0, & r > a \end{cases} \tag{13}$$

gives

$$\widetilde{\Phi}^{tm}(\zeta) = \frac{\mathrm{i}}{\kappa} \sum_{n=1}^{\infty} k_z b_n^m I_1(k_n, \zeta) . \tag{14}$$

Similarly substituting $\widetilde{\Phi}^{tm}(\zeta)$ into another boundary condition

$$\Phi^t(r,\phi,-d) = \Phi^d(r,\phi,-d), \qquad r < a \tag{15}$$

and manipulating (15) yields

$$\boxed{\sum_{n=1}^{\infty} \mathrm{i}k_z b_n^m I_2 = c_p^m I_1(k_p, k_p) \; .}$$

(16)

The far-zone scattered and transmitted fields at distances R_s and R_t are

$$\Phi^s(R_s, \theta_s, \phi) = \frac{k\cos\theta_s \exp(\mathrm{i}kR_s - \mathrm{i}\pi/2)}{R_s}$$

$$\cdot \sum_{m=-\infty}^{\infty} \exp[\mathrm{i}m(\phi - \pi/2)]\widetilde{\Phi}^{sm}(k\sin\theta_s)$$

(17)

$$\Phi^t(R_t, \theta_t, \phi) = \frac{k\cos\theta_t \exp(-\mathrm{i}kR_t + \mathrm{i}\pi/2)}{R_t} \exp(-\mathrm{i}kd\cos\theta_t)$$

$$\cdot \sum_{m=-\infty}^{\infty} \exp[\mathrm{i}m(\phi + \pi/2)]\widetilde{\Phi}^{tm}(k\sin\theta_t) \; .$$

(18)

The reflection coefficient ϱ (or transmission coefficient τ) is a ratio of the power reflected from (or transmitted through) an aperture to the power impinging on an aperture. They are

$$\varrho = -\frac{2}{a^2 k\cos\theta} \sum_{m=-\infty}^{\infty} \sum_{n=1}^{\infty} Im\Big\{ k_z^*(b_n^m\cos k_z d - c_n^m\sin k_z d)^*$$

$$\cdot [(b_n^m\sin k_z d + c_n^m\cos k_z d)I_1(k_n, k_n) - 2\mathrm{i}^m I_1(k\sin\theta, k_n)]\Big\}$$

(19)

$$\tau = \frac{2}{a^2 k\cos\theta} \sum_{m=-\infty}^{\infty} \sum_{n=1}^{\infty} Im\,(k_z^* c_n^m b_n^{m*})\,I_1(k_n, k_n) \; .$$

(20)

7.3 EM Scattering from a Circular Aperture in a Conducting Plane

Electromagnetic wave scattering from a circular aperture in a plane is an important canonical problem in diffraction theory [10-12]. A study of electromagnetic field penetration into a circular aperture is important for practical applications in electromagnetic compatibility and antenna radiation [13]. In this section we will analyze electromagnetic wave scattering from a circular aperture in a thick perfectly-conducting plane. For simplicity we normalize the fields $\boldsymbol{E}(t)$ and $\boldsymbol{H}(t)$ in a medium characterized by permittivity ϵ and permeability μ. Let's introduce the normalized fields $\boldsymbol{E}(t')$ and $\boldsymbol{H}(t')$ that are related by $\boldsymbol{E}(t) = \sqrt{\mu}\boldsymbol{E}(t')$, $\boldsymbol{H}(t) = -\mathrm{i}\sqrt{\epsilon}\boldsymbol{H}(t')$, and $t = \sqrt{\mu_0\epsilon_0}t'$. Maxwell's equations are then given in time domain by

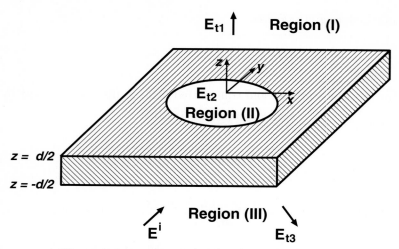

Fig. 7.3. A circular aperture in a conducting plane

$$\nabla \times \boldsymbol{E}(t') = \mathrm{i}n\frac{\partial \boldsymbol{H}(t')}{\partial t'} \tag{1}$$

$$\nabla \times \boldsymbol{H}(t') = \mathrm{i}n\frac{\partial \boldsymbol{E}(t')}{\partial t'} \tag{2}$$

where n is the refractive index of a medium. An electromagnetic wave $\boldsymbol{E}^i$ impinges from below on a circular aperture with radius a and thickness d in a thick perfectly-conducting plane. The wavenumbers in regions (I), (II), and (III) are $k_j = \omega n_j$, $(j = 1, 2, 3)$, respectively, where n_j is the corresponding medium refractive index. It is convenient to represent the transverse $\boldsymbol{E}_t$ and $\boldsymbol{H}_t$ fields in terms of two eigenvectors $\boldsymbol{e}_{jm}$ and $\boldsymbol{e}_{je}$ $(j = 1, 2, 3)$ that are transverse to the z-direction in circular cylindrical coordinates (r, ϕ, z). The eigenvectors $\boldsymbol{e}_{jm}$ and $\boldsymbol{e}_{je}$ are associated with the TM (transverse magnetic to the z-direction) and TE (transverse electric to the z-direction) waves [14]. In region (I) $(z > d/2)$ the transverse transmitted field is

$$\boldsymbol{E}_{t1} = \sum_{\nu=-\infty}^{\infty} \int_0^{\infty} \left[\widetilde{\varPi}_{tm}(\zeta)\boldsymbol{e}_{1m} + \widetilde{\varPi}_{te}(\zeta)\boldsymbol{e}_{1e} \right] \exp[\mathrm{i}\kappa_1(z - d/2)]\mathrm{d}\zeta \tag{3}$$

$$\boldsymbol{H}_{t1} = \sum_{\nu=-\infty}^{\infty} \int_0^{\infty} \left[\widetilde{\varPi}_{tm}(\zeta)\boldsymbol{h}_{1m} + \widetilde{\varPi}_{te}(\zeta)\boldsymbol{h}_{1e} \right] \exp[\mathrm{i}\kappa_1(z - d/2)]\mathrm{d}\zeta \tag{4}$$

where $\kappa_1 = \sqrt{\omega^2 n_1^2 - \zeta^2}$ and

$$\boldsymbol{e}_{1m} = \boldsymbol{h}_{1e} = \left[\hat{r}\mathrm{i}\kappa_1\zeta J_\nu'(\zeta r) - \hat{\phi}\frac{\kappa_1\nu}{r}J_\nu(\zeta r) \right] \mathrm{e}^{\mathrm{i}\nu\phi} \tag{5}$$

$$\boldsymbol{e}_{1e} = \boldsymbol{h}_{1m} = \left[\hat{r}\frac{\mathrm{i}\nu\omega n_1}{r}J_\nu(\zeta r) - \hat{\phi}\,\omega n_1\zeta J_\nu'(\zeta r) \right] \mathrm{e}^{\mathrm{i}\nu\phi} \tag{6}$$

and the prime denotes differentiation with respect to the argument. Note that $\hat{r}$ and $\hat{\phi}$ denote the unit vectors in cylindrical coordinates. In region (II) $(-d/2 < z < d/2$ and $r < a)$ the transverse field within the aperture is

$$
\boldsymbol{E}_{t2} = \sum_{\nu=-\infty}^{\infty} \sum_{j} \left\{ [A_{\nu j} \exp(\mathrm{i}\kappa_{\nu j} z) + B_{\nu j} \exp(-\mathrm{i}\kappa_{\nu j} z)] \boldsymbol{e}_{2m} \right.
$$
$$
\left. + [C_{\nu j} \exp(\mathrm{i}\kappa'_{\nu j} z) + D_{\nu j} \exp(-\mathrm{i}\kappa'_{\nu j} z)] \boldsymbol{e}_{2e} \right\} \tag{7}
$$

$$
\boldsymbol{H}_{t2} = \sum_{\nu=-\infty}^{\infty} \sum_{j} \left\{ [(A_{\nu j} \exp(\mathrm{i}\kappa_{\nu j} z) - B_{\nu j} \exp(-\mathrm{i}\kappa_{\nu j} z)] \boldsymbol{h}_{2m} \right.
$$
$$
\left. + [C_{\nu j} \exp(\mathrm{i}\kappa'_{\nu j} z) - D_{\nu j} \exp(-\mathrm{i}\kappa'_{\nu j} z)] \boldsymbol{h}_{2e} \right\} \tag{8}
$$

where $\kappa_{\nu j} = \sqrt{\omega^2 n_2^2 - (\chi_{\nu j}/a)^2}$, $\kappa'_{\nu j} = \sqrt{\omega^2 n_2^2 - (\chi'_{\nu j}/a)^2}$, and

$$
\boldsymbol{e}_{2m} = \left[\hat{r} \frac{\mathrm{i}\kappa_{\nu j} \chi_{\nu j}}{a} J'_\nu \left(\frac{\chi_{\nu j}}{a} r \right) - \hat{\phi} \frac{\kappa_{\nu j} \nu}{r} J_\nu \left(\frac{\chi_{\nu j}}{a} r \right) \right] \mathrm{e}^{\mathrm{i}\nu\phi} \tag{9}
$$

$$
\boldsymbol{e}_{2e} = \left[\hat{r} \frac{\mathrm{i}\nu\omega n_2}{r} J_\nu \left(\frac{\chi'_{\nu j}}{a} r \right) - \hat{\phi} \, \omega n_2 \frac{\chi'_{\nu j}}{a} J'_\nu \left(\frac{\chi'_{\nu j}}{a} r \right) \right] \mathrm{e}^{\mathrm{i}\nu\phi} \tag{10}
$$

$$
\boldsymbol{h}_{2m} = \left[\hat{r} \frac{\mathrm{i}\nu\omega n_2}{r} J_\nu \left(\frac{\chi_{\nu j}}{a} r \right) - \hat{\phi} \, \omega n_2 \frac{\chi_{\nu j}}{a} J'_\nu \left(\frac{\chi_{\nu j}}{a} r \right) \right] \mathrm{e}^{\mathrm{i}\nu\phi} \tag{11}
$$

$$
\boldsymbol{h}_{2e} = \left[\hat{r} \frac{\mathrm{i}\kappa'_{\nu j} \chi'_{\nu j}}{a} J'_\nu \left(\frac{\chi'_{\nu j}}{a} r \right) - \hat{\phi} \frac{\kappa'_{\nu j} \nu}{r} J_\nu \left(\frac{\chi'_{\nu j}}{a} r \right) \right] \mathrm{e}^{\mathrm{i}\nu\phi} . \tag{12}
$$

Note that $\chi_{\nu j}$ is the jth root of $J_\nu(\cdot) = 0$ and $\chi'_{\nu j}$ is the jth root of $J'_\nu(\cdot) = 0$. In region (III) $(z < -d/2)$ the total field consists of the incident, specularly reflected, and scattered components. For the TE wave incidence, the incident E-field takes the form of

$$
\boldsymbol{E}^i = \hat{y} \exp(\mathrm{i}k_3 x \sin\theta_i + \mathrm{i}k_3 z \cos\theta_i)
$$
$$
= \hat{y} \exp(\mathrm{i}k_3 z \cos\theta_i) \sum_{\nu=-\infty}^{\infty} \mathrm{i}^\nu J_\nu(k_3 r \sin\theta_i) \mathrm{e}^{\mathrm{i}\nu\phi} . \tag{13}
$$

For the TM wave incidence, the incident H-field $\boldsymbol{H}^i$ is represented by (13). The transverse scattered field in region (III) is

$$
\boldsymbol{E}_{t3} = \sum_{\nu=-\infty}^{\infty} \int_0^\infty [\tilde{\pi}_{tm}(\zeta) \boldsymbol{e}_{3m} + \tilde{\pi}_{te}(\zeta) \boldsymbol{e}_{3e}] \exp[-\mathrm{i}\kappa_3(z + d/2)] \mathrm{d}\zeta \tag{14}
$$

$$
\boldsymbol{H}_{t3} = -\sum_{\nu=-\infty}^{\infty} \int_0^\infty [\tilde{\pi}_{tm}(\zeta) \boldsymbol{h}_{3m} + \tilde{\pi}_{te}(\zeta) \boldsymbol{h}_{3e}] \exp[-\mathrm{i}\kappa_3(z + d/2)] \mathrm{d}\zeta \tag{15}
$$

where $\kappa_3 = \sqrt{\omega^2 n_3^2 - \zeta^2}$ and

$$\boldsymbol{e}_{3m} = \boldsymbol{h}_{3e} = \left[\hat{r}\ i\kappa_3\zeta J_\nu'(\zeta r) - \hat{\phi}\frac{\kappa_3\nu}{r}J_\nu(\zeta r)\right]e^{i\nu\phi} \tag{16}$$

$$\boldsymbol{e}_{3e} = \boldsymbol{h}_{3m} = \left[\hat{r}\frac{i\nu\omega n_3}{r}J_\nu(\zeta r) - \hat{\phi}\ \omega n_3\zeta J_\nu'(\zeta r)\right]e^{i\nu\phi}\ . \tag{17}$$

We are now in a position to determine the modal coefficients $A_{\nu j}$, $B_{\nu j}$, $C_{\nu j}$, and $D_{\nu j}$ by using the boundary conditions. The tangential E-field continuity at $z = d/2$ requires

$$\boldsymbol{E}_{t1}(r,\phi,d/2) = \begin{cases} \boldsymbol{E}_{t2}(r,\phi,d/2), & r < a \\ 0, & r > a\ . \end{cases} \tag{18}$$

In matching the boundary conditions on the field continuities (18), it is convenient to utilize the orthogonality property associated with the eigenvectors. Consider

$$\frac{1}{2\pi}\int_0^{2\pi}\int_0^\infty [\boldsymbol{E}_{t1} \times \boldsymbol{h}_{1m}^*(\zeta')] \cdot \hat{z}r\,dr\,d\phi$$

$$= \frac{1}{2\pi}\int_0^\infty \tilde{\Pi}_{tm}(\zeta)\int_0^{2\pi}\int_0^\infty [\boldsymbol{e}_{1m}(\zeta) \times \boldsymbol{h}_{1m}^*(\zeta')] \cdot \hat{z}r\,dr\,d\phi\,d\zeta$$

$$= -i\omega n_1\kappa_1\zeta\tilde{\Pi}_{tm}(\zeta)\big|_{\zeta\to\zeta'}\ . \tag{19}$$

A more detailed discussion on the eigenvector orthogonality property can be found in [14]. We define an operation $< \boldsymbol{a}|\boldsymbol{b} >$ as

$$< \boldsymbol{a}|\boldsymbol{b} > \equiv \frac{1}{2\pi}\int_0^{2\pi}\int_0^a (\boldsymbol{a} \times \boldsymbol{b}^*) \cdot \hat{z}r\,dr\,d\phi = -< \boldsymbol{b}|\boldsymbol{a} >^*\ . \tag{20}$$

Then

$$< \boldsymbol{E}_{t2}|\boldsymbol{h}_{1m}(\zeta') >$$

$$= \sum_j [A_{\nu j}\exp(i\kappa_{\nu j}d/2) + B_{\nu j}\exp(-i\kappa_{\nu j}d/2)] < \boldsymbol{e}_{2m}|\boldsymbol{h}_{1m}(\zeta') >$$

$$+ \sum_j [C_{\nu j}\exp(i\kappa_{\nu j}'d/2) + D_{\nu j}\exp(-i\kappa_{\nu j}'d/2)] < \boldsymbol{e}_{2e}|\boldsymbol{h}_{1m}(\zeta') >\ . \tag{21}$$

In view of (19) and (21), (18) becomes

$$-i\omega n_1\kappa_1\zeta\tilde{\Pi}_{tm}(\zeta) = \sum_j [A_{\nu j}\exp(i\kappa_{\nu j}d/2) + B_{\nu j}\exp(-i\kappa_{\nu j}d/2)]$$

$$\cdot \left[-i\omega n_1\kappa_{\nu j}\chi_{\nu j}J_\nu'(\chi_{\nu j})\frac{\zeta^2 J_\nu(\zeta a)}{\zeta^2 - (\chi_{\nu j}/a)^2}\right]$$

$$+ \sum_j [C_{\nu j}\exp(i\kappa_{\nu j}'d/2) + D_{\nu j}\exp(-i\kappa_{\nu j}'d/2)]$$

$$\cdot \left[-i\omega^2 n_1 n_2\nu J_\nu(\chi_{\nu j}')J_\nu(\zeta a)\right]\ . \tag{22}$$

Similarly applying the operation $< \boldsymbol{E}_1|\boldsymbol{h}_{1e}(\zeta') >$ to (18) gives

$$-\mathrm{i}\omega n_1 \kappa_1 \zeta \widetilde{\Pi}_{te}(\zeta) = \sum_j \left[C_{\nu j} \exp(\mathrm{i}\kappa'_{\nu j} d/2) + D_{\nu j} \exp(-\mathrm{i}\kappa'_{\nu j} d/2) \right]$$

$$\cdot \left[\mathrm{i}\omega n_2 \frac{(\chi'_{\nu j})^2}{a} J_\nu(\chi'_{\nu j}) \frac{\kappa_1 \zeta J'_\nu(\zeta a)}{\zeta^2 - (\chi'_{\nu j}/a)^2} \right] . \tag{23}$$

The tangential H-field continuity at $z = d/2$ and $r < a$ is

$$< \boldsymbol{H}_{t2} | \boldsymbol{e}_{2m} >$$
$$= \int_0^\infty \left[\widetilde{\Pi}_{tm}(\zeta) < \boldsymbol{h}_{1m} | \boldsymbol{e}_{2m} > + \widetilde{\Pi}_{te}(\zeta) < \boldsymbol{h}_{1e} | \boldsymbol{e}_{2m} > \right] \mathrm{d}\zeta \tag{24}$$

$$< \boldsymbol{H}_{t2} | \boldsymbol{e}_{2e} >$$
$$= \int_0^\infty \left[\widetilde{\Pi}_{tm}(\zeta) < \boldsymbol{h}_{1m} | \boldsymbol{e}_{2e} > + \widetilde{\Pi}_{te}(\zeta) < \boldsymbol{h}_{1e} | \boldsymbol{e}_{2e} > \right] \mathrm{d}\zeta . \tag{25}$$

Substituting $\widetilde{\Pi}_{tm}(\zeta)$ and $\widetilde{\Pi}_{te}(\zeta)$ into (24) and (25), and carrying out algebraic manipulation, we get

$$[A_{\nu q} \exp(\mathrm{i}\kappa_{\nu q} d/2) - B_{\nu q} \exp(-\mathrm{i}\kappa_{\nu q} d/2)]$$
$$\cdot \left\{ -\mathrm{i}\omega n_2 \kappa_{\nu q} \chi_{\nu q}^2 \frac{1}{2} [J'_\nu(\chi_{\nu q})]^2 \right\}$$
$$= \sum_p [A_{\nu p} \exp(\mathrm{i}\kappa_{\nu p} d/2) + B_{\nu p} \exp(-\mathrm{i}\kappa_{\nu p} d/2)]$$
$$\cdot [-\mathrm{i}\omega n_1 \kappa_{\nu p} \kappa_{\nu q} \chi_{\nu p} \chi_{\nu q} J'_\nu(\chi_{\nu p}) J'_\nu(\chi_{\nu q})] I_1(p,q)$$
$$+ \sum_p [C_{\nu p} \exp(\mathrm{i}\kappa'_{\nu p} d/2) + D_{\nu p} \exp(-\mathrm{i}\kappa'_{\nu p} d/2)]$$
$$\cdot [-\mathrm{i}\omega^2 n_1 n_2 \nu \kappa_{\nu q} \chi_{\nu q} J_\nu(\chi'_{\nu p}) J'_\nu(\chi_{\nu q})] I_2(q) \tag{26}$$

$$[C_{\nu q} \exp(\mathrm{i}\kappa'_{\nu q} d/2) - D_{\nu q} \exp(-\mathrm{i}\kappa'_{\nu q} d/2)]$$
$$\cdot \left[-\mathrm{i}\omega n_2 \kappa'_{\nu q} \frac{1}{2} (\chi'^2_{\nu q} - \nu^2) J^2_\nu(\chi'_{\nu q}) \right]$$
$$= \sum_p [A_{\nu p} \exp(\mathrm{i}\kappa_{\nu p} d/2) + B_{\nu p} \exp(-\mathrm{i}\kappa_{\nu p} d/2)]$$
$$\cdot [-\mathrm{i}\omega^2 n_1 n_2 \nu \kappa_{\nu p} \chi_{\nu p} J_\nu(\chi'_{\nu q}) J'_\nu(\chi_{\nu p})] I_2(p)$$
$$+ \sum_p [C_{\nu p} \exp(\mathrm{i}\kappa'_{\nu p} d/2) + D_{\nu p} \exp(-\mathrm{i}\kappa'_{\nu p} d/2)]$$
$$\cdot \left\{ [-\mathrm{i}\omega^3 n_1 n_2^2 \nu^2 J_\nu(\chi'_{\nu p}) J_\nu(\chi'_{\nu q})] I_3 \right.$$
$$\left. + \left[-\mathrm{i}\omega \frac{n_2^2}{n_1} \left(\frac{\chi'_{\nu p} \chi'_{\nu q}}{a} \right)^2 J_\nu(\chi'_{\nu p}) J_\nu(\chi'_{\nu q}) \right] I_4(p,q) \right\} \tag{27}$$

where

$$I_1(p,q) = \int_0^\infty \frac{\zeta^3 J_\nu^2(\zeta a)\mathrm{d}\zeta}{\kappa_1 \left[\zeta^2 - (\chi_{\nu p}/a)^2\right]\left[\zeta^2 - (\chi_{\nu q}/a)^2\right]} \tag{28}$$

$$I_2(p) = \int_0^\infty \frac{\zeta J_\nu^2(\zeta a)\mathrm{d}\zeta}{\kappa_1 \left[\zeta^2 - (\chi_{\nu p}/a)^2\right]} \tag{29}$$

$$I_3 = \int_0^\infty \frac{J_\nu^2(\zeta a)\mathrm{d}\zeta}{\kappa_1 \zeta} \tag{30}$$

$$I_4(p,q) = \int_0^\infty \frac{\kappa_1 \zeta {J_\nu'}^2(\zeta a)\mathrm{d}\zeta}{\left[\zeta^2 - (\chi_{\nu p}'/a)^2\right]\left[\zeta^2 - (\chi_{\nu q}'/a)^2\right]} . \tag{31}$$

It is possible to obtain another set of simultaneous equations for the modal coefficients $A_{\nu j}$, $B_{\nu j}$, $C_{\nu j}$, and $D_{\nu j}$ by using the boundary conditions at $z = -d/2$. The result is

$$
\begin{aligned}
&[A_{\nu q}\exp(-\mathrm{i}\kappa_{\nu q}d/2) - B_{\nu q}\exp(\mathrm{i}\kappa_{\nu q}d/2)] \\
&\cdot \left\{-\mathrm{i}\omega n_2 \kappa_{\nu q}\chi_{\nu q}^2 \frac{1}{2}[J_\nu'(\chi_{\nu q})]^2\right\} \\
&= \sum_p [A_{\nu p}\exp(-\mathrm{i}\kappa_{\nu p}d/2) + B_{\nu p}\exp(\mathrm{i}\kappa_{\nu p}d/2)] \\
&\cdot [\mathrm{i}\omega n_3 \kappa_{\nu p}\kappa_{\nu q}\chi_{\nu p}\chi_{\nu q}J_\nu'(\chi_{\nu p})J_\nu'(\chi_{\nu q})]\,\bar{I}_1(p,q) \\
&+ \sum_p [C_{\nu p}\exp(-\mathrm{i}\kappa_{\nu p}'d/2) + D_{\nu p}\exp(\mathrm{i}\kappa_{\nu p}'d/2)] \\
&\cdot [\mathrm{i}\omega^2 n_2 n_3 \nu \kappa_{\nu q}\chi_{\nu q}J_\nu'(\chi_{\nu q})J_\nu(\chi_{\nu p}')]\,\bar{I}_2(q) \\
&+ <(\boldsymbol{H}^i + \boldsymbol{H}^r)|e_{2m}>
\end{aligned}
\tag{32}
$$

$$\left[C_{\nu q} \exp(-\mathrm{i}\kappa'_{\nu q}d/2) - D_{\nu q}\exp(\mathrm{i}\kappa'_{\nu q}d/2)\right]$$

$$\cdot \left[-\mathrm{i}\omega n_2 \kappa'_{\nu q}\frac{1}{2}(\chi'^2_{\nu q} - \nu^2)J^2_\nu(\chi'_{\nu q})\right]$$

$$= \sum_p \left[A_{\nu p}\exp(-\mathrm{i}\kappa_{\nu p}d/2) + B_{\nu p}\exp(\mathrm{i}\kappa_{\nu p}d/2)\right]$$

$$\cdot \left[\mathrm{i}\omega^2 n_2 n_3 \nu \kappa_{\nu p}\chi_{\nu p} J'_\nu(\chi_{\nu p}) J_\nu(\chi'_{\nu q})\right] \bar{I}_2(p)$$

$$+ \sum_p \left[C_{\nu p}\exp(-\mathrm{i}\kappa'_{\nu p}d/2) + D_{\nu p}\exp(\mathrm{i}\kappa'_{\nu p}d/2)\right]$$

$$\cdot \left\{ \left[\mathrm{i}\omega^3 n_2^2 n_3 \nu^2 J_\nu(\chi'_{\nu p}) J_\nu(\chi'_{\nu q})\right] \bar{I}_3 \right.$$

$$\left. + \left[\mathrm{i}\omega \frac{n_2^2}{n_3}\left(\frac{\chi'_{\nu p}\chi'_{\nu q}}{a}\right)^2 J_\nu(\chi'_{\nu p}) J_\nu(\chi'_{\nu q})\right] \bar{I}_4(p,q)\right\}$$

$$+ < (\boldsymbol{H}^i + \boldsymbol{H}^r)|e_{2e} > \tag{33}$$

where $\bar{I}_1$ through $\bar{I}_4$ are obtained by replacing κ_1 in (28) through (31) with κ_3.

For the TE incidence, we note

$$< (\boldsymbol{H}^i + \boldsymbol{H}^r)|e_{2m} > = 0 \tag{34}$$

$$< (\boldsymbol{H}^i + \boldsymbol{H}^r)|e_{2e} > = -2\mathrm{i}^\nu \cos\theta_i \omega n_2 (\chi'^2_{\nu j}/a) J_\nu(\chi'_{\nu j})$$

$$\cdot \left[\frac{J'_\nu(k_3 a \sin\theta_i)}{(k_3\sin\theta_i)^2 - \left(\chi'_{\nu j}/a\right)^2}\right] \tag{35}$$

and for the TM incidence

$$< (\boldsymbol{H}^i + \boldsymbol{H}^r)|e_{2m} > = 2\mathrm{i}^\nu \kappa_{\nu j}\chi_{\nu j}k_3\sin\theta_i J'_\nu(\chi_{\nu j})$$

$$\cdot \left[\frac{J_\nu(k_3 a \sin\theta_i)}{(k_3\sin\theta_i)^2 - (\chi_{\nu j}/a)^2}\right] \tag{36}$$

$$< (\boldsymbol{H}^i + \boldsymbol{H}^r)|e_{2e} > = 2\mathrm{i}^\nu \frac{\omega n_2 \nu}{k_3 \sin\theta_i} J_\nu(\chi'_{\nu j}) J_\nu(k_3 a \sin\theta_i) . \tag{37}$$

The final formulation obtained in this section is shown to be somewhat analogous to the result in [10].

7.4 Acoustic Radiation from a Flanged Circular Cylinder [15]

A problem of acoustic wave radiation from a flanged circular cylinder was considered in [16]. In this section we will revisit the problem of an acoustic wave (velocity potential) $\Phi^i(r, z)$ radiating from a flanged circular waveguide

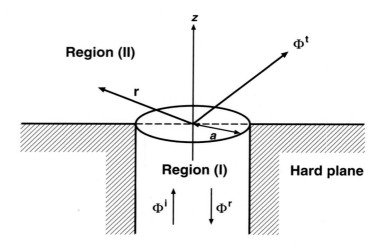

Fig. 7.4. A flanged circular cylinder

with radius a. The total wave in region (I) ($r < a$ and $z < 0$) is a sum of the incident and reflected waves $\Phi^i(r, z)$ and $\Phi^r(r, z)$

$$\Phi^i(r, z) = \frac{V_0}{k} e^{ikz} \tag{1}$$

$$\Phi^r(r, z) = \frac{V_0}{k} \sum_{m=0}^{\infty} A_m J_0(\beta_m r/a) e^{-i\sigma_m z} \tag{2}$$

where V_0 is the velocity amplitude of incident wave, k is the wavenumber, β_m is the roots of $J_1(\beta_m) = 0$, and $\sigma_m = \sqrt{k^2 - (\beta_m/a)^2}$. The transmitted field in region (II) ($z > 0$) is

$$\Phi^t(r, z) = \int_0^{\infty} \tilde{\Phi}^t(\zeta) \exp\left(iz\sqrt{k^2 - \zeta^2}\right) J_0(\zeta r)\zeta d\zeta . \tag{3}$$

The boundary condition at $z = 0$ requires

$$\frac{\partial \Phi^t(r, z)}{\partial z}\bigg|_{z=0} = \begin{cases} \partial\left[\Phi^i(r, z) + \Phi^r(r, z)\right]/\partial z\big|_{z=0}, & r < a \\ 0, & r > a . \end{cases} \tag{4}$$

Applying the Hankel transform to (4) gives

$$\sqrt{k^2 - \zeta^2}\tilde{\Phi}^t(\zeta) = \frac{V_0}{k} \int_0^a \left[k - \sum_{m=0}^{\infty} A_m \sigma_m J_0(\beta_m r/a)\right] J_0(\zeta r) r dr . \tag{5}$$

From (5), we get

$$\widetilde{\Phi}^t(\zeta) = \frac{V_0}{k\sqrt{k^2 - \zeta^2}} \left[\frac{ka}{\zeta} J_1(\zeta a) - \sum_{m=0}^{\infty} A_m \frac{a\sigma_m \zeta J_0(\beta_m) J_1(\zeta a)}{\zeta^2 - (\beta_m/a)^2} \right] . \quad (6)$$

Another boundary condition at $z = 0$ and $r < a$

$$\Phi^i(r, 0) + \Phi^r(r, 0) = \Phi^t(r, 0) \quad (7)$$

is rewritten as

$$\frac{V_0}{k} \left[1 + \sum_{m=0}^{\infty} A_m J_0(\beta_m r/a) \right] = \int_0^{\infty} \widetilde{\Phi}^t(\zeta) J_0(\zeta r) \zeta \mathrm{d}\zeta . \quad (8)$$

We substitute (6) into (8), multiply (8) by $r J_0(\beta_n r/a)$, and integrate with respect to r over $(0, a)$ for $n \geq 0$ to get

$$A_n + \frac{2}{J_0(\beta_n)} \sum_{m=0}^{\infty} A_m \sigma_m J_0(\beta_m) I_m = -\delta_{n0} + \frac{2k I_0}{J_0(\beta_n)} \quad (9)$$

where

$$I_m = \int_0^{\infty} \frac{\zeta^3 J_1^2(\zeta a)}{[\zeta^2 - (\beta_n/a)^2][\zeta^2 - (\beta_m/a)^2]\sqrt{k^2 - \zeta^2}} \mathrm{d}\zeta . \quad (10)$$

We note that (9) agrees with the result that is based on Morse's equation [16].

7.5 Acoustic Scattering from Circular Apertures in a Hard Plane [17]

Acoustic wave scattering from a single circular aperture in a thick plane has been studied in Sect. 7.2. In this section we will solve an acoustic scattering problem of two circular apertures in a thick hard plane. Consider an acoustic uniform plane wave impinging on two circular apertures in a thick hard plane. For simplicity, the wavenumbers in regions (I) through (IV) are assumed to be all identical with k. In region (I) ($z > 0$) an incident field Φ^i (velocity potential) impinges on two circular apertures. Regions (II) and (III) ($-h < z < 0$, $r < r_1$, and $r' < r_2$) denote two circular apertures and region (IV) ($z < -h$) denotes a half-space. In region (I) the total field consists of the incident, reflected, and scattered waves. The incident (Φ^i) and reflected (Φ^r) components are

$$\Phi^{i,r}(x, y, z) = \exp\left[ik \sin \theta_i (x \cos \phi_i + y \sin \phi_i) \mp ikz \cos \theta_i \right] . \quad (1)$$

Since $x = r \cos \phi - l/2 = r' \cos \phi' + l/2$ and $y = r \sin \phi = r' \sin \phi'$,

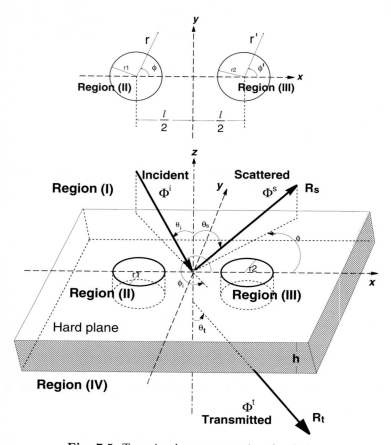

Fig. 7.5. Two circular apertures in a hard plane

$$\Phi^{i,r}(r,\phi,z) = \exp\left(\mp \mathrm{i}kz\cos\theta_i - \frac{\mathrm{i}kl\sin\theta_i}{2}\right)$$
$$\cdot \sum_{m=-\infty}^{\infty} \mathrm{i}^m J_m(kr\sin\theta_i)\exp[\mathrm{i}m(\phi-\phi_i)] \qquad (2)$$

$$\Phi^{i,r}(r',\phi',z) = \exp\left(\mp \mathrm{i}kz\cos\theta_i + \frac{\mathrm{i}kl\sin\theta_i}{2}\right)$$
$$\cdot \sum_{m=-\infty}^{\infty} \mathrm{i}^m J_m(kr'\sin\theta_i)\exp[\mathrm{i}m(\phi'-\phi_i)] \qquad (3)$$

where $J_m(\cdot)$ is the Bessel function of the first kind. In view of the superposition principle, the scattered wave in region (I) may be represented as a sum of two components

$$\Phi^s = \Phi^{s1}(r, \phi, z) + \Phi^{s2}(r', \phi', z) \tag{4}$$

where

$$\Phi^{s1}(r, \phi, z) = \sum_{m=-\infty}^{\infty} e^{im\phi} \int_0^\infty \tilde{\Phi}_m^{s1}(\zeta) J_m(\zeta r) e^{i\kappa z} \zeta d\zeta \tag{5}$$

$$\Phi^{s2}(r', \phi', z) = \sum_{m=-\infty}^{\infty} e^{im\phi'} \int_0^\infty \tilde{\Phi}_m^{s2}(\zeta) J_m(\zeta r') e^{i\kappa z} \zeta d\zeta \tag{6}$$

and $\kappa = \sqrt{k^2 - \zeta^2}$. In regions (II) and (III) the fields are, respectively, given as

$$\Phi^{II}(r, \phi, z) = \sum_{m=-\infty}^{\infty} e^{im\phi} \sum_{n=1}^{\infty}$$
$$[a_n^m \sin k_z^m(z + h) + b_n^m \cos k_z^m(z + h)] J_m(k_n^m r) \tag{7}$$

$$\Phi^{III}(r', \phi', z) = \sum_{m=-\infty}^{\infty} e^{im\phi'} \sum_{n=1}^{\infty}$$
$$[c_n^m \sin \bar{k}_z^m(z + h) + d_n^m \cos \bar{k}_z^m(z + h)] J_m(\bar{k}_n^m r') \tag{8}$$

where $k_z^m = \sqrt{k^2 - (k_n^m)^2}$, $J_m'(k_n^m r_1) = 0$; $\bar{k}_z^m = \sqrt{k^2 - (\bar{k}_n^m)^2}$, $J_m'(\bar{k}_n^m r_2) = 0$, and the prime $'$ denotes differentiation with respect to the argument. In region (IV) the transmitted wave is a sum of two components

$$\Phi^t = \Phi^{t1}(r, \phi, z) + \Phi^{t2}(r', \phi', z) \tag{9}$$

where

$$\Phi^{t1}(r, \phi, z) = \sum_{m=-\infty}^{\infty} e^{im\phi} \int_0^\infty \tilde{\Phi}_m^{t1}(\zeta) J_m(\zeta r) e^{-i\kappa(z+h)} \zeta d\zeta \tag{10}$$

$$\Phi^{t2}(r', \phi', z) = \sum_{m=-\infty}^{\infty} e^{im\phi'} \int_0^\infty \tilde{\Phi}_m^{t2}(\zeta) J_m(\zeta r') e^{-i\kappa(z+h)} \zeta d\zeta . \tag{11}$$

The boundary conditions at $z = 0$ require

$$\left. \frac{\partial \Phi^{s1}}{\partial z} \right|_{z=0} = \begin{cases} \left. \dfrac{\partial \Phi^{II}}{\partial z} \right|_{z=0}, & r < r_1 \\ 0, & r > r_1 \end{cases} \tag{12}$$

$$\left. \frac{\partial \Phi^{s2}}{\partial z} \right|_{z=0} = \begin{cases} \left. \dfrac{\partial \Phi^{III}}{\partial z} \right|_{z=0}, & r' < r_2 \\ 0, & r' > r_2 \end{cases} \tag{13}$$

$$\Phi^i(r, \phi, 0) + \Phi^r(r, \phi, 0) + \Phi^s(r, \phi, 0) = \Phi^{II}(r, \phi, 0), \qquad r < r_1 \tag{14}$$
$$\Phi^i(r', \phi', 0) + \Phi^r(r', \phi', 0) + \Phi^s(r', \phi', 0) = \Phi^{III}(r', \phi', 0), \quad r' < r_2 . \tag{15}$$

Applying the Hankel transforms to (12) and (13), respectively, yields

$$\widetilde{\Phi}_m^{s1}(\zeta) = \frac{-\mathrm{i}}{\kappa} \sum_{n=1}^{\infty} k_z^m \left(a_n^m \cos k_z^m h - b_n^m \sin k_z^m h\right) I_1^m(k_n^m, \zeta) \tag{16}$$

$$\widetilde{\Phi}_m^{s2}(\zeta) = \frac{-\mathrm{i}}{\kappa} \sum_{n=1}^{\infty} \bar{k}_z^m \left(c_n^m \cos \bar{k}_z^m h - d_n^m \sin \bar{k}_z^m h\right) I_2^m(\bar{k}_n^m, \zeta) \tag{17}$$

where for $\alpha \neq \beta$

$$I_{1,2}^m(\alpha, \beta) = \frac{r_{1,2}}{\alpha^2 - \beta^2} \Bigg[\alpha J_{m+1}(\alpha r_{1,2}) J_m(\beta r_{1,2})$$

$$- \beta J_m(\alpha r_{1,2}) J_{m+1}(\beta r_{1,2}) \Bigg] \tag{18}$$

otherwise

$$I_{1,2}^m(\alpha, \alpha) = \frac{r_{1,2}^2}{2} \left[J_m^2(\alpha r_{1,2}) - J_{m-1}(\alpha r_{1,2}) J_{m+1}(\alpha r_{1,2}) \right] . \tag{19}$$

Graf's addition theorem [18] gives

$$J_p(\zeta r') \mathrm{e}^{\mathrm{i}p\phi'} = \sum_{m=-\infty}^{\infty} J_{m-p}(\zeta l) J_m(\zeta r) \mathrm{e}^{\mathrm{i}m\phi} \tag{20}$$

$$J_p(\zeta r) \mathrm{e}^{\mathrm{i}p\phi} = \sum_{m=-\infty}^{\infty} J_{p-m}(\zeta l) J_m(\zeta r') \mathrm{e}^{\mathrm{i}m\phi'} . \tag{21}$$

Substituting (16), (17), and (20) into (14), multiplying (14) by $J_m(k_q^m r)r$, and integrating over $0 < r < r_1$, we get

$$\boxed{\begin{aligned}
& 2\exp\left(-\frac{\mathrm{i}kl}{2}\sin\theta_i - \mathrm{i}m\phi_i\right) \mathrm{i}^m I_2^m(k\sin\theta_i, k_q^m) \\
& - \sum_{n=1}^{\infty} \mathrm{i}k_z^m(a_n^m \cos k_z^m h - b_n^m \sin k_z^m h) I_{11} \\
& - \sum_{p=-\infty}^{\infty}\sum_{n=1}^{\infty} \mathrm{i}\bar{k}_z^p(c_n^p \cos \bar{k}_z^p h - d_n^p \sin \bar{k}_z^p h) I_{12} \\
& = \left(a_q^m \sin k_z^m h + b_q^m \cos k_z^m h\right) I_1^m(k_q^m, k_q^m)
\end{aligned}} \tag{22}$$

where

$$I_{11} = \int_0^{\infty} \kappa^{-1} I_1^m(k_n^m, \zeta) I_1^m(k_q^m, \zeta) \zeta \mathrm{d}\zeta \tag{23}$$

$$I_{12} = \int_0^{\infty} \kappa^{-1} J_{m-p}(\zeta l) I_1^m(k_q^m, \zeta) I_2^p(\bar{k}_n^p, \zeta) \zeta \mathrm{d}\zeta . \tag{24}$$

Similarly from (16), (17), (21), and (15), we obtain

$$\begin{aligned}
&2\exp\left(\frac{ikl}{2}\sin\theta_i - im\phi_i\right) i^m I_2^m(k\sin\theta_i, \bar{k}_q^m) \\
&- \sum_{p=-\infty}^{\infty}\sum_{n=1}^{\infty} ik_z^p(a_n^p\cos k_z^p h - b_n^p\sin k_z^p h)I_{21} \\
&- \sum_{n=1}^{\infty} i\bar{k}_z^m(c_n^m\cos\bar{k}_z^m h - d_n^m\sin\bar{k}_z^m h)I_{22} \\
&= \left(c_q^m\sin\bar{k}_z^m h + d_q^m\cos\bar{k}_z^m h\right) I_2^m(\bar{k}_q^m, \bar{k}_q^m)
\end{aligned} \tag{25}$$

where

$$I_{22} = \int_0^{\infty} \kappa^{-1} I_2^m(\bar{k}_n^m, \varsigma) I_2^m(\bar{k}_q^m, \varsigma)\varsigma d\varsigma \tag{26}$$

$$I_{21} = \int_0^{\infty} \kappa^{-1} J_{p-m}(\varsigma l) I_1^p(k_n^p, \varsigma) I_2^m(\bar{k}_q^m, \varsigma)\varsigma d\varsigma . \tag{27}$$

The additional boundary conditions at $z = -h$ require

$$\left.\frac{\partial \Phi^{t1}}{\partial z}\right|_{z=-h} = \begin{cases} \left.\dfrac{\partial \Phi^{II}}{\partial z}\right|_{z=-h}, & r < r_1 \\ 0, & r > r_1 \end{cases} \tag{28}$$

$$\left.\frac{\partial \Phi^{t2}}{\partial z}\right|_{z=-h} = \begin{cases} \left.\dfrac{\partial \Phi^{III}}{\partial z}\right|_{z=-h}, & r' < r_2 \\ 0, & r' > r_2 \end{cases} \tag{29}$$

$$\Phi^t(r, \phi, -h) = \Phi^{II}(r, \phi, -h), \qquad r < r_1 \tag{30}$$

$$\Phi^t(r', \phi', -h) = \Phi^{III}(r', \phi', -h), \quad r' < r_2 . \tag{31}$$

Applying the Hankel transforms to (28) and (29) gives respectively

$$\widetilde{\Phi}_m^{t1}(\varsigma) = \frac{i}{\kappa}\sum_{n=1}^{\infty} k_z^m a_n^m I_1^m(k_n^m, \varsigma) \tag{32}$$

$$\widetilde{\Phi}_m^{t2}(\varsigma) = \frac{i}{\kappa}\sum_{n=1}^{\infty} \bar{k}_z^m c_n^m I_2^m(\bar{k}_n^m, \varsigma) . \tag{33}$$

Substituting (20), (21), (32), and (33) into (30) and (31), we obtain

$$\sum_{n=1}^{\infty} ik_z^m a_n^m I_{11} + \sum_{p=-\infty}^{\infty}\sum_{n=1}^{\infty} i\bar{k}_z^p c_n^p I_{12} = b_q^m I_1^m(k_q^m, k_q^m) \tag{34}$$

$$\sum_{p=-\infty}^{\infty}\sum_{n=1}^{\infty} ik_z^p a_n^p I_{21} + \sum_{n=1}^{\infty} i\bar{k}_z^m c_n^m I_{22} = d_q^m I_2^m(\bar{k}_q^m, \bar{k}_q^m) . \tag{35}$$

The reflection coefficient ϱ_ν (or transmission coefficient τ_ν) is a ratio of the time-averaged power reflected from (or transmitted through) aperture ν to that impinging on aperture ν where ν is 1 or 2. Then the total reflection coefficient ϱ and transmission coefficient τ of two circular apertures are

$$\varrho = \varrho_1 + \varrho_2 \tag{36}$$

with

$$
\begin{aligned}
\varrho_1 = &-\frac{2}{r_1^2 k \cos\theta_i} \sum_{m=-\infty}^{\infty} \sum_{n=1}^{\infty} Im\Big\{ k_z^{m*}(a_n^m \cos k_z^m h - b_n^m \sin k_z^m h)^* \\
&\cdot\Big[(a_n^m \sin k_z^m h + b_n^m \cos k_z^m h) I_1^m(k_n^m, k_n^m) \\
&- 2i^m \exp\Big(-\frac{ikl}{2} \sin\theta_i - im\phi_i\Big) I_1^m(k\sin\theta_i, k_n^m)\Big]\Big\}
\end{aligned}
\tag{37}
$$

$$
\begin{aligned}
\varrho_2 = &-\frac{2}{r_2^2 k \cos\theta_i} \sum_{m=-\infty}^{\infty} \sum_{n=1}^{\infty} Im\Big\{ \bar{k}_z^{m*}(c_n^m \cos \bar{k}_z^m h - d_n^m \sin \bar{k}_z^m h)^* \\
&\cdot\Big[(c_n^m \sin \bar{k}_z^m h + d_n^m \cos \bar{k}_z^m h) I_2^m(\bar{k}_n^m, \bar{k}_n^m) \\
&- 2i^m \exp\Big(-\frac{ikl}{2} \sin\theta_i - im\phi_i\Big) I_2^m(k\sin\theta_i, \bar{k}_n^m)\Big]\Big\}
\end{aligned}
\tag{38}
$$

and

$$\tau = \tau_1 + \tau_2 \tag{39}$$

with

$$\tau_1 = \frac{2}{r_1^2 k \cos\theta_i} \sum_{m=-\infty}^{\infty} \sum_{n=1}^{\infty} Im(k_z^{m*} b_n^m a_n^{m*}) I_1^m(k_n^m, k_n^m) \tag{40}$$

$$\tau_2 = \frac{2}{r_2^2 k \cos\theta_i} \sum_{m=-\infty}^{\infty} \sum_{n=1}^{\infty} Im(\bar{k}_z^{m*} d_n^m c_n^{m*}) I_2^m(\bar{k}_n^m, \bar{k}_n^m) \ . \tag{41}$$

The far-zone scattered and transmitted fields at distances R_s and R_t are

$$
\begin{aligned}
&\Phi^s(R_s, \theta_s, \phi) \\
&= \frac{\exp(ikR_s - i\pi/4)\cos\theta_s}{R_s\sqrt{\sin\theta_s}} \sum_{m=-\infty}^{\infty} \sum_{p=-\infty}^{\infty} \\
&\Big\{ \exp(im\phi - ip\pi/2)\Big[-ie^{i\zeta l/2}\widetilde{\Phi}_p^{s1}(\zeta) + e^{-i\zeta l/2}\widetilde{\Phi}_p^{s2}(\zeta)\Big] \\
&+ (-1)^m \exp(im\phi + ip\pi/2)\Big[e^{-i\zeta l/2}\widetilde{\Phi}_p^{s1}(\zeta) - ie^{i\zeta l/2}\widetilde{\Phi}_p^{s2}(\zeta)\Big]\Big\}\Big|_{\zeta=k\sin\theta_s}
\end{aligned}
\tag{42}
$$

and

$$
\Phi^t(R_t, \theta_t, \phi)
$$

$$
= \frac{\exp(ikR_t - i\pi/4)\exp(-ikh\cos\theta_t)\cos\theta_t}{R_t\sqrt{\sin\theta_t}} \sum_{m=-\infty}^{\infty} \sum_{p=-\infty}^{\infty}
$$

$$
\left\{(-1)^m \exp(im\phi - ip\pi/2)\left[ie^{i\zeta l/2}\widetilde{\Phi}_p^{t1}(\zeta) - e^{-i\zeta l/2}\widetilde{\Phi}_p^{t2}(\zeta)\right]\right.
$$

$$
\left.+ \exp(im\phi + ip\pi/2)\left[-e^{-i\zeta l/2}\widetilde{\Phi}_p^{t1}(\zeta) + ie^{i\zeta l/2}\widetilde{\Phi}_p^{t2}(\zeta)\right]\right\}\bigg|_{\zeta=k\sin\theta_t} . \quad (43)
$$

7.6 Acoustic Radiation from Circular Cylinders in a Hard Plane [19]

A study of electromagnetic wave radiation from flanged circular cylinders is important for practical applications in microwave antenna array problems. Acoustic wave radiation from a single flanged circular cylinder has been studied in Sect. 7.4. In this section we will study acoustic wave radiation from multiple circular cylinders in a hard plane. A theoretical analysis given in this section is similar to that in Sect. 7.5. Consider an N number of circular cylinders ($\nu = 1, 2, 3, \ldots, N$) on an infinite hard plane where an acoustic wave emanates from the first flanged circular cylinder $\nu = 1$. In region (1) the total field (velocity potential) consists of the incident, reflected, and scattered waves, where the incident (Φ^i) and reflected (Φ^r) components are

$$
\Phi^{i,r}(x, y, z) = e^{\pm ikz} . \quad (1)
$$

Inside the νth cylinder the scattered component is represented as

$$
\Phi^{s\nu} = \sum_{m=-\infty}^{\infty} e^{im\phi_\nu} \sum_{n=1}^{\infty} b_{\nu n}^m \exp(-ik_\nu^m z) J_m(k_{\nu n}^m r_\nu) \quad (2)
$$

where $k_\nu^m = \sqrt{k^2 - (k_{\nu n}^m)^2}$, $J_m'(k_{\nu n}^m a_\nu) = 0$, and the prime $'$ denotes differentiation with respect to the argument. In region ($N + 1$) the transmitted wave, based on the superposition principle, is represented as

$$
\Phi^t = \sum_{\mu=1}^{N} \Phi^{t\mu}(r_\mu, \phi_\mu, z) \quad (3)
$$

$$
\Phi^{t\mu}(r_\mu, \phi_\mu, z) = \sum_{m=-\infty}^{\infty} e^{im\phi_\mu} \int_0^{\infty} \widetilde{\Phi}_m^{t\mu}(\zeta) J_m(\zeta r_\mu) e^{i\kappa z}\zeta d\zeta \quad (4)
$$

where $\kappa = \sqrt{k^2 - \zeta^2}$ and $\Phi^{t\mu}(r_\mu, \phi_\mu, z)$ is the wave transmitted through the μth cylinder.

The boundary conditions at $z = 0$ for the νth circular cylinder for $1 \le \nu \le N$ require

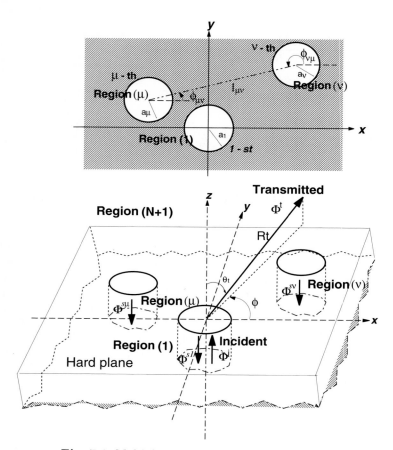

Fig. 7.6. Multiple circular cylinders in a hard plane

$$\left. \frac{\partial \Phi^{t\nu}}{\partial z} \right|_{z=0} = \begin{cases} \left. \dfrac{\partial \Phi^{s\nu}}{\partial z} \right|_{z=0}, & r_\nu < a_\nu \\[2mm] 0, & r_\nu > a_\nu \end{cases} \tag{5}$$

$$[\Phi^i(r_1,\phi_1,0) + \Phi^r(r_1,\phi_1,0)]\delta_{\nu 1} + \Phi^{s\nu}(r_\nu,\phi_\nu,0)$$
$$= \Phi^t(r,\phi,0), \qquad r_\nu < a_\nu \tag{6}$$

where δ_{ij} denotes the Kronecker delta. Applying the Hankel transform to (5) yields

$$\widetilde{\Phi}_m^{t\nu}(\zeta) = \frac{-\mathrm{i}}{\kappa} \sum_{n=1}^{\infty} k_\nu^m b_{\nu n}^m I_\nu^m(k_{\nu n}^m, \zeta) \tag{7}$$

where for $\alpha \neq \beta$

$$I_\nu^m(\alpha, \beta) = \frac{a_\nu}{\alpha^2 - \beta^2}$$
$$\cdot [\alpha J_{m+1}(\alpha a_\nu) J_m(\beta a_\nu) - \beta J_m(\alpha a_\nu) J_{m+1}(\beta a_\nu)] \qquad (8)$$

otherwise

$$I_\nu^m(\alpha, \alpha) = \frac{a_\nu^2}{2} \left[J_m^2(\alpha a_\nu) - J_{m-1}(\alpha a_\nu) J_{m+1}(\alpha a_\nu) \right] . \qquad (9)$$

Substituting (1), (2), and (3) into (6) yields

$$2\delta_{\nu 1} + \sum_{m=-\infty}^{\infty} e^{im\phi_\nu} \sum_{n=1}^{\infty} b_{\nu n}^m \exp(-ik_\nu^m z) J_m(k_{\nu n}^m r_\nu)$$

$$= \sum_{\mu=1}^{N} \Phi^{t\mu}(r_\mu, \phi_\mu, z) . \qquad (10)$$

Graf's addition theorem gives

$$J_p(\zeta r_\mu) e^{ip\phi_\mu} = \sum_{m=-\infty}^{\infty} J_{m-p}(\zeta l_{\nu\mu}) J_m(\zeta r_\nu) \exp[i(p-m)\phi_{\nu\mu}] e^{im\phi_\nu} . \quad (11)$$

We substitute (4), (7), and (11) into (10), multiply (10) by $J_m(k_{\nu q}^m r_\nu) r_\nu$, and integrate over $0 < r_\nu < a_\nu$ to obtain

$$\boxed{\begin{aligned} &\sum_{n=1}^{\infty} b_{\nu n}^m k_\nu^m I_1 + \sum_{\substack{\mu=1 \\ \mu \neq \nu}}^{N} \sum_{p=-\infty}^{\infty} \sum_{n=1}^{\infty} b_{\mu n}^p k_\mu^p \exp[i(p-m)\phi_{\nu\mu}] I_2 \\ &= -b_{\nu q}^m I_\nu(k_{\nu q}^m, k_{\nu q}^m) - a_1^2 \delta_{\nu 1} \delta_{m0} \delta_{q1} \end{aligned}} \qquad (12)$$

where

$$I_1 = \int_0^\infty \kappa^{-1} I_\nu^m(k_{\nu n}^m, \zeta) I_\nu^m(k_{\nu q}^m, \zeta) \zeta d\zeta \qquad (13)$$

$$I_2 = \int_0^\infty \kappa^{-1} J_{m-p}(\zeta l_{\nu\mu}) I_\nu^m(k_{\nu q}^m, \zeta) I_\mu^p(k_{\mu n}^p, \zeta) \zeta d\zeta . \qquad (14)$$

In case of single cylinder scattering ($a_1 \neq 0$ and $a_\nu = 0$ for $\nu \geq 2$), (12) reduces to (9) in Sect. 7.4.

The reflection coefficient C_{11} (self-coupling) is a ratio of the reflected power to the incident power and the coupling coefficient $C_{1\nu}$ (cross-coupling) is a ratio of the power coupled into the νth cylinder to the incident power. They are given as

$$C_{11} = \frac{2}{ka_1^2} \sum_{m=-\infty}^{\infty} \sum_{n=1}^{\infty} Re\left\{ k_1^{m*} |b_{1n}^m|^2 I_1^m(k_{1n}^m, k_{1n}^m) \right\}$$

$$+ 2Re(b_{11}^0) + 1 \tag{15}$$

$$C_{1\nu} = \frac{2}{ka_1^2} \sum_{m=-\infty}^{\infty} \sum_{n=1}^{\infty} Re\left\{ k_\nu^{m*} |b_{\nu n}^m|^2 I_\nu^m(k_{\nu n}^m, k_{\nu n}^m) \right\} . \tag{16}$$

The far-zone transmitted field at distance R_t is

$$\Phi^t(R_t, \theta_t, \phi)$$

$$= \frac{\exp(ikR_t - i\pi/4)\cos\theta_t}{R_t\sqrt{\sin\theta_t}}$$

$$\cdot \sum_{m=-\infty}^{\infty} e^{im\phi} \left\{ e^{-i\pi/2} \widetilde{\Phi}_m^{t1}(\zeta) + \sum_{\nu=2}^{N} \sum_{p=-\infty}^{\infty} \exp[i(p-m)\phi_{1\nu}] \right.$$

$$\cdot \left[\exp\left(-\frac{ip\pi}{2} - \frac{i\zeta l}{2}\right) - i(-1)^m \exp\left(\frac{ip\pi}{2} + \frac{i\zeta l}{2}\right) \right]$$

$$\left. \cdot \widetilde{\Phi}_p^{t\nu}(\zeta) \right\}\Bigg|_{\zeta=k\sin\theta_t} . \tag{17}$$

References for Chapter 7

1. J. H. Lee and H. J. Eom, "Electrostatic potential through a circular aperture in a thick conducting plane," *IEEE Trans. Microwave Theory Tech.*, vol. 81, no. 12, pp. 341-343, Feb. 1996.
2. R. L. Gluckstern, R. Li, and R. K. Cooper, "Electric polarizability and magnetic susceptibility of small holes in a thin screen," *IEEE Trans. Microwave Theory Tech.*, vol. 38, no. 2, pp. 186-191, Feb. 1990.
3. R. L. Gluckstern and J. A. Diamond, "Penetration of fields through a circular hole in a wall of finite thickness," *IEEE Trans. Microwave Theory Tech*, vol. 39, no. 2, pp. 274-279, Feb. 1991.
4. J. G. Lee and H. J. Eom, "Magnetostatic potential distribution through a circular aperture in a thick conducting plane," *IEEE Trans. Electromagn. Compat.*, vol. 40, no. 2, pp. 97-99, 1998.
5. K. F. Casey, "Low-frequency electromagnetic penetration of loaded apertures," *IEEE Trans. Electromagn. Compat.*, vol. 23, no. 4, pp. 367-377, Nov. 1981.
6. K. H. Jun and H. J. Eom, "Acoustic scattering from a circular aperture in a thick hard screen," *J. Acoust. Soc. Am.*, vol. 98, no. 4, pp. 2324-2327, Oct. 1995.
7. S. N. Karp and J. B. Keller, "Multiple diffraction by an aperture in a hard screen," *Optica Acta*, vol. 8, pp. 61-72, Jan. 1961.
8. C. J. Bouwkamp, "Theoretical and numerical treatment of diffraction through a circular hole," *IEEE Trans. Antennas Propagat.*, vol. 18, no. 2, pp. 152-176, March 1970.
9. G. P. Wilson and W. W. Soroka, "Approximation to the diffraction of sound by a circular aperture in a rigid wall of finite thickness," *J. Acoust. Soc. Am.*, vol. 37, no. 2, pp. 286-297, Feb. 1965.

10. A. Roberts, "Electromagnetic theory of diffraction by a circular aperture in a thick, perfectly conducting screen," *J. Opt. Soc. Am. A.*, vol. 4, no. 10, pp. 1970-1983, Oct. 1987.

11. L. J. Palumbo and A. M. Platzck, "Diffraction by a circular aperture: a new approach," *J. Opt. Soc. Am. A.*, vol. 4, no. 5, pp. 839-842, May 1987.

12. K. Hongo, "Diffraction of an electromagnetic plane wave by circular disk and circular hole," *IEICE Trans. Electron.*, vol. E80-C, no. 11, pp. 1360-1366, Nov. 1997.

13. W. T. Cathey, Jr., "Approximate expressions for field penetration through circular apertures," *IEEE Trans. Electromagn. Compat.*, vol. 25, no. 3, pp. 339-345, Aug. 1983.

14. R. E. Collin, *Field Theory of Guided Waves*, New York, IEEE Press, Second Edition, Chapter 5 and pp. 121-123, 1991.

15. H. J. Eom, T. J. Park, and S. Kozaki, "A series solution for acoustic radiation from a flanged circular pipe," *Acustica*, vol. 80, no. 3, pp. 315-316, May/June 1994.

16. W. E. Zorumski, "Generalized radiation impedances and reflection coefficients of circular and annular ducts," *J. Acoust. Soc. Am.*, vol. 54, no. 6, pp. 1667-1673, 1973.

17. J. S. Seo, H. J. Eom, and H. S. Lee, "Acoustic scattering from two circular apertures in a thick hard plane," *J. Acoust. Soc. Am.*, vol. 107, no. 5, Pt. 1, pp. 2338-2343, May 2000.

18. M. Abramowitz and I. A. Stegun, *Handbook of Mathematical Functions*, New York, Dover Publications, p. 363, 1965.

19. J. S. Seo and H. J. Eom, "Acoustic scattering from flanged circular cylinders," *Acustica*, vol. 86, no. 5, pp. 780-783, Sept./Oct. 2000.

8. Annular Aperture in a Plane

8.1 Static Potential Through an Annular Aperture in a Plane

The polarizability of various aperture shapes in a conducting plane finds practical applications in low-frequency microwave scattering and penetration problems [1]. For instance, when an aperture size is small compared to an incident wavelength, the polarizability is a useful concept to estimate a field penetration into apertures. The behavior of polarizability for an annular aperture in a conducting plane has been well studied in [2-3] based on the variational method. In the present section we will revisit the problem of polarizability of an annular aperture in a thick conducting plane and analyze its static potential distribution through an annular aperture.

8.1.1 Electrostatic Distribution [4,5]

In this subsection we will investigate an electrostatic potential distribution through an annular aperture with a floating inner conductor. An electrostatic potential $\Phi^i(r, z)$ is applied to an annular aperture in a thick conducting plane. A conducting plane ($r > b$ and $|z| < d/2$) is at zero potential and the inner conductor ($r < a$ and $|z| < d/2$) is electrically floated with potential V and charge Q. Regions (I), (II), and (III) are an upper half-space ($z > d/2$) with permittivity ϵ_1, an annular aperture (radii: a and b, depth: d) with ϵ_2, and a lower half-space ($z < -d/2$) with ϵ_3. In region (I) the total electrostatic potential is a sum of the incident and scattered components

$$\Phi^i(r, z) = E_0(z - d/2) \tag{1}$$

$$\Phi^s(r, z) = \int_0^\infty \widetilde{\Phi}^s(\zeta) J_0(\zeta r) \exp[-\zeta(z - d/2)]\zeta \mathrm{d}\zeta . \tag{2}$$

The electrostatic potential in region (II) is assumed to be

$$\Phi^d(r, z) = V\frac{\ln b/r}{\ln b/a} + \sum_{n=1}^\infty (a_n \mathrm{e}^{k_n z} + b_n \mathrm{e}^{-k_n z}) R(k_n r) \tag{3}$$

where

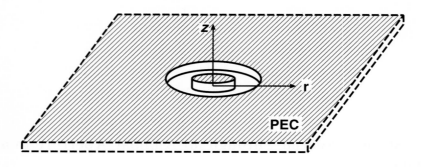

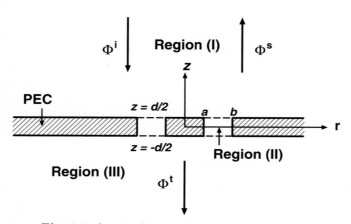

Fig. 8.1. An annular aperture in a conducting plane

$$R(k_n r) = J_0(k_n r) - \frac{J_0(k_n a)}{N_0(k_n a)} N_0(k_n r) \tag{4}$$

and a constant k_n is determined by $R(k_n b) = 0$. The transmitted electrostatic potential in region (III) is

$$\Phi^t(r, z) = \int_0^\infty \widetilde{\Phi}^t(\zeta) J_0(\zeta r) \exp[\zeta(z + d/2)]\zeta \mathrm{d}\zeta . \tag{5}$$

The boundary condition on the field continuity at $z = d/2$ (Dirichlet boundary condition) requires

$$\Phi^i(r, d/2) + \Phi^s(r, d/2) = \begin{cases} V, & 0 < r < a \\ \Phi^d(r, d/2), & a < r < b \\ 0, & \text{otherwise} . \end{cases} \tag{6}$$

Taking the Hankel transform of (6) yields

$$\widetilde{\Phi}^s(\zeta) = \frac{V}{\ln b/a}\frac{J_0(\zeta a) - J_0(\zeta b)}{\zeta^2} + \sum_{n=1}^{\infty}\left(a_n e^{k_n d/2} + b_n e^{-k_n d/2}\right)\Xi_n(\zeta) \quad (7)$$

where

$$\Xi_n(\zeta) = \frac{2}{\pi}\left[\frac{J_0(\zeta a)}{N_0(k_n a)} - \frac{J_0(\zeta b)}{N_0(k_n b)}\right]\frac{1}{\zeta^2 - k_n^2} \cdot \quad (8)$$

The boundary condition on the field continuity (Neumann boundary condition) requires

$$\epsilon_1\left(\frac{\partial\Phi^i}{\partial z}\bigg|_{z=0} + \frac{\partial\Phi^s}{\partial z}\bigg|_{z=0}\right) = \epsilon_2\frac{\partial\Phi^d}{\partial z}\bigg|_{z=0}, \quad a < r < b. \quad (9)$$

We substitute $\widetilde{\Phi}^s(\zeta)$ into (9), multiply (9) by $rR(k_p r)$, and integrate with respect to r from a to b to obtain

$$\begin{aligned}
&\epsilon_1 E_0 L_p - \frac{\epsilon_1 V}{\ln b/a}I_2 - \epsilon_1\sum_{n=1}^{\infty}\left(a_n e^{k_n d/2} + b_n e^{-k_n d/2}\right)I_1 \\
&= \epsilon_2 k_p\left(a_p e^{k_p d/2} - b_p e^{-k_p d/2}\right)[A(b) - A(a)]
\end{aligned} \quad (10)$$

where

$$A(r) = \frac{2}{\pi^2 k_p^2}\frac{1}{N_0^2(k_p r)} \quad (11)$$

$$I_1 = \int_0^{\infty}\Xi_n(\zeta)\Xi_p(\zeta)\zeta^2 d\zeta \quad (12)$$

$$I_2 = \int_0^{\infty}[J_0(\zeta a) - J_0(\zeta b)]\Xi_p(\zeta)d\zeta \quad (13)$$

$$L_n = \int_a^b R(k_n r)r dr = \frac{2}{\pi k_n^2}\left[\frac{1}{N_0(k_n b)} - \frac{1}{N_0(k_n a)}\right]\cdot \quad (14)$$

Similarly from the boundary conditions at $z = -d/2$, we obtain

$$\begin{aligned}
&\frac{\epsilon_3 V}{\ln b/a}I_2 + \epsilon_3\sum_{n=1}^{\infty}\left(a_n e^{-k_n d/2} + b_n e^{k_n d/2}\right)I_1 \\
&= \epsilon_2 k_p\left(a_p e^{-k_p d/2} - b_p e^{k_p d/2}\right)[A(b) - A(a)]\cdot
\end{aligned} \quad (15)$$

The total charge Q on the inner conductor must be conserved since the inner conductor is electrically floated. In order to determine a relationship between the charge Q and the potential V, we apply Gauss's law over the surface of the inner conductor. The relationship is

$$\begin{aligned}
Q &= \frac{2\pi V}{\ln b/a}[\epsilon_2 d + (\epsilon_1 + \epsilon_3)a\gamma] \\
&\quad -\pi\epsilon_1 a^2 E_0 + 2\pi a\sum_{n=1}^{\infty}(a_n c_1 + b_n c_2)
\end{aligned} \quad (16)$$

where

$$c_1 = I_3 \left(\epsilon_1 e^{k_n d/2} + \epsilon_3 e^{-k_n d/2} \right) - 2\epsilon_2 R'(k_n a) \sinh(k_n d/2) \qquad (17)$$

$$c_2 = I_3 \left(\epsilon_1 e^{-k_n d/2} + \epsilon_3 e^{k_n d/2} \right) - 2\epsilon_2 R'(k_n a) \sinh(k_n d/2) \qquad (18)$$

$$I_3 = \int_0^\infty J_1(\zeta a) \Xi_n(\zeta) \zeta d\zeta \qquad (19)$$

$$\gamma = \int_0^\infty [J_0(\zeta a) - J_0(\zeta b)] \frac{J_1(\zeta a)}{\zeta} d\zeta . \qquad (20)$$

We assume that Q is linearly proportional to V, and $Q = 0$ when $E_0 = V = 0$. By setting E_0 to zero and V to unity, we solve (10) and (15) for a_n and b_n. Substituting a_n and b_n into (16) yields the capacitance C

$$C \equiv \frac{\partial Q}{\partial V} = \frac{2\pi}{\ln b/a} [\epsilon_2 d + (\epsilon_1 + \epsilon_3) a \gamma] + 2\pi a \sum_{n=1}^\infty (a_n c_1 + b_n c_2) . \qquad (21)$$

The electric polarizability $\chi_e(z)$ is

$$\chi_e(z) \equiv 4\pi \int_a^b \Phi^d(r, z) r dr$$

$$= \frac{\pi V(b^2 - a^2)}{\ln b/a} + 4\pi \sum_{n=1}^\infty \left(a_n e^{k_n z} + b_n e^{-k_n z} \right) L_n . \qquad (22)$$

8.1.2 Magnetostatic Distribution [4]

Consider a problem of magnetostatic distribution through an annular aperture in a thick conducting plane. Regions (I) ($z > 0$), (II) ($-d/2 < z < d/2$ and $a < r < b$), and (III) ($z < -d/2$) denote an upper half-space, an annular aperture, and a lower half-space, respectively. In region (I) the total magnetostatic potential consists of the incident and scattered components. An incident magnetostatic potential is

$$\Phi^i(r, \phi, z) = x = r \cos \phi . \qquad (23)$$

The scattered magnetostatic potential in region (I) takes the form of

$$\Phi^s(r, \phi, z) = \cos \phi \int_0^\infty \widetilde{\Phi}^s(\zeta) J_1(\zeta r) e^{-\zeta(z-d/2)} \zeta d\zeta . \qquad (24)$$

In region (II) the magnetostatic potential is

$$\Phi^d(r, \phi, z)$$

$$= \sum_{m=1}^\infty [a_m \sinh k_m(z + d/2) + b_m \cosh k_m(z + d/2)] R(k_m r) \cos \phi \qquad (25)$$

where

$$R(k_m r) = J_1(k_m r) - \frac{J_1'(k_m a)}{N_1'(k_m a)} N_1(k_m r) \ . \tag{26}$$

Note that the prime denotes differentiation with respect to the argument and a constant k_m is given by the condition $\left. \dfrac{\partial \Phi^d}{\partial r} \right|_{r=b} = 0$. In region (III) the transmitted magnetostatic potential is

$$\Phi^t(r, \phi, z) = \cos\phi \int_0^\infty \widetilde{\Phi}^t(\zeta) J_1(\zeta r) e^{\zeta(z+d/2)} \zeta d\zeta \ . \tag{27}$$

The boundary condition at $z = d/2$ requires

$$\left. \frac{\partial \Phi^i}{\partial z} \right|_{z=d/2} + \left. \frac{\partial \Phi^s}{\partial z} \right|_{z=d/2} = \begin{cases} \left. \dfrac{\partial \Phi^d}{\partial z} \right|_{z=d/2}, & a < r < b \\[2mm] 0, & \text{otherwise} \ . \end{cases} \tag{28}$$

Applying the Hankel transform to (28) yields

$$\widetilde{\Phi}^s(\zeta) = - \sum_{m=1}^\infty (a_m \cosh k_m d + b_m \sinh k_m d) k_m \Xi_m(\zeta) \tag{29}$$

where

$$\Xi_m(\zeta) = \frac{2}{\pi k_m} \left[\frac{J_1'(\zeta a)}{N_1'(k_m a)} - \frac{J_1'(\zeta b)}{N_1'(k_m b)} \right] \frac{1}{\zeta^2 - k_m^2} \ . \tag{30}$$

An additional boundary condition at $z = d/2$ for $a < r < b$ requires

$$\left. \Phi^i \right|_{z=d/2} + \left. \Phi^s \right|_{z=d/2} = \left. \Phi^d \right|_{z=d/2} \ . \tag{31}$$

It is expedient to apply to (31) the orthogonality property

$$\int_a^b R(k_m r) R(k_n r) r dr = \delta_{mn} \left[A(b) - A(a) \right] \tag{32}$$

where

$$A(r) = \frac{2}{\pi^2 k_n^2} \left[1 - \frac{1}{(k_n r)^2} \right] \frac{1}{N_1'(k_n r)^2} \ . \tag{33}$$

Multiplying (31) by $R(k_n r) r$ and integrating with respect to r over (a, b), we get

$$\boxed{\begin{aligned} L_n - \sum_{m=1}^\infty (a_m \cosh k_m d + b_m \sinh k_m d) k_m I \\ = (a_n \sinh k_n d + b_n \cosh k_n d) \left[A(b) - A(a) \right] \end{aligned}} \tag{34}$$

where

$$I = \int_0^\infty \Xi_n(\zeta)\Xi_m(\zeta)\zeta^2 \mathrm{d}\zeta \tag{35}$$

$$L_n = \int_a^b r^2 R(k_n r)\mathrm{d}r = \frac{2}{\pi k_n^3}\left[\frac{1}{N_1'(k_n b)} - \frac{1}{N_1'(k_n a)}\right] . \tag{36}$$

From the boundary condition at $z = -d/2$, we obtain

$$\boxed{\sum_{m=1}^\infty a_m k_m I = b_n\left[A(b) - A(a)\right] .} \tag{37}$$

The magnetic polarizability $\chi_m(z)$ is shown to be

$$\chi_m(z) \equiv \pi \int_a^b r^2 \frac{\partial \Phi^d}{\partial z}\mathrm{d}r$$

$$= \pi \sum_{n=1}^\infty \left[a_n \cosh k_n(z + d/2) + b_n \sinh k_n(z + d/2)\right] k_n L_n . \tag{38}$$

8.2 EM Radiation from a Coaxial Line into a Parallel-Plate Waveguide [6]

Electromagnetic wave radiation from a coaxial line into a half-space or a parallel-plate waveguide was studied for material permittivity characterization and antenna feed application [7-9]. In this section we will analyze electromagnetic radiation from a coaxial line into a parallel-plate waveguide. Assume that an incident TEM wave emanates from a flanged coaxial line. Due to a symmetry of the problem geometry, the H-field has only a ϕ-component. In region (I) ($a < r < b$ and $z < 0$) the incident and reflected H-fields are

$$H_\phi^i(r, z) = M_0 \frac{\mathrm{e}^{\mathrm{i}\beta_1 z}}{r} \tag{1}$$

$$H_\phi^r(r, z) = c_0 M_0 \frac{\mathrm{e}^{-\mathrm{i}\beta_1 z}}{r} + \sum_{n=1}^\infty c_n M_n R_n(r)\mathrm{e}^{-\mathrm{i}k_{zn} z} \tag{2}$$

where $\beta_1 = \omega\sqrt{\mu\epsilon_1}$, $k_{zn} = \beta_1\sqrt{1 - (k_n/\beta_1)^2}$, $R_n(r) = J_1(k_n r)N_0(k_n b) - N_1(k_n r)J_0(k_n b)$, $M_0 = 1/\sqrt{\ln(b/a)}$, and $M_n = \pi k_n/\sqrt{2 - 2J_0^2(k_n b)/J_0^2(k_n a)}$. Note that an eigenvalue k_n is given by $J_0(k_n a)N_0(k_n b) - N_0(k_n a)J_0(k_n b) = 0$. The transmitted H-field in region (II) ($z > 0$) is

$$H_\phi^t(r, z) = \int_0^\infty \tilde{H}(\zeta)\left(\mathrm{e}^{\mathrm{i}\kappa z} + \mathrm{e}^{2\mathrm{i}\kappa h}\mathrm{e}^{-\mathrm{i}\kappa z}\right) J_1(\zeta r)\zeta \mathrm{d}\zeta \tag{3}$$

where $\kappa = \sqrt{\beta_2^2 - \zeta^2}$ and the wavenumber is $\beta_2 = \omega\sqrt{\mu\epsilon_2} = 2\pi/\lambda$.

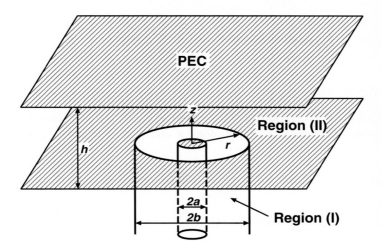

Fig. 8.2. A coaxial line radiating into a parallel-plate waveguide

The tangential E-field continuity at $z = 0$ requires

$$E_r^t(r,0) = \begin{cases} E_r^i(r,0) + E_r^r(r,0), & b < r < a \\ 0, & \text{otherwise} . \end{cases} \tag{4}$$

Applying the Hankel transform to (4) yields

$$\tilde{H}(\zeta) = \frac{\epsilon_2}{\epsilon_1} \frac{1}{\kappa(1 - e^{2i\kappa h})} \left[(1 - c_0)\beta_1 f_0(\zeta) - \sum_{n=1}^{\infty} c_n k_{zn} f_n(\zeta) \right] \tag{5}$$

where

$$f_0(\zeta) = -\frac{M_0 \left[J_0(b\zeta) - J_0(a\zeta) \right]}{\zeta} \tag{6}$$

$$f_n(\zeta) = \frac{2M_n\zeta \left[J_0(\zeta b) J_0(k_n a) - J_0(\zeta a) J_0(k_n b) \right]}{\pi k_n J_0(k_n a)(k_n^2 - \zeta^2)} . \tag{7}$$

The tangential H-field continuity at $z = 0$ requires

$$H_\phi^t(r,0) = H_\phi^i(r,0) + H_\phi^r(r,0), \qquad a < r < b . \tag{8}$$

We multiply (8) by $r M_p R_p(r) dr$ and integrate from a to b to obtain

$$\boxed{C = (U - A)^{-1}\Gamma} \tag{9}$$

where C is a column vector of c_p, U is a unit matrix, and the elements of matrices A and Γ are for $p \geq 0$

$$a_{p0} = \int_0^\infty \beta_1 g(\zeta) f_0(\zeta) f_p(\zeta) \zeta \mathrm{d}\zeta \tag{10}$$

$$a_{pn} = \int_0^\infty k_{zn} g(\zeta) f_n(\zeta) f_p(\zeta) \zeta \mathrm{d}\zeta, \quad n \geq 1 \tag{11}$$

$$\gamma_p = -a_{p0} - \delta_{p0} \tag{12}$$

$$g(\zeta) = -\mathrm{i} \frac{\epsilon_2}{\epsilon_1} \frac{\cot(\kappa h)}{\kappa}. \tag{13}$$

It is convenient to transform the integral a_{pn} into fast convergent series by performing the residue calculus along a contour path in the ζ-plane. The results are

$$a_{00} = -\mathrm{i}\sqrt{\frac{\epsilon_2}{\epsilon_1}} \cot \beta_2 h - \frac{\beta_1}{2} \sum_{m=0}^\infty \pi M_0^2 \alpha_m \frac{\Lambda_{00}(\xi_m)}{h \xi_m^2} \tag{14}$$

$$a_{0n} = \frac{\pi k_{zn}}{2h} \sum_{m=0}^\infty \alpha_m \phi_n(\xi_m) \Lambda_{10}(\xi_m) \tag{15}$$

$$a_{p0} = \frac{\pi \beta_1}{2h} \sum_{m=0}^\infty \alpha_m \phi_p(\xi_m) \Lambda_{01}(\xi_m) \tag{16}$$

$$a_{pn} = \frac{k_{zn} M_n^2 \cot\left(\sqrt{\beta_2^2 - k_n^2} h\right)}{\pi k_n J_0^2(k_n a) \sqrt{\beta_2^2 - k_n^2}} \Lambda'_{11}(k_n) \delta_{pn}$$
$$- \frac{\pi k_{zn}}{2h} \sum_{m=0}^\infty \alpha_m \psi_{pn}(\xi_m) \Lambda_{11}(\xi_m) \tag{17}$$

where $\Lambda'_{11}(k_n)$ means $\left. \dfrac{\mathrm{d}\Lambda_{11}(\zeta)}{\mathrm{d}\zeta} \right|_{\zeta = k_n}$ and

$$\Lambda_{st}(\zeta) = \Big[J_0(sak_n) J_0(tak_p) J_0(b\zeta) H_0^{(1)}(b\zeta)$$
$$- J_0(sak_n) J_0(tbk_p) J_0(a\zeta) H_0^{(1)}(b\zeta)$$
$$- J_0(sbk_n) J_0(tak_p) J_0(a\zeta) H_0^{(1)}(b\zeta)$$
$$+ J_0(sbk_n) J_0(tbk_p) J_0(a\zeta) H_0^{(1)}(a\zeta) \Big] \frac{\epsilon_2}{\epsilon_1}, \quad s, t = 1, 0 \tag{18}$$

$$\alpha_m = \begin{cases} 1, & m = 0 \\ 2, & m = 1, 2, 3, \ldots \end{cases} \tag{19}$$

$$\xi_m = \sqrt{\beta_2^2 - (m\pi/h)^2}, \quad m = 0, 1, 2, \ldots \tag{20}$$

$$\phi_n(\zeta) = \frac{2 M_0 M_n}{\pi k_n J_0(k_n a)(k_n^2 - \zeta^2)} \tag{21}$$

$$\psi_{pn}(\zeta) = \frac{4 M_p M_n \zeta^2}{\pi^2 k_p k_n J_0(k_p a) J_0(k_n a)(k_p^2 - \zeta^2)(k_n^2 - \zeta^2)}. \tag{22}$$

Performing the residue calculus, we represent the transmitted H-field in rapidly-convergent series

$$H_\phi^t(r,z) = B_1 \sum_{m=0}^{\infty} \begin{cases} \dfrac{\alpha_m}{\sqrt{1-\left(\dfrac{m\pi}{\beta_2 h}\right)^2}} \cos\dfrac{m\pi}{h}z \\ \quad \cdot [J_0(\xi_m b) - J_0(\xi_m a)]H_1^{(1)}(\xi_m r) \\[2em] \dfrac{\alpha_m}{\sqrt{1-\left(\dfrac{m\pi}{\beta_2 h}\right)^2}} \cos\dfrac{m\pi}{h}z \\ \quad \cdot [J_1(\xi_m r)H_0^{(1)}(\xi_m b) - H_1^{(1)}(\xi_m r)J_0(\xi_m a)] \\ \quad - \dfrac{2hi}{\pi r}\dfrac{\cos\beta_2(h-z)}{\sin\beta_2 h}\delta_{m0} \\[2em] \dfrac{\alpha_m}{\sqrt{1-\left(\dfrac{m\pi}{\beta_2 h}\right)^2}} \cos\dfrac{m\pi}{h}z\, J_1(\xi_m r) \\ \quad \cdot \left[H_0^{(1)}(\xi_m b) - H_0^{(1)}(\xi_m a)\right] \end{cases}$$

$$-\sum_{n=1}^{\infty}\sum_{m=0}^{\infty} B_2 \frac{\cos(m\pi/h)z}{(k_n^2 - \xi_m^2)}$$
$$\begin{cases} [J_0(\xi_m b)J_0(k_n a) - J_0(\xi_m a)J_0(k_n b)]H_1^{(1)}(\xi_m r) \\[0.8em] J_1(\xi_m r)J_0(k_n a)H_0^{(1)}(\xi_m b) - H_1^{(1)}(\xi_m r)J_0(\xi_m a)J_0(k_n b) \\[0.8em] J_1(\xi_m r)[H_0^{(1)}(\xi_m b)J_0(k_n a) - H_0^{(1)}(\xi_m a)J_0(k_n b)] \end{cases}$$

$$-\sum_{n=1}^{\infty} \frac{\epsilon_2 c_n k_{zn} M_n}{\epsilon_1}\frac{i\cos\kappa_n(h-z)}{\kappa_n \sin\kappa_n h}$$
$$\begin{cases} 0, & r \geq b \\[0.8em] N_0(k_n b)J_1(k_n r) - J_0(k_n b)N_1(k_n r), & a \leq r \leq b \\[0.8em] 0, & r \leq a \end{cases} \tag{23}$$

where $B_1 = -\sqrt{\dfrac{\epsilon_2}{\mu_0}}\dfrac{\pi V_0}{2h\ln(b/a)}$, $B_2 = \dfrac{2c_n\epsilon_2 k_{zn}\xi_m M_n\alpha_m}{\epsilon_1 k_n J_0(k_n a)h}$, $V_0 = \dfrac{\beta_1(1-c_0)}{\omega\epsilon_1 M_0}$, and $\kappa_n = \sqrt{k_2^2 - k_n^2}$.

Let's consider a coaxial line radiating into a half-space. When $h \to \infty$, the solution (9)-(12) is applicable with $g(\zeta) = -\epsilon_2/(\epsilon_1\kappa)$, and it is expedient to transform the integrals into fast convergent forms by performing a branch-cut integration. We choose a similar deformed contour path as shown in Fig. 1.2 with a branch cut associated with a branch point $\zeta = \beta_2$. The contour integration results give

$$a_{00} = -\sqrt{\frac{\epsilon_2}{\epsilon_1}} + \beta_1 \int_0^\infty \frac{M_0^2}{\kappa \zeta} \Lambda_{00}(\zeta) \mathrm{i} dv \tag{24}$$

$$a_{0n} = -k_{zn} \int_0^\infty \frac{\zeta}{\kappa} \phi_n(\zeta) \Lambda_{10}(\zeta) \mathrm{i} dv \tag{25}$$

$$a_{p0} = -\beta_1 \int_0^\infty \frac{\zeta}{\kappa} \phi_p(\zeta) \Lambda_{01}(\zeta) \mathrm{i} dv \tag{26}$$

$$a_{pn} = -\frac{\mathrm{i}\epsilon_2 M_n^2}{\epsilon_1 \pi k_n J_0^2(k_n a)} \Lambda_{11}'(k_n) \delta_{pn}$$
$$+ k_{zn} \int_0^\infty \frac{\zeta}{\kappa} \psi_{pn}(\zeta) \Lambda_{11}(\zeta) \mathrm{i} dv \tag{27}$$

where $\zeta = \beta_2 + \mathrm{i}v$.

8.3 EM Radiation from a Coaxial Line into a Dielectric Slab [10]

An open-ended coaxial line radiating into a half-space and a stratified dielectric medium was studied in [11-16]. In this section we will analyze the electromagnetic wave radiation from a flanged coaxial line into a dielectric slab that is displaced from an open-ended coaxial line aperture. A theoretical analysis given in this section is in continuation of Sect. 8.2. Assume that a TEM wave radiates from region (I) ($a < r < b$ and $z < 0$). The incident (TEM) and reflected (TM$_{0n}$ waves) H-fields in region (I) are

$$H_\phi^i(r,z) = M_0 \frac{\mathrm{e}^{\mathrm{i}\beta z}}{r} \tag{1}$$

$$H_\phi^r(r,z) = c_0 M_0 \frac{\mathrm{e}^{-\mathrm{i}\beta z}}{r} + \sum_{n=1}^\infty c_n M_n R_n(r) \mathrm{e}^{-\mathrm{i}\beta_n z} \tag{2}$$

where $\beta = \omega\sqrt{\mu\epsilon_1}$, $\beta_n = \sqrt{\beta^2 - \lambda_n^2}$, $M_0 = \dfrac{1}{\sqrt{\ln(b/a)}}$, $M_n = \dfrac{\pi\lambda_n}{\sqrt{2 - 2\dfrac{J_0^2(\lambda_n b)}{J_0^2(\lambda_n a)}}}$,

and $R_n(r) = J_1(\lambda_n r)N_0(\lambda_n b) - N_1(\lambda_n r)J_0(\lambda_n b)$. The eigenvalue λ_n is given by $J_0(\lambda_n a)N_0(\lambda_n b) - N_0(\lambda_n a)J_0(\lambda_n b) = 0$. Regions (II), (III), and (IV) represent a background (wavenumber: $k_2 = \omega\sqrt{\mu\epsilon_2}$), a dielectric slab of thickness

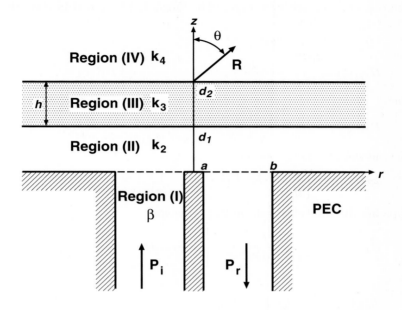

Fig. 8.3. A flanged coaxial line radiating into a dielectric slab

h ($k_3 = \omega\sqrt{\mu\epsilon_3}$), and a half-space ($k_4 = \omega\sqrt{\mu\epsilon_4}$), respectively. The H-field in region (II) ($0 < z < d_1$) is

$$H_\phi^{II}(r, z) = \int_0^\infty \left[\widetilde{H}_{II}^+(\zeta)\mathrm{e}^{\mathrm{i}\kappa_2 z} + \widetilde{H}_{II}^-(\zeta)\mathrm{e}^{-\mathrm{i}\kappa_2 z} \right] \zeta J_1(\zeta r)\mathrm{d}\zeta \tag{3}$$

where $\kappa_2 = \sqrt{k_2^2 - \zeta^2}$. In region (III) ($d_1 < z < d_2$) the H-field is

$$H_\phi^{III}(r, z) = \int_0^\infty \left[\widetilde{H}_{III}^+(\zeta)\mathrm{e}^{\mathrm{i}\kappa_3 z} + \widetilde{H}_{III}^-(\zeta)\mathrm{e}^{-\mathrm{i}\kappa_3 z} \right] \zeta J_1(\zeta r)\mathrm{d}\zeta \tag{4}$$

where $\kappa_3 = \sqrt{k_3^2 - \zeta^2}$. In region (IV) ($z > d_2$) the H-field is

$$H_\phi^{IV}(r, z) = \int_0^\infty \widetilde{H}_{IV}(\zeta)\exp[\mathrm{i}\kappa_4(z - d_2)]\zeta J_1(\zeta r)\mathrm{d}\zeta \tag{5}$$

where $\kappa_4 = \sqrt{k_4^2 - \zeta^2}$.

The tangential E- and H-field continuities at $z = d_2$ give

$$\widetilde{H}_{III}^-(\zeta) = \left(\frac{\eta_3 - \eta_4}{\eta_3 + \eta_4} \right) \mathrm{e}^{\mathrm{i}2\kappa_3 d_2} \widetilde{H}_{III}^+(\zeta) \tag{6}$$

$$\widetilde{H}_{IV}(\zeta) = \left(\frac{2\eta_3}{\eta_3 + \eta_4} \right) \mathrm{e}^{\mathrm{i}\kappa_3 d_2} \widetilde{H}_{III}^+(\zeta) \tag{7}$$

where $\eta_3 = \kappa_3/\epsilon_3$ and $\eta_4 = \kappa_4/\epsilon_4$. The tangential E- and H-field continuities at $z = d_1$ similarly give

$$\widetilde{H}_{II}^-(\zeta) = \Gamma e^{i2\kappa_2 d_1}\widetilde{H}_{II}^+(\zeta) \tag{8}$$

$$\widetilde{H}_{III}^+(\zeta) = \left[\frac{(1+\Gamma)(\eta_3 + \eta_4)e^{i\kappa_2 d_1 - i\kappa_3 d_2}}{(\eta_3 + \eta_4)e^{-i\kappa_3 h} + (\eta_3 - \eta_4)e^{i\kappa_3 h}}\right]\widetilde{H}_{II}^+(\zeta) \tag{9}$$

where

$$\Gamma = \frac{(\eta_2 - \eta_3)(\eta_3 + \eta_4)e^{-i\kappa_3 h} + (\eta_2 + \eta_3)(\eta_3 - \eta_4)e^{i\kappa_3 h}}{(\eta_2 + \eta_3)(\eta_3 + \eta_4)e^{-i\kappa_3 h} + (\eta_2 - \eta_3)(\eta_3 - \eta_4)e^{i\kappa_3 h}} . \tag{10}$$

The tangential E-field continuity at $z = 0$ is

$$E_r^{II}(r,0) = \begin{cases} E_r^i(r,0) + E_r^r(r,0), & a < r < b \\ 0, & \text{otherwise} . \end{cases} \tag{11}$$

Applying the Hankel transform to (11) yields

$$\widetilde{H}_{II}^+(\zeta) = \frac{\epsilon_2}{\epsilon_1 \kappa_2 (1 - \Gamma e^{i2\kappa_2 d_1})}\left[(1 - c_0)\beta f_0(\zeta) - \sum_{n=1}^{\infty} c_n \beta_n f_n(\zeta)\right] \tag{12}$$

where

$$f_0(\zeta) = -\frac{M_0}{\zeta}[J_0(\zeta b) - J_0(\zeta a)] \tag{13}$$

$$f_n(\zeta) = \frac{2M_n\zeta[J_0(\zeta b)J_0(\lambda_n a) - J_0(\zeta a)J_0(\lambda_n b)]}{\pi\lambda_n(\lambda_n^2 - \zeta^2)J_0(\lambda_n a)} . \tag{14}$$

The tangential H-field continuity at $z = 0$ requires

$$H_\phi^{II}(r,0) = H_\phi^i(r,0) + H_\phi^r(r,0), \quad a < r < b . \tag{15}$$

Substituting (8) and (12) into (15), multiplying (15) by $rM_mR_m(r)dr$, and integrating from a to b, we get a matrix equation

$$\boxed{C = (U - A)^{-1}Q} \tag{16}$$

where C is a column vector of c_m, U is a unit matrix, and the elements of matrices A and Q for $m \geq 0$ are

$$a_{m0} = -\int_0^\infty \frac{\epsilon_2\beta}{\epsilon_1\kappa_2}\frac{g_1(\zeta)}{g_2(\zeta)}f_m(\zeta)f_0(\zeta)\zeta d\zeta \tag{17}$$

$$a_{mn} = -\int_0^\infty \frac{\epsilon_2\beta_n}{\epsilon_1\kappa_2}\frac{g_1(\zeta)}{g_2(\zeta)}f_m(\zeta)f_n(\zeta)\zeta d\zeta \tag{18}$$

$$q_m = -\delta_{m0} - a_{m0} \tag{19}$$

with

$$g_1(\zeta) = \eta_2\eta_3\cos(\kappa_3 h)\cos(\kappa_2 d_1) - \eta_3^2\sin(\kappa_3 h)\sin(\kappa_2 d_1)$$
$$\quad -i[\eta_2\eta_4\sin(\kappa_3 h)\cos(\kappa_2 d_1) + \eta_3\eta_4\cos(\kappa_3 h)\sin(\kappa_2 d_1)] \tag{20}$$

$$g_2(\zeta) = \eta_3\eta_4\cos(\kappa_3 h)\cos(\kappa_2 d_1) - \eta_2\eta_4\sin(\kappa_3 h)\sin(\kappa_2 d_1)$$
$$\quad -i[\eta_3^2\sin(\kappa_3 h)\cos(\kappa_2 d_1) + \eta_2\eta_3\cos(\kappa_3 h)\sin(\kappa_2 d_1)] . \tag{21}$$

It is convenient to transform a_{mn} into numerically-efficient forms by per-

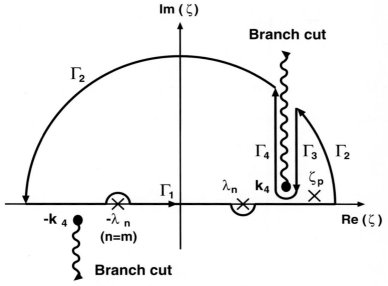

Fig. 8.4. Contour path in the ζ-plane

forming a contour integral in the complex ζ-plane. The results are

$$a_{00} = \sqrt{\frac{\epsilon_2}{\epsilon_1}} \frac{g_1(\zeta)}{g_2(\zeta)} \delta_{00} - \sum_{\zeta_p} \mathrm{i}\pi \frac{\beta}{\kappa_2} \frac{g_1(\zeta)}{g_2'(\zeta)} \frac{M_0^2}{\zeta} \Lambda_{00}(\zeta) \bigg|_{\zeta=\zeta_p}$$

$$+\frac{\mathrm{i}}{2} \int_0^\infty \beta \frac{M_0^2}{\zeta} \Lambda_{00}(\zeta) \bigg|_{\zeta=k_4+\mathrm{i}v} G(v)\mathrm{d}v \qquad (22)$$

$$a_{m0} = \sum_{\zeta_p} \mathrm{i}\pi \frac{\beta}{\kappa_2} \frac{g_1(\zeta)}{g_2'(\zeta)} \phi_m(\zeta) \Lambda_{10}(\zeta) \bigg|_{\zeta=\zeta_p}$$

$$-\frac{\mathrm{i}}{2} \int_0^\infty \beta \phi_m(\zeta) \Lambda_{10}(\zeta) \bigg|_{\zeta=k_4+\mathrm{i}v} G(v)\mathrm{d}v \qquad (23)$$

$$a_{0n} = \sum_{\zeta_p} \mathrm{i}\pi \frac{\beta_n}{\kappa_2} \frac{g_1(\zeta)}{g_2'(\zeta)} \phi_n(\zeta) \Lambda_{01}(\zeta) \bigg|_{\zeta=\zeta_p}$$

$$-\frac{\mathrm{i}}{2} \int_0^\infty \beta_n \phi_n(\zeta) \Lambda_{01}(\zeta) \bigg|_{\zeta=k_4+\mathrm{i}v} G(v)\mathrm{d}v \qquad (24)$$

$$a_{mn} = -\frac{i}{\pi}\frac{\beta_n}{\kappa_2}\frac{g_1(\zeta)}{g_2(\zeta)}\frac{M_m M_n \Lambda_{11}'(\zeta)}{\lambda_m J_0(\lambda_m a)J_0(\lambda_n a)}\bigg|_{\zeta=\lambda_n}\delta_{mn}$$

$$-\sum_{\zeta_p}i\pi\frac{\beta_n}{\kappa_2}\frac{g_1(\zeta)}{g_2'(\zeta)}\psi_{mn}(\zeta)\Lambda_{11}(\zeta)\bigg|_{\zeta=\zeta_p}$$

$$+\frac{i}{2}\int_0^\infty \beta_n\psi_{mn}(\zeta)\Lambda_{11}(\zeta)\bigg|_{\zeta=k_4+iv}G(v)dv \tag{25}$$

where the prime $'$ denotes differentiation with respect to ζ and

$$\Lambda_{ts}(\zeta) = \frac{\epsilon_2}{\epsilon_1}\bigg[J_0(sa\lambda_n)J_0(ta\lambda_m)J_0(b\zeta)H_0^{(1)}(b\zeta)$$

$$-J_0(sa\lambda_n)J_0(tb\lambda_m)J_0(a\zeta)H_0^{(1)}(b\zeta)$$

$$-J_0(sb\lambda_n)J_0(ta\lambda_m)J_0(a\zeta)H_0^{(1)}(b\zeta)$$

$$+J_0(sb\lambda_n)J_0(tb\lambda_m)J_0(a\zeta)H_0^{(1)}(a\zeta)\bigg], \qquad s,t=1,0 \tag{26}$$

$$\phi_n(\zeta) = \frac{2M_0 M_n \zeta}{\pi\lambda_n J_0(\lambda_n a)(\lambda_n^2 - \zeta^2)} \tag{27}$$

$$\psi_{mn}(\zeta) = \frac{4M_m M_n \zeta^3}{\pi^2 \lambda_m \lambda_n J_0(\lambda_m a)J_0(\lambda_n a)(\lambda_m^2 - \zeta^2)(\lambda_n^2 - \zeta^2)} \tag{28}$$

$$G(v) = \frac{g_1(\zeta)}{\kappa_2 g_2(\zeta)}\bigg|_{\zeta=k_4+iv} - \frac{g_1(\zeta)}{\kappa_2 g_2(\zeta)}\bigg|_{\kappa_4\to-\kappa_4,\zeta=k_4+iv}. \tag{29}$$

The first term in a_{mn} is a residue contribution at $\zeta = \pm\lambda_m$ and the second term represents residue contributions at $\zeta = \zeta_p$, which are zeros of $g_2(\zeta)$. The third term in a_{mn} is a branch-cut integration along Γ_3 and Γ_4 associated with a branch-point at $\zeta = k_4$. In low-frequency limit, a single-mode solution ($a_{mn} = 0$ for $m,n \geq 1$) is given by

$$c_0 \simeq -\frac{1+a_{00}}{1-a_{00}}, \qquad c_n \simeq 0, \qquad n \geq 1. \tag{30}$$

The far-zone field in region (IV) is

$$H_\phi^{IV}(R,\theta) \sim -\frac{\eta_3 k_4 \cos\theta}{\epsilon_1 g_2(\zeta)}$$

$$\cdot\bigg[(1-c_0)\beta f_0(\zeta) - \sum_{n=1}^\infty c_n\beta_n f_n(\zeta)\bigg]\bigg|_{\zeta=k_4\sin\theta}\frac{e^{ik_4 R}}{R} \tag{31}$$

where $R = \sqrt{r^2 + z^2}$.

Next we consider two special cases corresponding to the half-space and attached-slab scattering problems.

When $k_2 = k_3 = k_4$ and $d_2 = 0$ (half-space case), we obtain

$$a_{m0} = -\int_0^\infty \frac{\epsilon_4 \beta}{\epsilon_1 \kappa_4} f_m(\zeta) f_0(\zeta) \zeta J_1(\zeta r) \mathrm{d}\zeta \tag{32}$$

$$a_{mn} = -\int_0^\infty \frac{\epsilon_4 \beta_n}{\epsilon_1 \kappa_4} f_m(\zeta) f_n(\zeta) \zeta J_1(\zeta r) \mathrm{d}\zeta \tag{33}$$

$$H_\phi^{IV}(R,\theta) \sim \frac{\epsilon_4}{\epsilon_1} \left(F_0 + \sum_{n=1}^\infty F_n \right) \frac{e^{ik_4 R}}{R} \tag{34}$$

where

$$F_0 = (1 - c_0)\beta M_0 \left[\frac{J_0(k_4 b \sin\theta) - J_0(k_4 a \sin\theta)}{k_4 \sin\theta} \right] \tag{35}$$

$$
\begin{aligned}
F_n = {} & 2c_n \beta_n M_n k_4 \sin\theta \\
& \cdot \left[\frac{J_0(k_4 b \sin\theta) J_0(\lambda_n a) - J_0(k_4 a \sin\theta) J_0(\lambda_n b)}{\pi \lambda_n (\lambda_n^2 - k_4^2 \sin^2\theta) J_0(\lambda_n a)} \right] .
\end{aligned}
\tag{36}
$$

When $k_2 = k_3$, $d_1 = 0$, and $k_3 > k_4$ (attached-slab case), we get

$$
\begin{aligned}
a_{m0} = {} & -\int_0^\infty \left(\frac{\epsilon_3 \beta}{\epsilon_1 \kappa_3} \frac{\eta_3 \cos\kappa_3 d_2 - i\eta_4 \sin\kappa_3 d_2}{\eta_4 \cos\kappa_3 d_2 - i\eta_3 \sin\kappa_3 d_2} \right) \\
& \cdot f_m(\zeta) f_0(\zeta) \zeta J_1(\zeta r) \mathrm{d}\zeta
\end{aligned}
\tag{37}
$$

$$
\begin{aligned}
a_{mn} = {} & -\int_0^\infty \left(\frac{\epsilon_3 \beta_n}{\epsilon_1 \kappa_3} \frac{\eta_3 \cos\kappa_3 d_2 - i\eta_4 \sin\kappa_3 d_2}{\eta_4 \cos\kappa_3 d_2 - i\eta_3 \sin\kappa_3 d_2} \right) \\
& \cdot f_m(\zeta) f_n(\zeta) \zeta J_1(\zeta r) \mathrm{d}\zeta
\end{aligned}
\tag{38}
$$

and

$$
\begin{aligned}
H_\phi^{IV}(R,\theta) \sim {} & \frac{\epsilon_4}{\epsilon_1} \frac{k_4 \cos\theta}{(k_4 \cos\kappa_3 d_2 - i\epsilon_4/\epsilon_3 \kappa_3 \sin\kappa_3 d_2)} \bigg|_{\zeta = k_4 \sin\theta} \\
& \cdot \left(F_0 + \sum_{n=1}^\infty F_n \right) \frac{e^{ik_4 R}}{R} .
\end{aligned}
\tag{39}
$$

8.4 EM Radiation from a Monopole into a Parallel-Plate Waveguide [17]

A study of electromagnetic wave radiation from a coaxially-fed monopole antenna into a parallel-plate waveguide is important for practical applications in antenna feeder [18,19]. In this section we will present a scattering analysis for a coaxially-fed monopole antenna radiating into a parallel-plate waveguide. When an incident TEM wave radiates from a coaxial line, the total field in region (III) ($a_1 < r < a_2$ and permittivity: ϵ_3) consists of the incident, reflected, and scattered components

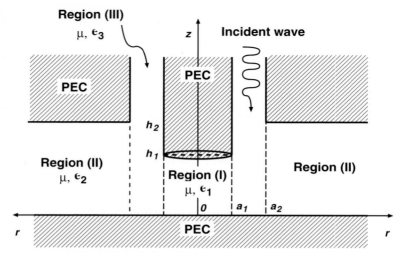

Fig. 8.5. A monopole antenna radiating into a parallel-plate waveguide

$$H_\phi^i = -\frac{e^{-ik_3 z}}{\eta r} \tag{1}$$

$$H_\phi^r = -\frac{e^{ik_3 z}}{\eta r} \tag{2}$$

$$E_z^{III}(r, z) = \frac{2}{i\omega\epsilon_3\pi} \int_0^\infty R(\kappa r)\cos(\zeta z)d\zeta . \tag{3}$$

In regions (I) ($0 < r < a_1$ and permittivity: ϵ_1) and (II) ($r > a_2$ and permittivity: ϵ_2) the scattered fields are respectively

$$E_z^I(r, z) = \frac{i}{\omega\epsilon_1} \sum_{m=0}^\infty p_m \xi_{1m}\cos(h_{1m}z)J_0(\xi_{1m}r) \tag{4}$$

$$E_z^{II}(r, z) = \frac{i}{\omega\epsilon_2} \sum_{m=0}^\infty q_m \xi_{2m}\cos(h_{2m}z)H_0^{(1)}(\xi_{2m}r) \tag{5}$$

where $h_{pm} = m\pi/h_p$, $\xi_{pm} = \sqrt{k_p^2 - h_{pm}^2}$, $\kappa = \sqrt{k_3^2 - \zeta^2}$, $k_p = \omega\sqrt{\mu\epsilon_p}$, $\eta = \sqrt{\mu/\epsilon_3}$, $R(\kappa r) = J_0(\kappa r)\widetilde{E}^+(\zeta) - N_0(\kappa r)\widetilde{E}^-(\zeta)$, $H_0^{(1)}(\cdot) = J_0(\cdot) + iN_0(\cdot)$, $J_0(\cdot)$ is the 0th order Bessel function, and $N_0(\cdot)$ is the 0th order Neumann function.

Applying the Fourier cosine transform to the E_z continuity at $r = a_1$, $E_z^i(a_1, z) + E_z^r(a_1, z) + E_z^{III}(a_1, z) = E_z^I(a_1, z)$, yields

$$R(\kappa a_1) = -\sum_{m=0}^\infty p_m \frac{\epsilon_3}{\epsilon_1}\xi_{1m}J_0(\xi_{1m}a_1)\Xi_m^1(\zeta) . \tag{6}$$

Similarly from the E_z continuity at $r = a_2$, we obtain

$$R(\kappa a_2) = - \sum_{m=0}^{\infty} q_m \frac{\epsilon_3}{\epsilon_2} \xi_{2m} H_0^{(1)}(\xi_{2m}a_2) \Xi_m^2(\zeta) \tag{7}$$

where

$$\Xi_m^p(\zeta) = \frac{(-1)^m \zeta \sin(\zeta h_p)}{\zeta^2 - h_{pm}^2}. \tag{8}$$

We multiply the H_ϕ continuity at $r = a_1$ for $0 < z < h_1$ by $\cos(n\pi/h_1)z$ and integrate to get

$$\boxed{I_1 + p_n \frac{h_1}{2} J_1(\xi_{1n}a_1)\varepsilon_n = -\frac{2}{\eta} \frac{\Xi_n^1(k_3)}{a_1}, \qquad n = 0, 1, \ldots} \tag{9}$$

where

$$I_1 = -\frac{2}{\pi} \int_0^{\infty} \frac{1}{\kappa} R'(\kappa a_1) \Xi_n^1(\zeta) \mathrm{d}\zeta \tag{10}$$

$\varepsilon_0 = 2$, $\varepsilon_n = 1$ $(n = 1, 2, \ldots)$, and $R'(\cdot) = \mathrm{d}R(\cdot)/\mathrm{d}(\cdot)$. Utilizing a residue calculus, it is possible to transform I_1 into a rapidly-convergent series

$$I_1 = \sum_{m=0}^{\infty} (-1)^{m+n} \Big[p_m \frac{\epsilon_3}{\epsilon_1} \xi_{1m} J_0(\xi_{1m}a_1) \overline{I}_1$$
$$+ q_m \frac{\epsilon_3}{\epsilon_2} \xi_{2m} H_0^{(1)}(\xi_{2m}a_2) \overline{I}_{12} \Big]. \tag{11}$$

Similarly using the H_ϕ continuity at $r = a_2$ for $0 < z < h_2$, we get

$$\boxed{I_2 + q_n \frac{h_2}{2} H_1^{(1)}(\xi_{2n}a_2)\varepsilon_n = -\frac{2}{\eta} \frac{\Xi_n^2(k_3)}{a_2}, \qquad n = 0, 1, \ldots} \tag{12}$$

where

$$I_2 = -\frac{2}{\pi} \int_0^{\infty} \frac{1}{\kappa} R'(\kappa a_2) \Xi_n^2(\zeta) \mathrm{d}\zeta$$
$$= \sum_{m=0}^{\infty} (-1)^{m+n} \Big[q_m \frac{\epsilon_3}{\epsilon_2} \xi_{2m} H_0^{(1)}(\xi_{2m}a_2) \overline{I}_2$$
$$- p_m \frac{\epsilon_3}{\epsilon_1} \xi_{1m} J_0(\xi_{1m}a_1) \overline{I}_{21} \Big] \tag{13}$$

$$\overline{I}_1 = \frac{h_1}{2} \frac{\Delta_1 \delta_{mn} \varepsilon_m}{\kappa_{1m} \Delta(\kappa_{1m})} + \frac{ik_3}{2a_1 \ln(a_2/a_1)} \frac{1 - \exp(i2k_3 h_1)}{(k_3^2 - h_{1m}^2)(k_3^2 - h_{1n}^2)}$$
$$- \frac{i}{a_1} \sum_{v=1}^{\infty} \frac{\zeta_v}{[1 - J_0^2(\kappa_v a_1)/J_0^2(\kappa_v a_2)]}$$
$$\cdot \frac{1 - \exp(i2\zeta_v h_1)}{(\zeta_v^2 - h_{1m}^2)(\zeta_v^2 - h_{1n}^2)} \tag{14}$$

$$\overline{I}_2 = \frac{h_2}{2} \frac{\Delta_2 \delta_{mn}\varepsilon_m}{\kappa_{2m}\Delta(\kappa_{2m})} - \frac{ik_3}{2a_2\ln(a_2/a_1)} \frac{1 - \exp(i2k_3h_2)}{(k_3^2 - h_{2m}^2)(k_3^2 - h_{2n}^2)}$$

$$- \frac{i}{a_2}\sum_{v=1}^{\infty} \frac{\zeta_v}{[1 - J_0^2(\kappa_v a_2)/J_0^2(\kappa_v a_1)]}$$

$$\cdot \frac{1 - \exp(i2\zeta_v h_2)}{(\zeta_v^2 - h_{2m}^2)(\zeta_v^2 - h_{2n}^2)} \tag{15}$$

$$\overline{I}_{pq} = X_{pq} - \frac{ik_3}{2a_p\ln(a_2/a_1)} \frac{\exp(ik_3|h_2 - h_1|) - \exp[ik_3(h_2 + h_1)]}{(k_3^2 - h_{pn}^2)(k_3^2 - h_{qm}^2)}$$

$$- \frac{i}{a_p}\sum_{v=1}^{\infty} \frac{\zeta_v}{[J_0(\kappa_v a_1)/J_0(\kappa_v a_2) - J_0(\kappa_v a_2)/J_0(\kappa_v a_1)]}$$

$$\cdot \frac{\exp(i\zeta_v|h_2 - h_1|) - \exp[i\zeta_v(h_2 + h_1)]}{(\zeta_v^2 - h_{pn}^2)(\zeta_v^2 - h_{qm}^2)} \tag{16}$$

$$X_{12} = \begin{cases} \dfrac{2(-1)^m}{\pi a_p\kappa_{2m}^2\Delta(\kappa_{2m})} \dfrac{h_{2m}\sin\left(\dfrac{h_1}{h_2}m\pi\right)}{h_{2m}^2 - h_{1n}^2}, & h_2 > h_1 \\[20pt] \dfrac{2(-1)^n}{\pi a_p\kappa_{1n}^2\Delta(\kappa_{1n})} \dfrac{h_{1n}\sin\left(\dfrac{h_2}{h_1}n\pi\right)}{h_{1n}^2 - h_{2m}^2}, & h_2 < h_1 \end{cases} \tag{17}$$

δ_{mn} is the Kronecker delta, $\Delta(\kappa) = J_0(\kappa a_1)N_0(\kappa a_2) - J_0(\kappa a_2)N_0(\kappa a_1)$, $\Delta_1 = J_0(\kappa_{1m}a_2)N_1(\kappa_{1m}a_1) - J_1(\kappa_{1m}a_1)N_0(\kappa_{1m}a_2)$, $\Delta_2 = J_1(\kappa_{2m}a_2)N_0(\kappa_{2m}a_1) - J_0(\kappa_{2m}a_1)N_1(\kappa_{2m}a_2)$, $\kappa_{pm} = \sqrt{k_3^2 - h_{pm}^2}$, κ_v is determined by $\Delta(\kappa_v) = 0$, and $\zeta_v = \sqrt{k_3^2 - \kappa_v^2}$.

The reflected plus scattered TEM wave at $z = \infty$ is shown to be

$$H_\phi^r + H_\phi^{III} = -\frac{\exp(ik_3 z)}{\eta r}(1 + L_0 - M_0) \tag{18}$$

where

$$L_0 = \frac{\eta i}{k_3\ln(a_2/a_1)}\sum_{m=0}^{\infty} p_m\frac{\epsilon_3}{\epsilon_1}\xi_{1m}J_0(\xi_{1m}a_1)\Xi_m^1(k_3) \tag{19}$$

$$M_0 = \frac{\eta i}{k_3\ln(a_2/a_1)}\sum_{m=0}^{\infty} q_m\frac{\epsilon_3}{\epsilon_2}\xi_{2m}H_0^{(1)}(\xi_{2m}a_2)\Xi_m^2(k_3) . \tag{20}$$

References for Chapter 8

1. R. E. Collin, *Field Theory of Guided Waves*, New York; McGraw-Hill, 1960.

2. A. Zolotov and V. P. Kazantsev, "An analytic solution of the problem of the polarizability of a circular ring aperture in an unbounded planar screen of zero thickness obtained by the variational method," *Soviet J. Commun. Technol. Electron.*, vol. 37, no. 4, pp. 103–105, May 1992.
3. S. S. Kurennoy, "Polarizabilities of an annular cut in the wall of an arbitrary thickness," *IEEE Trans. Microwave Theory Tech.*, vol. 44, no. 7, pp. 1109–1114, July 1996.
4. H. S. Lee and H. J. Eom, "Polarizabilities of an annular aperture in a thick conducting plane," *J. Electromagn. Waves Appl.*, vol. 12, pp. 269–279, Feb. 1998.
5. H. S. Lee and H. J. Eom, "Potential distribution through an annular aperture with a floating inner conductor," *IEEE Trans. Microwave Theory Tech.*, vol. 47, no. 3, pp. 372-374, 1999.
6. J. H. Lee, H. J. Eom, and K. H. Jun, "Reflection of a coaxial line radiating into a parallel plate," *IEEE Microwave Guided Wave Lett.*, vol. 6, pp. 135-137, Mar. 1996.
7. D. C. Chang, "Input admittance and complete near-field distribution of an annular aperture antenna driven by a coaxial line," *IEEE Trans. Antennas Propagat.*, vol. 18, no. 5, pp. 610-616, Sept. 1970.
8. J. R. Mosig, J. E. Besson, M. Gex-Fabry, and F. E. Gardiol, "Reflection of an open-ended coaxial line and application to nondestructive measurement of materials," *IEEE Trans. Instrum. Meas.*, vol. 30, no. 1, pp. 46-51, March 1981.
9. B. Tomasic and A. Hessel, "Electric and magnetic current sources in the parallel plate waveguide," *IEEE Trans. Antennas Propagat.*, vol. 45, no. 11, pp. 1307-1310, Nov. 1987.
10. Y. C. Noh and H. J. Eom, "Radiation from a flanged coaxial line into a dielectric slab," *IEEE Trans. Microwave Theory Tech.*, vol. 47, no. 11, pp. 2158-2161, Nov. 1999.
11. M. Stuchly and S. Stuchly, "Coaxial line reflection methods for measuring dielectric properties of biological substances at radio and microwave frequencies - A review," *IEEE Trans. Instrum. Meas.*, vol. 29, pp. 176-183, Sept. 1980.
12. D. Misra, "A quasi-static analysis of the open-ended coaxial line," *IEEE Trans. Microwave Theory Tech.*, vol. 35, pp. 925-928, Oct. 1987.
13. E. Burdette, F. Cain, and J. Seals, "In vivo probe measurement technique for determining dielectric properties at VHF through microwave frequencies," *IEEE Trans. Microwave Theory Tech.*, vol. 28, pp. 414-426, Apr., 1980.
14. S. Fan, K. Staebel, and S. Stuchly, "Static analysis of an open-ended coaxial line terminated by layered media," *IEEE Trans. Instrum. Meas.*, vol. 39, pp. 435-437, Apr. 1990.
15. L. Anderson, G. Gajda, and S. Stuchly, "Analysis of an open-ended coaxial line sensor in layered dielectrics," *IEEE Trans. Instrum. Meas.*, vol. 35, pp. 13-18, Mar. 1986.
16. S. Bakhtiari, S. I. Ganchev, and R. Zoughi, "Analysis of radiation from an open-ended coaxial line into stratified dielectrics," *IEEE Trans. Microwave Theory Tech.*, vol. 42, pp. 1261-1267, July 1994.
17. H. J. Eom, Y. H. Cho, and M. S. Kwon, "Monopole antenna radiation into a parallel-plate waveguide," *IEEE Trans. Antennas Propagat.*, vol. 48, no. 7, pp. 1142-1144, July 2000.
18. A. G. Williamson, "Radial-line/coaxial-line junctions: analysis and equivalent circuits," *Int. J. Electron.*, vol. 58, no. 1, pp. 91-104, 1985.
19. Z. Shen and R. H. MacPhie, "Modal expansion analysis of monopole antennas driven from a coaxial line," *Radio Sci.*, vol. 31, no. 5, pp. 1037-1046, Sept.-Oct. 1996.

9. Circumferential Apertures on a Circular Cylinder

9.1 EM Radiation from an Aperture on a Shorted Coaxial Line [1]

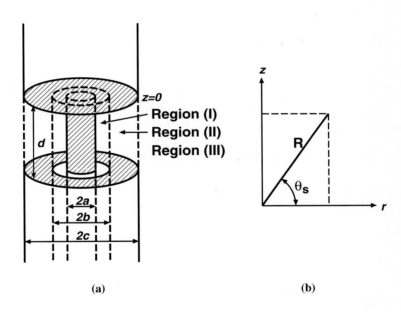

(a)   (b)

Fig. 9.1. A circumferential aperture on a shorted coaxial line

9.1.1 Field Analysis

Electromagnetic scattering and radiation from a circumferential aperture on a shorted coaxial line (dielectric-filled edge-slot antenna, DFES antenna) was considered in [2]. In this section we will revisit the problem of DFES antenna radiation. Consider radiation from a shorted coaxial line whose outer conductor is circumferentially removed and replaced by a dielectric medium with permittivity $\epsilon = \epsilon_0 \epsilon_r$. An incident TEM wave propagates from below inside a coaxial line that is shorted at $z = 0$. The field in region (I) $(a < r < b)$ consists of the incident and scattered waves

$$E_{rI}^i(r, z) = \frac{e^{ikz}}{r} \tag{1}$$

$$E_{rI}(r, z) = -\frac{e^{-ikz}}{r} - \frac{2}{\pi} \int_0^\infty \frac{\zeta}{\kappa} \widetilde{E}_I(\zeta) R'(\kappa r) \sin(\zeta z) d\zeta \tag{2}$$

$$E_{zI}(r, z) = \frac{2}{\pi} \int_0^\infty \widetilde{E}_I(\zeta) R(\kappa r) \cos(\zeta z) d\zeta \tag{3}$$

where $\kappa = \sqrt{k^2 - \zeta^2}$, $k = \omega\sqrt{\mu\epsilon} = 2\pi/\lambda$, $R(\kappa r) = J_0(\kappa r)N_0(\kappa a) - N_0(\kappa r)J_0(\kappa a)$, and $R'(\kappa r) = dR(\kappa r)/d(\kappa r)$. The total field in region (II) $(b < r < c)$ of permittivity ϵ is

$$E_{zII}(r, z) = \sum_{m=0}^\infty R_0(\kappa_m r) \cos(a_m z) \tag{4}$$

where $R_0(\kappa_m r) = p_m \frac{H_0^{(1)}(\kappa_m r)}{H_1^{(1)}(\kappa_m b)} + q_m \frac{H_0^{(2)}(\kappa_m r)}{H_1^{(2)}(\kappa_m c)}$, $\kappa_m = \sqrt{k^2 - a_m^2}$, and $a_m = m\pi/d$. The scattered field in region (III) $(c < r)$ of permittivity ϵ_0 takes the form of

$$E_{zIII}(r, z) = \frac{1}{2\pi} \int_{-\infty}^\infty \widetilde{E}_{III}(\zeta) H_0^{(1)}(\xi r) e^{-i\zeta z} d\zeta \tag{5}$$

where $\xi = \sqrt{k_0^2 - \zeta^2}$ and $k_0 = \omega\sqrt{\mu\epsilon_0}$.

The tangential E-field continuity condition at $r = b$ requires

$$E_{zI}(b, z) = \begin{cases} E_{zII}(b, z), & -d < z < 0 \\ 0, & \text{otherwise} . \end{cases} \tag{6}$$

Applying the Fourier cosine transform to (6) yields

$$\widetilde{E}_I(\zeta) = \sum_{m=0}^\infty \frac{R_0(\kappa_m b)}{R(\kappa b)} \Xi_m(\zeta) \tag{7}$$

where

$$\Xi_m(\zeta) = \frac{\zeta(-1)^m \sin(\zeta d)}{(\zeta^2 - a_m^2)} . \tag{8}$$

The H_ϕ field continuity at $r = b$ for $-d < z < 0$ is

$$H_{\phi I}(b, z) + H_{\phi I}^i(b, z) = H_{\phi II}(b, z) . \tag{9}$$

We multiply (9) by $\cos(a_s z)$ $(s = 0, 1, 2, \dots)$ and integrate with respect to z from $-d$ to 0 to obtain

$$\sum_{m=0}^{\infty} \left[\frac{H_0^{(1)}(\kappa_m b)}{H_1^{(1)}(\kappa_m b)} I_1 + \frac{d}{2} \frac{\varepsilon_m}{\kappa_m} \delta_{ms} \right] p_m$$

$$+ \sum_{m=0}^{\infty} \left[\frac{H_0^{(2)}(\kappa_m b)}{H_1^{(2)}(\kappa_m b)} I_1 + \frac{d}{2} \frac{\varepsilon_m}{\kappa_m} \frac{H_1^{(2)}(\kappa_m b)}{H_1^{(2)}(\kappa_m c)} \delta_{ms} \right] q_m$$

$$= \frac{2\mathrm{i}}{kb} \Xi_s(k) \tag{10}$$

where

$$I_1 = \frac{2}{\pi} \int_0^{\infty} \frac{R'(\kappa b)}{\kappa R(\kappa b)} \Xi_s(\zeta) \Xi_m(\zeta) \mathrm{d}\zeta . \tag{11}$$

Using the residue calculus, we get

$$I_1 = \left. \frac{dR'(\kappa b)}{2\kappa R(\kappa b)} \right|_{\zeta = a_m} \varepsilon_m \delta_{ms}$$

$$- \sum_{n=1}^{\infty} \left. \frac{\mathrm{i}\zeta(-1)^{m+s}(1 - \mathrm{e}^{2\mathrm{i}\zeta d})}{b[1 - J_0^2(\kappa b)/J_0^2(\kappa a)](\zeta^2 - a_m^2)(\zeta^2 - a_s^2)} \right|_{\zeta = \zeta_n}$$

$$- \frac{\mathrm{i}k(-1)^{m+s}(1 - \mathrm{e}^{2ikd})}{2b \ln (b/a)(k^2 - a_m^2)(k^2 - a_s^2)} \tag{12}$$

where ζ_n is given by $R(\kappa b)|_{\zeta = \zeta_n} = 0$.

The E_z field continuity at $r = c$ is written as

$$E_{zIII}(c, z) = \begin{cases} E_{zII}(c, z), & -d < z < 0 \\ 0, & \text{otherwise} . \end{cases} \tag{13}$$

Applying the Fourier transform to (13) yields

$$\widetilde{E}_{III}(\zeta) = \sum_{m=0}^{\infty} \frac{R_0(\kappa_m c)}{H_0^{(1)}(\xi c)} G_m(\zeta) \tag{14}$$

where

$$G_m(\zeta) = \frac{-\mathrm{i}\zeta[1 - (-1)^m \mathrm{e}^{-\mathrm{i}\zeta d}]}{\zeta^2 - a_m^2} . \tag{15}$$

The H_ϕ field continuity at $r = c$ is given by

$$H_{\phi II}(c, z) = H_{\phi III}(c, z) . \tag{16}$$

We multiply (16) by $\cos(a_s z)$ and integrate with respect to z from $-d$ to 0 to get

$$\sum_{m=0}^{\infty} \left[\frac{H_0^{(1)}(\kappa_m c)}{H_1^{(1)}(\kappa_m b)} I_2 - \epsilon_r \frac{d}{2} \frac{\varepsilon_m}{\kappa_m} \frac{H_1^{(1)}(\kappa_m c)}{H_1^{(1)}(\kappa_m b)} \delta_{ms} \right] p_m$$

$$+ \sum_{m=0}^{\infty} \left[\frac{H_0^{(2)}(\kappa_m c)}{H_1^{(2)}(\kappa_m c)} I_2 - \epsilon_r \frac{d}{2} \frac{\varepsilon_m}{\kappa_m} \delta_{ms} \right] q_m = 0 \tag{17}$$

where

$$I_2 = \frac{1}{2\pi} \int_{-\infty}^{\infty} -\frac{H_0^{(1)\prime}(\xi c)}{\xi H_0^{(1)}(\xi c)} G_m(\zeta) G_s(-\zeta) \mathrm{d}\zeta \ . \tag{18}$$

It is convenient to transform I_2 into a numerically-efficient form by using the residue calculus. The evaluation of I_2 is summarized in Subsect. 9.1.2 Appendix.

The scattered field at $z = -\infty$ in region (I) is

$$H_{\phi I}(r, -\infty) = (1 + L_0) \frac{e^{-ikz}}{\eta r} + \sum_{n=1}^{\infty} L_n(\zeta) R'(\kappa r) e^{i\zeta z} \bigg|_{\zeta = -\zeta_n} \tag{19}$$

where η is the intrinsic impedance $\sqrt{\mu/\epsilon}$, the eigenvalue ζ_n is given by $R(\kappa b)|_{\zeta = \zeta_n} = 0$, and

$$L_0 = \frac{k \sin(kd)}{\ln(b/a)} \sum_{m=0}^{\infty} \frac{(-1)^m R_0(\kappa_m b)}{(k^2 - a_m^2)} \tag{20}$$

$$L_n(\zeta) = \frac{-2\omega\epsilon \sin(\zeta d)}{bR'(\kappa b)[1 - J_0^2(\kappa b)/J_0^2(\kappa a)]} \sum_{m=0}^{\infty} \frac{(-1)^m R_0(\kappa_m b)}{(\zeta^2 - a_m^2)} \ . \tag{21}$$

The reflected power at $z = -\infty$ is

$$P_r = \frac{1}{2} Re \left\{ \int_0^{2\pi} \int_a^b E_{rI}(r, z) H_{\phi I}^*(r, z) r \mathrm{d}r \mathrm{d}\phi \right\} \tag{22}$$

$$= \frac{\pi}{\eta} \ln(b/a) |1 + L_0|^2 \ . \tag{23}$$

The far-zone radiation field at distance R ($r = R \cos\theta_s$ and $z = R \sin\theta_s$) is

$$H_{\phi III}(R, \theta_s)$$
$$= \frac{e^{ik_0 R}}{R} \sum_{m=0}^{\infty} \frac{\omega\epsilon_0 R_0(\kappa_m c) \tan\theta_s [(-1)^m \exp(ik_0 d \sin\theta_s) - 1]}{\pi H_0^{(1)}(k_0 c \cos\theta_s)(k_0^2 \sin^2\theta_s - a_m^2)} \ . \tag{24}$$

9.1.2 Appendix

Consider

$$I_2 = \frac{1}{2\pi} \int_{-\infty}^{\infty} \frac{H_1^{(1)}(\xi c)\zeta^2 [1 - (-1)^m e^{-i\zeta d}][1 - (-1)^s e^{i\zeta d}]}{\xi H_0^{(1)}(\xi c)(\zeta^2 - a_m^2)(\zeta^2 - a_s^2)} \mathrm{d}\zeta \ . \tag{25}$$

When $m + s$ is odd, $I_2 = 0$. When $m + s$ is even, I_2 is rewritten as

$$I_2 = \frac{1}{2\pi} \int_{-\infty}^{\infty} \frac{2H_1^{(1)}(\xi c)\zeta^2 [1 - (-1)^m e^{i\zeta d}]}{\xi H_0^{(1)}(\xi c)(\zeta^2 - a_m^2)(\zeta^2 - a_s^2)} \mathrm{d}\zeta \ . \tag{26}$$

Let's evaluate I_2 along the contour path in the complex ζ-plane as shown in Fig. 1.2 of Sect. 1.1. The integrand has a pair of branch points corresponding to $\xi = 0$ and two simple poles at $\zeta = \pm a_m$ when $m = s$. Integrating along the deformed contour path $\Gamma_1, \Gamma_2, \Gamma_3$, and Γ_4 gives

$$I_2 = \frac{d}{2} \frac{H_1^{(1)}(\xi c)}{\xi H_0^{(1)}(\xi c)} \varepsilon_m \delta_{ms} \bigg|_{\zeta=a_m} + (I_3 + I_4) \tag{27}$$

where

$$I_3 = \frac{4(-1)^m}{\pi^2} \int_0^\infty$$
$$\frac{(1+iv)^2 e^{ik_0 d} e^{-k_0 v d}}{k_0 c [g(v)]^2 [(1+iv)^2 - (a_m/k_0)^2][(1+iv)^2 - (a_s/k_0)^2]}$$
$$\cdot \frac{1}{J_0^2[g(v)c] + N_0^2[g(v)c]} dv \tag{28}$$

$$I_4 = \frac{-4}{\pi^2} \int_0^\infty$$
$$\frac{(1+iv)^2}{k_0 c [g(v)]^2 [(1+iv)^2 - (a_m/k_0)^2][(1+iv)^2 - (a_s/k_0)^2]}$$
$$\cdot \frac{1}{J_0^2[g(v)c] + N_0^2[g(v)c]} dv \tag{29}$$

$$g(v) = k_0 \sqrt{v(-2i + v)} . \tag{30}$$

Note that $\varepsilon_m \dfrac{d}{2} \dfrac{H_1^{(1)}(\xi c)}{\xi H_0^{(1)}(\xi c)} \bigg|_{\zeta=a_m} \sim O(1/k_0)$ and $(I_3 + I_4) \sim O(1/k_0^2)$; hence

$$I_2 \approx \varepsilon_m \delta_{ms} \frac{d}{2} \frac{H_1^{(1)}(\xi c)}{\xi H_0^{(1)}(\xi c)} \bigg|_{\zeta=a_m} \quad \text{in high-frequency limit } (k_0 \to \infty).$$

9.2 EM Radiation from Apertures on a Shorted Coaxial Line [3]

In Sect. 9.1 the radiation characteristics of a dielectric-filled edge-slot (DFES) antenna were studied. In the present section we will investigate a multiple dielectric-filled edge-slot antenna that is made by circumferentially and multiply removing the outer conductor of a shorted coaxial line. Consider a TEM wave exciting a multiple dielectric-filled edge-slot antenna consisting of an N number of periodic circumferential slots on the outer conductor of a coaxial line. The field representations in regions (I) and (III) are the same as those in Sect. 9.1. The field in region (II) is represented as

$$H_{\phi II}(r, z) = \sum_{n=0}^{N-1} \sum_{m=0}^{\infty} \frac{i\omega\epsilon}{\kappa_m} R_0'(\kappa_m r) \cos a_m(z + nT)$$
$$\cdot [u(z + d + nT) - u(z + nT)] \tag{1}$$

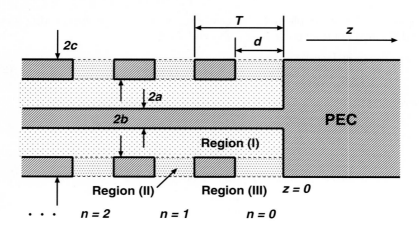

Fig. 9.2 Multiple circumferential apertures on a shorted coaxial line

where $R_0(\kappa_m r) = p_m^n \dfrac{H_0^{(1)}(\kappa_m r)}{H_1^{(1)}(\kappa_m b)} + q_m^n \dfrac{H_0^{(2)}(\kappa_m r)}{H_1^{(2)}(\kappa_m c)}$, $\kappa_m = \sqrt{k^2 - a_m^2}$, $k = \omega\sqrt{\mu\epsilon_0\epsilon_r}$, $a_m = m\pi/d$, and $u(\cdot)$ is a unit step function.

A procedure of applying the boundary conditions is somewhat similar to the dielectric-filled edge-slot antenna problem considered in Sect. 9.1, thus leading to

$$
\sum_{n=0}^{N-1}\sum_{m=0}^{\infty}\left[\frac{H_0^{(1)}(\kappa_m b)}{H_1^{(1)}(\kappa_m b)}X_{sm}^{pn} + \frac{d}{2}\frac{\varepsilon_m}{\kappa_m}\delta_{ms}\delta_{np}\right]p_m^n
$$
$$
+\sum_{n=0}^{N-1}\sum_{m=0}^{\infty}\left[\frac{H_0^{(2)}(\kappa_m b)}{H_1^{(2)}(\kappa_m c)}X_{sm}^{pn} + \frac{d}{2}\frac{\varepsilon_m}{\kappa_m}\frac{H_1^{(2)}(\kappa_m b)}{H_1^{(2)}(\kappa_m c)}\delta_{ms}\delta_{np}\right]q_m^n
$$
$$
= \frac{2\mathrm{i}}{kb}\Xi_s^p(k)
$$
(2)

$$
\sum_{n=0}^{N-1}\sum_{m=0}^{\infty}\left[\frac{H_0^{(1)}(\kappa_m c)}{H_1^{(1)}(\kappa_m b)}Y_{sm}^{pn} - \epsilon_r\frac{d}{2}\frac{\varepsilon_m}{\kappa_m}\frac{H_1^{(1)}(\kappa_m c)}{H_1^{(1)}(\kappa_m b)}\delta_{ms}\delta_{np}\right]p_m^n
$$
$$
+\sum_{n=0}^{N-1}\sum_{m=0}^{\infty}\left[\frac{H_0^{(2)}(\kappa_m c)}{H_1^{(2)}(\kappa_m c)}Y_{sm}^{pn} - \epsilon_r\frac{d}{2}\frac{\varepsilon_m}{\kappa_m}\delta_{ms}\delta_{np}\right]q_m^n = 0
$$
(3)

where δ_{ms} is the Kronecker delta, $\varepsilon_m = 2$ ($m = 0$), 1 ($m = 1, 2, \ldots$), and

$$X_{sm}^{pn} = \frac{2}{\pi} \int_0^\infty \frac{R'(\kappa b)}{\kappa R(\kappa b)} \Xi_s^p(\zeta) \Xi_m^n(\zeta) \mathrm{d}\zeta \tag{4}$$

$$Y_{sm}^{pn} = \frac{1}{2\pi} \int_{-\infty}^\infty \frac{H_1^{(1)}(\xi c)}{\xi H_0^{(1)}(\xi c)} G_m^n(\zeta) G_s^p(-\zeta) \mathrm{d}\zeta \tag{5}$$

$$\Xi_m^n(\zeta) = \frac{1}{2} \left[G_m^n(\zeta) + G_m^n(-\zeta) \right] \tag{6}$$

$$G_m^n(\zeta) = \frac{-\mathrm{i}\zeta[1 - (-1)^m \mathrm{e}^{-\mathrm{i}\zeta d}]}{\zeta^2 - a_m^2} \, \mathrm{e}^{-\mathrm{i}\zeta nT} \, . \tag{7}$$

Note that $\xi = \sqrt{k_0^2 - \zeta^2}$ and $k_0 = \omega\sqrt{\mu\epsilon_0}$ is the wavenumber in region (III). When $N = 1$, (2) and (3) reduce to the single-aperture case considered in Sect. 9.1.

The reflection coefficient Γ_{in} from a feeding point at $z = -(d + NT)$ is related to

$$\Gamma_{in} = \left. \frac{E_{rI}(r,z)}{E_{rI}^i(r,z)} \right|_{z=-(d+NT)} = -(1 + L_0^n) \exp[2\mathrm{i}k(d + NT)] \tag{8}$$

$$L_0^n = \sum_{n=0}^{N-1} \sum_{m=0}^\infty \frac{kR_0(\kappa_m b)}{\ln(b/a)(k^2 - a_m^2)}$$
$$\cdot [(-1)^m \sin k(d + nT) - \sin(knT)] \, . \tag{9}$$

The far-zone radiation field at distance R is

$$H_{\phi III}(R, \theta_s) = \sum_{n=0}^{N-1} \sum_{m=0}^\infty \frac{\mathrm{i}R_0(\kappa_m c) G_m^n(-k_0 \sin\theta_s)}{\pi \cos\theta_s H_0^{(1)}(k_0 c \cos\theta_s)} \frac{\mathrm{e}^{\mathrm{i}k_0 R}}{\eta_0 R} \tag{10}$$

where $\eta_0 = \sqrt{\mu/\epsilon_0}$ is the intrinsic impedance of free-space and the observation point is at $z = R\sin\theta_s$ and $r = R\cos\theta_s$.

9.3 EM Radiation from Apertures on a Coaxial Line [4]

Electromagnetic wave radiation from slots on a coaxial line is of practical interest in antenna and electromagnetic interference problems [5]. In this section we will analyze electromagnetic wave radiation from a finite N number of circumferential periodic slots (apertures) on a coaxial line. A scattering analysis in this section is an extension of the single-aperture problem discussed in Sect. 9.1. An incident TEM wave is assumed to propagate along a slotted coaxial line. In region (I) ($a < r < b$ and wavenumber: k) the incident and scattered H-fields are

$$H_{\phi I}^i(r, z) = \frac{\mathrm{e}^{\mathrm{i}kz}}{\eta r} \tag{1}$$

$$H_{\phi I}(r, z) = \frac{\mathrm{i}\omega\epsilon}{2\pi} \int_{-\infty}^\infty \frac{1}{\kappa} \widetilde{E}_I(\zeta) R'(\kappa r) \mathrm{e}^{-\mathrm{i}\zeta z} \mathrm{d}\zeta \, . \tag{2}$$

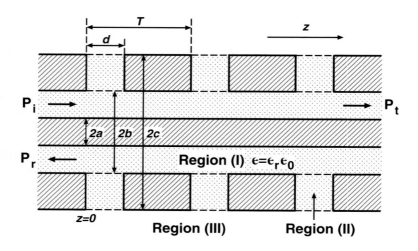

Fig. 9.3. Multiple apertures on a coaxial line

In region (III) $(c < r$ and wavenumber: $k_0)$ the scattered H-field is

$$H_{\phi III}(r, z) = \frac{-i\omega\epsilon_0}{2\pi} \int_{-\infty}^{\infty} \frac{1}{\xi} \widetilde{E}_{III}(\zeta) H_1^{(1)}(\xi r) e^{-i\zeta z} d\zeta \tag{3}$$

where $\kappa = \sqrt{k^2 - \zeta^2}$, $\eta = \sqrt{\mu/\epsilon}$, $R(\kappa r) = J_0(\kappa r)N_0(\kappa a) - N_0(\kappa r)J_0(\kappa a)$, $k = \omega\sqrt{\mu\epsilon} = \omega\sqrt{\mu\epsilon_r\epsilon_0}$, $R'(\kappa r) = dR(\kappa r)/d(\kappa r)$, $\xi = \sqrt{k_0^2 - \zeta^2}$, and $k_0 = \omega\sqrt{\mu\epsilon_0}$. In region (II) $(b < r < c$ and wavenumber: $k)$ the H-field is

$$H_{\phi II}(r, z) = \sum_{n=0}^{N-1} \sum_{m=0}^{\infty} \frac{i\omega\epsilon}{\kappa_m} R_0'(\kappa_m r) \cos a_m (z - nT)$$
$$\cdot [u(z - nT) - u(z - d - nT)] \tag{4}$$

where $R_0(\kappa_m r) = p_m^n \dfrac{H_0^{(1)}(\kappa_m r)}{H_1^{(1)}(\kappa_m b)} + q_m^n \dfrac{H_0^{(2)}(\kappa_m r)}{H_1^{(2)}(\kappa_m c)}$, $\kappa_m = \sqrt{k^2 - a_m^2}$, $a_m = m\pi/d$, and $u(.)$ is a unit step function.

The procedure for matching the boundary conditions is somewhat similar to that in Sect. 9.1. By enforcing the boundary conditions on the tangential E and H-field continuities at $r = b$ and c, we obtain

$$\sum_{n=0}^{N-1}\sum_{m=0}^{\infty}\left[\frac{H_0^{(1)}(\kappa_m b)}{H_1^{(1)}(\kappa_m b)}I_1 + \frac{d}{2}\frac{\varepsilon_m}{\kappa_m}\delta_{ml}\delta_{np}\right]p_m^n$$

$$+\sum_{n=0}^{N-1}\sum_{m=0}^{\infty}\left[\frac{H_0^{(2)}(\kappa_m b)}{H_1^{(2)}(\kappa_m c)}I_1 + \frac{d}{2}\frac{\varepsilon_m}{\kappa_m}\frac{H_1^{(2)}(\kappa_m b)}{H_1^{(2)}(\kappa_m c)}\delta_{ml}\delta_{np}\right]q_m^n$$

$$= \frac{i}{kb}G_l^p(k) \tag{5}$$

$$\sum_{n=0}^{N-1}\sum_{m=0}^{\infty}\left[\frac{H_0^{(1)}(\kappa_m c)}{H_1^{(1)}(\kappa_m b)}I_2 - \epsilon_r\frac{d}{2}\frac{\varepsilon_m}{\kappa_m}\frac{H_1^{(1)}(\kappa_m c)}{H_1^{(1)}(\kappa_m b)}\delta_{ml}\delta_{np}\right]p_m^n$$

$$+\sum_{n=0}^{N-1}\sum_{m=0}^{\infty}\left[\frac{H_0^{(2)}(\kappa_m c)}{H_1^{(2)}(\kappa_m c)}I_2 - \epsilon_r\frac{d}{2}\frac{\varepsilon_m}{\kappa_m}\delta_{ml}\delta_{np}\right]q_m^n = 0 \tag{6}$$

where

$$I_1 = \frac{1}{2\pi}\int_{-\infty}^{\infty}\frac{R'(\kappa b)}{\kappa R(\kappa b)}G_m^n(\zeta)G_l^p(-\zeta)\mathrm{d}\zeta \tag{7}$$

$$I_2 = \frac{1}{2\pi}\int_{-\infty}^{\infty}\frac{H_1^{(1)}(\xi c)}{\xi H_0^{(1)}(\xi c)}G_m^n(\zeta)G_l^p(-\zeta)\mathrm{d}\zeta \tag{8}$$

$$G_m^n(\zeta) = \frac{-i\zeta[(-1)^m e^{i\zeta d}-1]}{\zeta^2 - a_m^2}e^{i\zeta nT}\,. \tag{9}$$

It is convenient to transform I_1 and I_2 into numerically-efficient forms by utilizing the residue calculus.

The scattered field at $z = \pm\infty$ in region (I) is

$$H_{\phi I}(r, \pm\infty) = L_0^{\pm}\frac{e^{\pm ikz}}{\eta r} + \left.\sum_{j=1}^{\infty}L_j^{\pm}(\zeta)R'(\kappa r)e^{\pm i\zeta z}\right|_{\zeta=\zeta_j} \tag{10}$$

where ζ_j is determined by $R(\kappa b) = 0$ and

$$L_0^{\pm} = \sum_{n=0}^{N-1}\sum_{m=0}^{\infty}\frac{R_0(\kappa_m b)G_m^n(\mp k)}{2\ln(b/a)} \tag{11}$$

$$L_j^{\pm}(\zeta) = \mp\sum_{n=0}^{N-1}\sum_{m=0}^{\infty}\frac{i\omega\epsilon R_0(\kappa_m b)[1-(-1)^m e^{\mp i\zeta d}]}{bR'(\kappa b)[1-J_0^2(\kappa b)/J_0^2(\kappa a)](\zeta^2-a_m^2)}e^{\mp i\zeta nT}\,. \tag{12}$$

The reflection (P_r/P_i) and transmission (P_t/P_i) coefficients are $|L_0^-|^2$ and $|1+L_0^+|^2$, respectively. The far-zone radiation field at distance R ($z = R\cos\theta$ and $r = R\sin\theta$) is

$$H_{\phi III}(R, \theta)$$

$$= \sum_{n=0}^{N-1} \sum_{m=0}^{\infty} \frac{e^{ik_0 R}}{\eta_0 R}$$

$$\cdot \frac{k_0 R_0(\kappa_m c)\left[1 - (-1)^m \exp(-ik_0 d \cos \theta)\right] \exp(-ik_0 n T \cos \theta)}{\pi \tan \theta \; H_0^{(1)}(k_0 c \sin \theta)\left[(k_0 \cos \theta)^2 - a_m^2\right]} \tag{13}$$

where $\eta_0 = \sqrt{\mu/\epsilon_0}$ is the intrinsic impedance of free-space.

9.4 EM Radiation from Apertures on a Coaxial Line with a Cover [6]

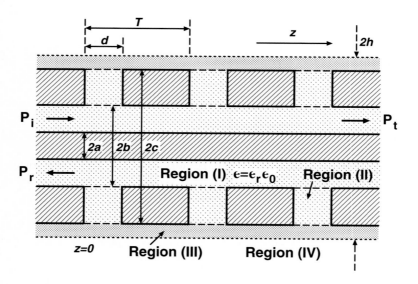

Fig. 9.4. Multiple apertures on a coaxial line with a dielectric cover

In Sect. 9.3 we have analyzed electromagnetic wave radiation from apertures on a coaxial line. In this section we will present an analysis of radiation from multiple circumferential apertures (slots) on a coaxial line covered with a dielectric layer. A similar scattering analysis for apertures on a dielectric-covered coaxial line was performed in [7–9]. A TEM wave is incident from left along a slotted coaxial line with a dielectric cover. Regions (I), (II), (III), and (IV) denote a coaxial line interior ($a < r < b$), apertures in the outer conductor ($b < r < c$), a dielectric layer ($c < r < h$), and a free space, respectively. The incident field components are assumed to be $E^i_{rI}(r,z) = \dfrac{e^{ik_1 z}}{r}$ and $H^i_{\phi I}(r,z) = \dfrac{e^{ik_1 z}}{\eta_1 r}$ where $k_1 = \omega\sqrt{\mu\epsilon_0\epsilon_{r1}}$ and $\eta_1 = \sqrt{\mu/(\epsilon_0\epsilon_{r1})}$. The scattered fields in regions (I), (II), (III), and (IV) are

$$H_{\phi I}(r,z) = \frac{i\omega\epsilon_1}{2\pi}\int_{-\infty}^{\infty}\frac{1}{\kappa_1}\widetilde{E}_I(\zeta)R'(\kappa_1 r)e^{-i\zeta z}d\zeta \tag{1}$$

$$H_{\phi II}(r,z) = \sum_{n=0}^{N-1}\sum_{m=0}^{\infty}\frac{i\omega\epsilon_2}{\kappa_m^0}R_0'(\kappa_m^0 r)\cos\frac{m\pi}{d}(z-nT)$$
$$\cdot\,[u(z-nT)-u(z-d-nT)] \tag{2}$$

$$H_{\phi III}(r,z) = \frac{-i\omega\epsilon_3}{2\pi}\int_{-\infty}^{\infty}\frac{1}{\kappa_3}\left[\widetilde{E}^+_{III}(\zeta)H_1^{(1)}(\kappa_3 r)+\widetilde{E}^-_{III}(\zeta)H_1^{(2)}(\kappa_3 r)\right]$$
$$\cdot\,e^{-i\zeta z}d\zeta \tag{3}$$

$$H_{\phi IV}(r,z) = \frac{-i\omega\epsilon_4}{2\pi}\int_{-\infty}^{\infty}\frac{1}{\kappa_4}\widetilde{E}_{IV}(\zeta)H_1^{(1)}(\kappa_4 r)e^{-i\zeta z}d\zeta \tag{4}$$

where $R(\kappa_1 r) = J_0(\kappa_1 r)N_0(\kappa_1 a) - N_0(\kappa_1 r)J_0(\kappa_1 a)$, $\kappa_i = \sqrt{k_i^2 - \zeta^2}$, $k_i = \omega\sqrt{\mu\epsilon_0\epsilon_{ri}}$ for $i = 1,2,3,4$. Note $R_0(\kappa_m^0 r) = p_m^n\dfrac{H_0^{(1)}(\kappa_m^0 r)}{H_1^{(1)}(\kappa_m^0 b)} + q_m^n\dfrac{H_0^{(2)}(\kappa_m^0 r)}{H_1^{(2)}(\kappa_m^0 c)}$, $\kappa_m^0 = \sqrt{k_2^2 - (m\pi/d)^2}$, and $u(\cdot)$ is a unit step function. The prime denotes differentiation with respect to an argument. Note that d denotes an aperture width and T is a distance between two adjacent apertures.

Following the Fourier transform procedure of matching the boundary conditions as was done in Sect. 9.3, we obtain

$$\sum_{n=0}^{N-1}\sum_{m=0}^{\infty}\left[\frac{H_0^{(1)}(\kappa_m^0 b)}{H_1^{(1)}(\kappa_m^0 b)}\varPhi_{ml}^{np} + \frac{\epsilon_{r2}}{\epsilon_{r1}}\frac{d}{2}\frac{\epsilon_m}{\kappa_m^0}\delta_{ml}\delta_{np}\right]p_m^n$$
$$+\sum_{n=0}^{N-1}\sum_{m=0}^{\infty}\left[\frac{H_0^{(2)}(\kappa_m^0 b)}{H_1^{(2)}(\kappa_m^0 c)}\varPhi_{ml}^{np} + \frac{\epsilon_{r2}}{\epsilon_{r1}}\frac{d}{2}\frac{\epsilon_m}{\kappa_m^0}\frac{H_1^{(2)}(\kappa_m^0 b)}{H_1^{(2)}(\kappa_m^0 c)}\delta_{ml}\delta_{np}\right]q_m^n$$
$$= \frac{i}{k_1 b}G_l^p(k_1) \tag{5}$$

$$\sum_{n=0}^{N-1}\sum_{m=0}^{\infty}\left[\frac{H_0^{(1)}(\kappa_m^0 c)}{H_1^{(1)}(\kappa_m^0 b)}\varPsi_{ml}^{np} - \frac{\epsilon_{r2}}{\epsilon_{r3}}\frac{d}{2}\frac{\epsilon_m}{\kappa_m^0}\frac{H_1^{(1)}(\kappa_m^0 c)}{H_1^{(1)}(\kappa_m^0 b)}\delta_{ml}\delta_{np}\right]p_m^n$$
$$+\sum_{n=0}^{N-1}\sum_{m=0}^{\infty}\left[\frac{H_0^{(2)}(\kappa_m^0 c)}{H_1^{(2)}(\kappa_m^0 c)}\varPsi_{ml}^{np} - \frac{\epsilon_{r2}}{\epsilon_{r3}}\frac{d}{2}\frac{\epsilon_m}{\kappa_m^0}\delta_{ml}\delta_{np}\right]q_m^n = 0 \tag{6}$$

where $\varepsilon_m = 2\ (m = 0),\ 1\ (m = 1, 2, \dots)$, δ_{ml} is the Kronecker delta, and

$$\Phi_{ml}^{np} = \frac{1}{2\pi} \int_{-\infty}^{\infty} \frac{R'(\kappa_1 b)}{\kappa_1 R(\kappa_1 b)} G_m^n(\zeta) G_l^p(-\zeta) d\zeta \tag{7}$$

$$\Psi_{ml}^{np} = \frac{1}{2\pi} \int_{-\infty}^{\infty} \frac{A(\zeta) R_1^{(1)}(\kappa_3 h) + R_1^{(1)'}(\kappa_3 h)}{\kappa_3 \left[A(\zeta) R_1^{(0)}(\kappa_3 h) + R_1^{(0)'}(\kappa_3 h) \right]}$$
$$\cdot G_m^n(\zeta) G_l^p(-\zeta) d\zeta \tag{8}$$

$$R_1^{(i)}(\kappa_3 r) = H_0^{(1)}(\kappa_3 r) H_i^{(2)}(\kappa_3 c) - H_0^{(2)}(\kappa_3 r) H_i^{(1)}(\kappa_3 c), \quad i = 0, 1 \tag{9}$$

$$A(\zeta) = \frac{\epsilon_{r4} \kappa_3 H_1^{(1)}(\kappa_4 h)}{\epsilon_{r3} \kappa_4 H_0^{(1)}(\kappa_4 h)} \tag{10}$$

$$G_m^n(\zeta) = \frac{-i\zeta[(-1)^m e^{i\zeta d} - 1]}{\zeta^2 - (m\pi/d)^2} e^{i\zeta nT} . \tag{11}$$

The scattered fields at $z = \pm\infty$ in regions (I) and (III) are

$$H_{\phi I}(r, \pm\infty) = \sum_{n=0}^{N-1} \sum_{m=0}^{\infty} \frac{R_0(\kappa_m^0 b) G_m^n(\mp k_1)}{2\ln(b/a)} \frac{e^{\pm ik_1 z}}{\eta_1 r}$$
$$+ \sum_{j=1}^{\infty} L_j^{\pm}(\zeta) R'(\kappa_1 r) e^{\pm i\zeta z} \Big|_{\zeta = \zeta_j} \tag{12}$$

$$H_{\phi III}(r, \pm\infty) = \pm\omega\epsilon_3 \sum_{n=0}^{N-1} \sum_{m=0}^{\infty} \sum_{s=1}^{\infty} \frac{R_0(\kappa_m^0 c)}{\frac{\partial \Delta}{\partial \zeta}}$$
$$\cdot T(\kappa_3 r) G_m^n(\zeta) e^{-i\zeta z} \Big|_{\zeta = \mp\zeta_s} \tag{13}$$

where ζ_j and ζ_s are determined by $R(\kappa_1 b) = 0$ and $\Delta = 0$, respectively, and

$$L_j^{\pm}(\zeta) = \sum_{n=0}^{N-1} \sum_{m=0}^{\infty} \frac{\omega\epsilon_1 R_0(\kappa_m^0 b) G_m^n(\mp\zeta)}{b\zeta R'(\kappa_1 b)[1 - J_0^2(\kappa_1 b)/J_0^2(\kappa_1 a)]} \tag{14}$$

$$\Delta = \kappa_3 \left[A(\zeta) R_1^{(0)}(\kappa_3 h) + R_1^{(0)'}(\kappa_3 h) \right] \tag{15}$$

$$T(\kappa_3 r) = [H_1^{(1)}(\kappa_3 h) - A(\zeta) H_0^{(1)}(\kappa_3 h)] H_1^{(2)}(\kappa_3 r)$$
$$- [H_1^{(2)}(\kappa_3 h) - A(\zeta) H_0^{(2)}(\kappa_3 h)] H_1^{(1)}(\kappa_3 r) . \tag{16}$$

The solution at $\zeta = \pm\zeta_s$ represents a contribution from the surface wave propagating in the $\pm z$ directions. The reflection coefficient at $z = 0$ is

$$\Gamma_{in} = \sum_{n=0}^{N-1} \sum_{m=0}^{\infty} \frac{R_0(\kappa_m^0 b) G_m^n(k_1)}{2\ln(b/a)} . \tag{17}$$

The far-zone radiation field at distance R ($z = R\cos\theta$ and $r = R\sin\theta$) is

$$H_{\phi IV}(R,\theta) = \sum_{n=0}^{N-1} \sum_{m=0}^{\infty} \frac{2R_0(\kappa_m^0 c)G_m^n(-k_4\cos\theta)}{\pi^2 g(\theta)\tan\theta H_0^{(1)}(k_4 h\sin\theta)} \frac{e^{ik_4 R}}{\eta_4 R}$$

$$\cdot \left[A(-k_4\cos\theta)R_1^{(0)}(g(\theta)) + R_1^{(0)'}(g(\theta)) \right]^{-1} \tag{18}$$

where $g(\theta) = h\sqrt{k_3^2 - k_4^2\cos^2\theta}$ and $\eta_4 = \sqrt{\mu/(\epsilon_0\epsilon_{r4})}$ is the intrinsic impedance of free space.

9.5 EM Radiation from Apertures on a Circular Cylinder [10]

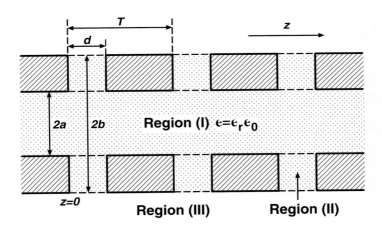

Fig. 9.5. Multiple apertures on a conducting circular cylinder

Electromagnetic wave radiation from circumferential apertures (slots) on a conducting cylinder was studied in [11-16] and applied to antenna and interference problems. A study of radiation from multiple apertures on a cylinder is useful for the design of microwave and millimeter wave array antenna of cylindrical structure. In this section we will analyze radiation from a finite (N) number of circumferential apertures on a conducting cylinder.

9.5.1 TE Radiation

Consider electromagnetic wave radiation from periodic circumferential apertures on a thick circular conducting cylinder. The incident TE_{01} wave is assumed to propagate along the cylinder. The scattered E-field has no variation in the ϕ-direction due to a circular symmetry of the problem geometry. In region (I) ($r < a$) the total E-field is a sum of the incident and scattered components

$$E^i_{\phi I}(r, z) = \frac{i\omega\mu}{k_c} J_1(k_c r) e^{i\beta_z z} \tag{1}$$

$$E_{\phi I}(r, z) = \frac{i\omega\mu}{2\pi} \int_{-\infty}^{\infty} \frac{1}{\kappa} \widetilde{E}_I(\zeta) J_1(\kappa r) e^{-i\zeta z} d\zeta \tag{2}$$

where $k_c = 3.832/a$, $\beta_z = \sqrt{\beta^2 - k_c^2}$, $\beta = \omega\sqrt{\mu\epsilon_r\epsilon_0} = 2\pi/\lambda$, and $\kappa = \sqrt{\beta^2 - \zeta^2}$. In region (II) ($a < r < b$) the E-field is

$$E_{\phi II}(r, z) = \sum_{n=0}^{N-1} \sum_{k=1}^{\infty} \frac{-i\omega\mu}{\kappa_k} Q'(\kappa_k r) \sin a_k(z - nT)$$
$$\cdot [u(z - nT) - u(z - d - nT)] \tag{3}$$

where $Q(\kappa_k r) = r_k^n \dfrac{H_0^{(1)}(\kappa_k r)}{H_1^{(1)}(\kappa_k a)} + s_k^n \dfrac{H_0^{(2)}(\kappa_k r)}{H_1^{(2)}(\kappa_k b)}$ and the prime denotes differentiation with respect to the argument $\kappa_k r$. Furthermore note that $\kappa_k = \sqrt{\beta^2 - a_k^2}$, $a_k = k\pi/d$, and $u(\cdot)$ is a unit step function. In region (III) ($r > b$) the E-field is

$$E_{\phi III}(r, z) = \frac{i\omega\mu}{2\pi} \int_{-\infty}^{\infty} \frac{1}{\xi} \widetilde{E}_{III}(\zeta) H_1^{(1)}(\xi r) e^{-i\zeta z} d\zeta \tag{4}$$

where $\xi = \sqrt{\beta_0^2 - \zeta^2}$ and $\beta_0 = \omega\sqrt{\mu\epsilon_0}$.

The E_ϕ field continuity at $r = a$ is given by

$$E_{\phi I}(a, z) = \begin{cases} E_{\phi II}(a, z), & nT < z < d + nT \\ 0, & \text{otherwise} . \end{cases} \tag{5}$$

Taking the Fourier transform of (5) yields

$$\widetilde{E}_I(\zeta) = -\sum_{n=0}^{N-1} \sum_{k=1}^{\infty} \frac{Q'(\kappa_k a)}{\kappa_k} \frac{\kappa \Xi_k^n(\zeta)}{J_1(\kappa a)} \tag{6}$$

where

$$\Xi_k^n(\zeta) = \frac{a_k[(-1)^k e^{i\zeta d} - 1]}{\zeta^2 - a_k^2} e^{i\zeta nT} . \tag{7}$$

We multiply the H_z field continuity, $H^i_{zI}(a, z) + H_{zI}(a, z) = H_{zII}(a, z)$, by $\sin a_l(z - pT)$ and integrate with respect to z from pT to $d + pT$ to get

$$\sum_{n=0}^{N-1} \sum_{k=1}^{\infty} \left[\frac{I_1}{\kappa_k} - \frac{d}{2} \frac{H_0^{(1)}(\kappa_k a)}{H_1^{(1)}(\kappa_k a)} \delta_{kl} \delta_{np} \right] r_k^n$$

$$+ \sum_{n=0}^{N-1} \sum_{k=1}^{\infty} \left[\frac{H_1^{(2)}(\kappa_k a)}{H_1^{(2)}(\kappa_k b)} \frac{I_1}{\kappa_k} - \frac{d}{2} \frac{H_0^{(2)}(\kappa_k a)}{H_1^{(2)}(\kappa_k b)} \delta_{kl} \delta_{np} \right] s_k^n$$

$$= -J_0(k_c a) \Xi_l^p(\beta_z) \tag{8}$$

where

$$I_1 = \frac{1}{2\pi} \int_{-\infty}^{\infty} \frac{\kappa J_0(\kappa a)}{J_1(\kappa a)} \Xi_k^n(\zeta) \Xi_l^p(-\zeta) \mathrm{d}\zeta . \tag{9}$$

Using the residue calculus, we transform I_1 into a fast convergent series

$$I_1 = \frac{d}{2} \frac{\kappa J_0(\kappa a)}{J_1(\kappa a)} \delta_{kl} \delta_{np} \bigg|_{\zeta=a_k}$$

$$- a_k a_l \sum_{j=1}^{\infty} \mathrm{i}\kappa^2 \frac{A_1}{a\zeta(\zeta^2 - a_k^2)(\zeta^2 - a_l^2)} \bigg|_{\zeta=\zeta_j} \tag{10}$$

$$A_1 = [(-1)^{k+l} + 1] \exp(\mathrm{i}\zeta|n - p|T)$$

$$- (-1)^k \exp[\mathrm{i}\zeta|d + (n - p)T|]$$

$$- (-1)^l \exp[\mathrm{i}\zeta|(n - p)T - d|] \tag{11}$$

where ζ_j are the roots of $J_1(\kappa a)|_{\zeta=\zeta_j} = 0$. In addition, the tangential field continuities at $r = b$ result in

$$\sum_{n=0}^{N-1} \sum_{k=1}^{\infty} \left[\frac{H_1^{(1)}(\kappa_k b)}{H_1^{(1)}(\kappa_k a)} \frac{I_2}{\kappa_k} - \frac{d}{2} \frac{H_0^{(1)}(\kappa_k b)}{H_1^{(1)}(\kappa_k a)} \delta_{kl} \delta_{np} \right] r_k^n$$

$$+ \sum_{n=0}^{N-1} \sum_{k=1}^{\infty} \left[\frac{I_2}{\kappa_k} - \frac{d}{2} \frac{H_0^{(2)}(\kappa_k b)}{H_1^{(2)}(\kappa_k b)} \delta_{kl} \delta_{np} \right] s_k^n = 0 \tag{12}$$

where

$$I_2 = \frac{1}{2\pi} \int_{-\infty}^{\infty} \frac{\xi H_0^{(1)}(\xi b)}{H_1^{(1)}(\xi b)} \Xi_k^n(\zeta) \Xi_l^p(-\zeta) \mathrm{d}\zeta . \tag{13}$$

Using the contour integral technique shown in Fig. 1.2 of Sect. 1.1, we transform I_2 into a numerically-efficient integral

$$I_2 = \frac{d}{2} \frac{\xi H_0^{(1)}(\xi b)}{H_1^{(1)}(\xi b)} \delta_{kl} \delta_{np} \bigg|_{\zeta=a_k} - a_k a_l \Big\{ [1 + (-1)^{k+l}] I_3(gT)$$

$$- (-1)^k I_3(d + gT) - (-1)^l I_3(-d + gT) \Big\} \tag{14}$$

where $g = n - p$ and

$$I_3(x) = -\frac{2}{\pi^2} \int_0^\infty \frac{e^{i\beta_0|x|}e^{-\beta_0 v|x|}}{\beta_0^3 b[(1+iv)^2 - (a_k/\beta_0)^2][(1+iv)^2 - (a_l/\beta_0)^2]}$$
$$\cdot \frac{1}{J_1^2[\beta_0 b\sqrt{v(-2i+v)}] + N_1^2[\beta_0 b\sqrt{v(-2i+v)}]} dv \ . \tag{15}$$

The scattered field at $z = \pm\infty$ in region (I) is

$$E_{\phi I}(r, \pm\infty) = \sum_{j=1}^\infty L_j^\pm(\zeta)\frac{i\omega\mu}{\kappa}J_1(\kappa r)e^{\pm i\zeta z}\bigg|_{\zeta=\zeta_j} \tag{16}$$

where

$$L_j^\pm(\zeta) = \sum_{n=0}^{N-1}\sum_{k=1}^\infty \frac{ia_k Q'(\kappa_k a)}{\kappa_k a}\frac{\kappa^2[(-1)^k e^{\mp i\zeta d} - 1]}{\zeta J_0(\kappa a)(\zeta^2 - a_k^2)}e^{\mp i\zeta nT} \tag{17}$$

and ζ_j is determined by $J_1(\kappa a)|_{\zeta=\zeta_j} = 0$. The reflection ($\varrho$), transmission ($\tau$), and scattering ($\sigma$) coefficients are

$$\tau = \frac{P_t}{P_i}$$
$$= |1 + L_j^+(\zeta_j)|_{\zeta_j=\beta_z}^2 + \sum_{\zeta_j\neq\beta_z}\frac{k_c^2 Re(\zeta)J_0^2(\kappa a)}{\kappa^2 Re(\beta_z)J_0^2(k_c a)}|L_j^+(\zeta)|^2\bigg|_{\zeta=\zeta_j} \tag{18}$$

$$\sigma = \frac{P_s}{P_i} \tag{19}$$

$$\varrho = \frac{P_r}{P_i} = \sum_{j=1}^\infty \frac{k_c^2 Re(\zeta)J_0^2(\kappa a)}{\kappa^2 Re(\beta_z)J_0^2(k_c a)}|L_j^-(\zeta)|^2\bigg|_{\zeta=\zeta_j} \tag{20}$$

where P_i, P_t, P_s, and P_r are the incident, transmitted, scattered (radiated through apertures), and reflected powers, respectively, with

$$P_i = \frac{\pi a^2\omega\mu}{2k_c^2}J_0^2(k_c a)Re(\beta_z) \tag{21}$$

$$P_s = \frac{\pi bd}{2}Re\left\{\sum_{n=0}^{N-1}\sum_{k=1}^\infty \frac{-i\omega\mu}{\kappa_k}Q'(\kappa_k b)Q^*(\kappa_k b)\right\} \ . \tag{22}$$

The power conservation stipulates $\tau + \sigma + \varrho = 1$. The far-zone radiation field at distance R ($z = R\cos\theta$ and $r = R\sin\theta$) is

$$E_{\phi III}(R, \theta)$$
$$= \frac{e^{i\beta_0 R}}{R}\sum_{n=0}^{N-1}\sum_{k=1}^\infty \frac{Q'(\kappa_k b)}{\pi\kappa_k}$$
$$\cdot\left\{\frac{i\omega\mu a_k[(-1)^k\exp(-i\beta_0 d\cos\theta) - 1]\exp(-i\beta_0 nT\cos\theta)}{H_1^{(1)}(\beta_0 b\sin\theta)[(\beta_0\cos\theta)^2 - a_k^2]}\right\} \ . \tag{23}$$

9.5.2 TM Radiation

Consider an incident TM_{01} wave traveling along a circular cylinder. The H-field in region (I) $(r < a)$ consists of

$$H_{\phi I}^i(r,z) = \frac{-i\omega\epsilon}{k_c} J_1(k_c r) e^{i\beta_z z} \tag{24}$$

$$H_{\phi I}(r,z) = \frac{-i\omega\epsilon}{2\pi} \int_{-\infty}^{\infty} \frac{1}{\kappa} \widetilde{E}_I(\zeta) J_1(\kappa r) e^{-i\zeta z} d\zeta \tag{25}$$

where $k_c = 2.405/a$, $\beta_z = \sqrt{\beta^2 - k_c^2}$, $\beta = \omega\sqrt{\mu\epsilon_r\epsilon_0}$, and $\kappa = \sqrt{\beta^2 - \zeta^2}$. In region (II) $(a < r < b)$ the H-field is

$$H_{\phi II}(r,z) = \sum_{n=0}^{N-1} \sum_{m=0}^{\infty} \frac{i\omega\epsilon}{\kappa_m} R_0'(\kappa_m r) \cos a_m(z - nT)$$
$$\cdot [u(z - nT) - u(z - d - nT)] \tag{26}$$

where $R_0(\kappa_m r) = p_m^n \dfrac{H_0^{(1)}(\kappa_m r)}{H_1^{(1)}(\kappa_m a)} + q_m^n \dfrac{H_0^{(2)}(\kappa_m r)}{H_1^{(2)}(\kappa_m b)}$, $\kappa_m = \sqrt{\beta^2 - a_m^2}$, and $a_m = m\pi/d$. In region (III) $(r > b)$ the scattered field takes the form of

$$H_{\phi III}(r,z) = \frac{-i\omega\epsilon_0}{2\pi} \int_{-\infty}^{\infty} \frac{1}{\xi} \widetilde{E}_{III}(\zeta) H_1^{(1)}(\xi r) e^{-i\zeta z} d\zeta \tag{27}$$

where $\xi = \sqrt{\beta_0^2 - \zeta^2}$ and $\beta_0 = \omega\sqrt{\mu\epsilon_0}$.

The E_z and H_ϕ field continuities at $r = a$ between regions (I) and (II) yield

$$\sum_{n=0}^{N-1} \sum_{m=0}^{\infty} \left[\frac{H_0^{(1)}(\kappa_m a)}{H_1^{(1)}(\kappa_m a)} \bar{I}_1 - \frac{d}{2} \frac{\varepsilon_m}{\kappa_m} \delta_{ms}\delta_{np} \right] p_m^n$$
$$+ \sum_{n=0}^{N-1} \sum_{m=0}^{\infty} \left[\frac{H_0^{(2)}(\kappa_m a)}{H_1^{(2)}(\kappa_m b)} \bar{I}_1 - \frac{d}{2} \frac{\varepsilon_m}{\kappa_m} \frac{H_1^{(2)}(\kappa_m a)}{H_1^{(2)}(\kappa_m b)} \delta_{ms}\delta_{np} \right] q_m^n$$
$$= -\frac{J_1(k_c a)}{k_c} G_s^p(\beta_z) \tag{28}$$

where

$$\bar{I}_1 = \frac{1}{2\pi} \int_{-\infty}^{\infty} \frac{J_1(\kappa a)}{\kappa J_0(\kappa a)} G_m^n(\zeta) G_s^p(-\zeta) d\zeta \tag{29}$$

$$G_m^n(\zeta) = \frac{-i\zeta[(-1)^m e^{i\zeta d} - 1]}{\zeta^2 - a_m^2} e^{i\zeta nT} . \tag{30}$$

Using the residue calculus, we convert $\bar{I}_1$ into a fast convergent series

$$\bar{I}_1 = \frac{d}{2} \frac{J_1(\kappa a)}{\kappa J_0(\kappa a)} \varepsilon_m \delta_{ms} \delta_{np} \bigg|_{\zeta=a_m}$$

$$+ \sum_{j=1}^{\infty} i\zeta \frac{A_1}{a(\zeta^2 - a_m^2)(\zeta^2 - a_s^2)} \bigg|_{\zeta=\zeta_j} \tag{31}$$

$$A_1 = \left[(-1)^{m+s} + 1\right] \exp(i\zeta|n - p|T)$$
$$- (-1)^m \exp\left[i\zeta|d + (n - p)T|\right]$$
$$- (-1)^s \exp\left[i\zeta|(n - p)T - d|\right] \tag{32}$$

where ζ_j are the roots of $J_0(\kappa a)|_{\zeta=\zeta_j} = 0$. In addition, the E_z and H_ϕ field continuities at $r = b$ between regions (II) and (III) yield

$$\boxed{\begin{aligned}
&\sum_{n=0}^{N-1} \sum_{m=0}^{\infty} \left[\frac{H_0^{(1)}(\kappa_m b)}{H_1^{(1)}(\kappa_m a)} \bar{I}_2 - \epsilon_r \frac{d}{2} \frac{\varepsilon_m}{\kappa_m} \frac{H_1^{(1)}(\kappa_m b)}{H_1^{(1)}(\kappa_m a)} \delta_{ms} \delta_{np} \right] p_m^n \\
&+ \sum_{n=0}^{N-1} \sum_{m=0}^{\infty} \left[\frac{H_0^{(2)}(\kappa_m b)}{H_1^{(2)}(\kappa_m b)} \bar{I}_2 - \epsilon_r \frac{d}{2} \frac{\varepsilon_m}{\kappa_m} \delta_{ms} \delta_{np} \right] q_m^n = 0
\end{aligned}} \tag{33}$$

where

$$\bar{I}_2 = \frac{1}{2\pi} \int_{-\infty}^{\infty} \frac{H_1^{(1)}(\xi b)}{\xi H_0^{(1)}(\xi b)} G_m^n(\zeta) G_s^p(-\zeta) d\zeta . \tag{34}$$

We transform $\bar{I}_2$ into a numerically-efficient integral

$$\bar{I}_2 = \frac{d}{2} \frac{H_1^{(1)}(\xi b)}{\xi H_0^{(1)}(\xi b)} \varepsilon_m \delta_{ms} \delta_{np} \bigg|_{\zeta=a_m} - \left\{ [1 + (-1)^{m+s}] \bar{I}_3(gT) \right.$$

$$\left. - (-1)^m \bar{I}_3(d + gT) - (-1)^s \bar{I}_3(-d + gT) \right\} \tag{35}$$

where $g = n - p$ and

$$\bar{I}_3(x) = \frac{2}{\pi^2} \int_0^{\infty}$$

$$\frac{(1 + iv)^2 e^{i\beta_0|x|} e^{-\beta_0 v|x|}}{\beta_0^3 v(-2i + v) b[(1 + iv)^2 - (a_m/\beta_0)^2][(1 + iv)^2 - (a_s/\beta_0)^2]}$$

$$\cdot \frac{1}{J_0^2[\beta_0 b\sqrt{v(-2i + v)}] + N_0^2[\beta_0 b\sqrt{v(-2i + v)}]} dv . \tag{36}$$

The scattered field at $z = \pm\infty$ in region (I) is

$$H_{\phi I}(r, \pm\infty) = \sum_{j=1}^{\infty} L_j^{\pm}(\zeta) \frac{-i\omega\epsilon}{\kappa} J_1(\kappa r) e^{\pm i\zeta z} \bigg|_{\zeta=\zeta_j} \tag{37}$$

where

$$L_j^{\pm}(\zeta) = \mp \sum_{n=0}^{N-1} \sum_{m=0}^{\infty} \frac{R_0(\kappa_m a)}{a} \frac{\kappa[(-1)^m e^{\mp i\zeta d} - 1]}{J_1(\kappa a)(\zeta^2 - a_m^2)} e^{\mp i\zeta nT} \tag{38}$$

and ζ_j is determined by $J_0(\kappa a)|_{\zeta=\zeta_j} = 0$. The reflection ($\varrho$), transmission ($\tau$), and scattering ($\sigma$) coefficients are

$$\tau = \frac{P_t}{P_i}$$

$$= |1 + L_j^+(\zeta_j)|^2_{\zeta_j=\beta_z} + \sum_{\zeta_j \neq \beta_z} \frac{k_c^2 Re(\zeta) J_1^2(\kappa a)}{\kappa^2 Re(\beta_z) J_1^2(k_c a)} |L_j^+(\zeta)|^2 \Bigg|_{\zeta=\zeta_j} \tag{39}$$

$$\sigma = \frac{P_s}{P_i} \tag{40}$$

$$\varrho = \frac{P_r}{P_i} = \sum_{j=1}^{\infty} \frac{k_c^2 Re(\zeta) J_1^2(\kappa a)}{\kappa^2 Re(\beta_z) J_1^2(k_c a)} |L_j^-(\zeta)|^2 \Bigg|_{\zeta=\zeta_j} \tag{41}$$

where

$$P_i = \frac{\pi a^2 \omega \epsilon}{2 k_c^2} J_1^2(k_c a) Re(\beta_z) \tag{42}$$

$$P_s = \frac{\pi b d}{2} Re \left\{ \sum_{n=0}^{N-1} \sum_{m=0}^{\infty} \frac{i\omega\epsilon\varepsilon_m}{\kappa_m^*} R_0(\kappa_m b) R_0'(\kappa_m b)^* \right\} . \tag{43}$$

The far-zone radiation field at distance R ($z = R\cos\theta$ and $r = R\sin\theta$) is

$$H_{\phi III}(R,\theta)$$

$$= \frac{e^{i\beta_0 R}}{R} \sum_{n=0}^{N-1} \sum_{m=0}^{\infty} \frac{R_0(\kappa_m b)}{\pi \tan\theta}$$

$$\cdot \left\{ \frac{\omega\epsilon_0 [1 - (-1)^m \exp(-i\beta_0 d \cos\theta)] \exp(-i\beta_0 nT\cos\theta)}{H_0^{(1)}(\beta_0 b \sin\theta) [(\beta_0 \cos\theta)^2 - a_m^2]} \right\} . \tag{44}$$

References for Chapter 9

1. J. K. Park and H. J. Eom, "Fourier transform analysis of dielectric-filled edge-slot antenna," *Radio Sci.*, vol. 32, no. 6, pp. 2149-2154, Nov./Dec. 1997. Correction to "Fourier transform analysis of dielectric-filled edge-slot antenna," vol. 33, no. 3, May-June 1998.
2. D. L. Sengupta and L. F. Martins-Camelo, "Theory of dielectric-filled edge-slot antenna," *IEEE Trans. Antennas Propagat.*, vol. 28, no. 4, pp. 481-490, July 1980.
3. J. K. Park and H. J. Eom, "Multiple dielectric-filled edge-slot antenna," *Microwave Opt. Technol. Lett.*, vol. 21, no. 5, pp. 366-367, June 1999.
4. J. K. Park and H. J. Eom, "Radiation from multiple circumferential slots on coaxial cable," *Microwave Opt. Technol. Lett.*, vol. 26, no. 3, pp. 160-162, Aug. 2000.
5. J. F. Kiang, "Radiation properties of circumferential slots on a coaxial cable," *IEEE Trans. Microwave Theory Tech.*, vol. 45, no. 1, pp. 102-107, Jan. 1997.

6. J. K. Park and H. J. Eom, "Radiation from multiple circumferential slots on coaxial cable with a dielectric or plasma layer," *J. Electromagn. Waves Appl.*, vol. 14, no. 3, pp. 359-368, 2000.
7. W. J. Dewar and J. C. Beal, "Coaxial-slot surface wave launchers," *IEEE Trans. Microwave Theory Tech.*, vol. 18, no. 8, pp. 449-455, Aug. 1970.
8. J. R. Wait and D. A. Hill, "On the electromagnetic field of a dielectric coated coaxial cable with an interrupted shield," *IEEE Trans. Antennas Propagat.*, vol. 23, no. 4, pp. 470-479, July 1975.
9. J. R. Wait and D. A. Hill, "Electromagnetic fields of a dielectric coated coaxial cable with an interrupted shield - Quasi-static approach," *IEEE Trans. Antennas Propagat.*, vol. 23, no. 4, pp. 679-682, Sept. 1975.
10. J. K. Park and H. J. Eom, "Radiation from multiple circumferential slots on a conducting circular cylinder," *IEEE Trans. Antennas Propagat.*, vol. 47, no. 2, pp. 287-292, Feb. 1999.
11. C. H. Papas, "Radiation from a transverse slot in an infinite cylinder," *J. Math. Phys.*, vol. 28, pp. 227-236, Jan. 1950.
12. J. R. Wait, *Electromagnetic Radiation from Cylindrical Structures*, Elmsford, NY: Pergamon, 1959.
13. D. C. Chang, "Equivalent-circuit representation and characteristics of a radiating cylinder driven through a circumferential slot," *IEEE Trans. Antennas Propagat.*, vol. 21, no. 6, pp. 792-796, Nov. 1973.
14. S. Papatheodorou, J. R. Mautz, and R. F. Harrington, "The aperture admittance of a circumferential slot in a circular cylinder," *IEEE Trans. Antennas Propagat.*, vol. 40, no. 2, pp. 240-244, Feb. 1992.
15. C. M. Knop and L. F. Libelo, "On the leakage radiation from a circumferentially-slotted cylinder and its application to the EMI produced by TEM-coaxial rotary joints," *IEEE Trans. Electromagn. Compat.*, vol. 37, no. 4, pp. 583-589, Nov. 1995.
16. S. Xu and X. Wu, "Millimeter-wave omnidirectional dielectric rod metallic grating antenna," *IEEE Trans. Antennas Propagat.*, vol. 44, no. 1, pp. 74-79, Jan. 1996.

A. Appendix

A.1 Vector Potentials and Field Representations

Maxwell's equations in a time-harmonic case are

$$\nabla \times \boldsymbol{E} = i\omega\boldsymbol{B} - \boldsymbol{M} \tag{1}$$
$$\nabla \times \boldsymbol{H} = -i\omega\boldsymbol{D} + \boldsymbol{J} \tag{2}$$
$$\nabla \cdot \boldsymbol{D} = \rho_e \tag{3}$$
$$\nabla \cdot \boldsymbol{B} = \rho_m \ . \tag{4}$$

Based on the superposition principle, $\boldsymbol{E}$ and $\boldsymbol{H}$ are decomposed into

$$\boldsymbol{E} = \boldsymbol{E}_e + \boldsymbol{E}_m \tag{5}$$
$$\boldsymbol{H} = \boldsymbol{H}_e + \boldsymbol{H}_m \tag{6}$$

where $\boldsymbol{E}_e$ and $\boldsymbol{H}_e$ are due to the electric current density $\boldsymbol{J}$, and $\boldsymbol{E}_m$ and $\boldsymbol{H}_m$ are due to the magnetic current density $\boldsymbol{M}$, respectively. Maxwell's equations for $\boldsymbol{E}_e$ and $\boldsymbol{H}_e$ are therefore

$$\nabla \times \boldsymbol{E}_e = i\omega\mu\boldsymbol{H}_e \tag{7}$$
$$\nabla \times \boldsymbol{H}_e = -i\omega\epsilon\boldsymbol{E}_e + \boldsymbol{J} \tag{8}$$
$$\nabla \cdot \boldsymbol{D}_e = \rho_e \tag{9}$$
$$\nabla \cdot \boldsymbol{B}_e = 0 \ . \tag{10}$$

Introducing the magnetic vector potential $\boldsymbol{A}$ such as

$$\boldsymbol{B}_e = \nabla \times \boldsymbol{A} \tag{11}$$

and the Lorentz condition $\nabla \cdot \boldsymbol{A} = i\omega\mu\epsilon\phi_e$, we get

$$(\nabla^2 + k^2)\boldsymbol{A} = -\mu\boldsymbol{J} \ . \tag{12}$$

Note that ϕ_e is the electric scalar potential and $k=\omega\sqrt{\mu\epsilon}$ is the wavenumber. Similarly to determine $\boldsymbol{E}_m$ and $\boldsymbol{H}_m$, we introduce the electric vector potential $\boldsymbol{F}$ where $\boldsymbol{D}_m = -\nabla \times \boldsymbol{F}$. We then obtain

$$(\nabla^2 + k^2)\boldsymbol{F} = -\epsilon\boldsymbol{M} \ . \tag{13}$$

Hence the total $\boldsymbol{E}$ and $\boldsymbol{H}$ are

$$\boldsymbol{E} = \boldsymbol{E}_e + \boldsymbol{E}_m = \mathrm{i}\omega\boldsymbol{A} + \frac{\mathrm{i}}{\omega\mu\epsilon}\nabla(\nabla\cdot\boldsymbol{A}) - \frac{1}{\epsilon}\nabla\times\boldsymbol{F} \tag{14}$$

$$\boldsymbol{H} = \boldsymbol{H}_e + \boldsymbol{H}_m = \frac{1}{\mu}\nabla\times\boldsymbol{A} + \mathrm{i}\omega\boldsymbol{F} + \frac{\mathrm{i}}{\omega\mu\epsilon}\nabla(\nabla\cdot\boldsymbol{F})\,. \tag{15}$$

Consider an electromagnetic wave propagation in the z-direction along an infinitely long waveguide with a uniform cross section. The vector potentials

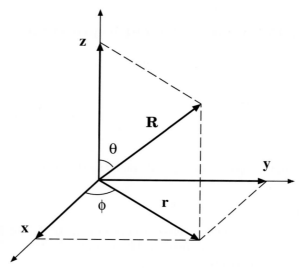

Fig. A.1. Orthogonal coordinate system

in a source-free region of cylindrical waveguide satisfy the wave equations

$$(\nabla^2 + k^2)\boldsymbol{A} = 0 \tag{16}$$
$$(\nabla^2 + k^2)\boldsymbol{F} = 0\,. \tag{17}$$

If we choose $\boldsymbol{F} = \hat{z}F_z(x,y,z)$ and $\boldsymbol{A} = 0$, then $E_z(x,y,z) = 0$ and this type is called the TE wave (transverse electric to the wave propagation direction z). Similarly when $\boldsymbol{A} = \hat{z}A_z(x,y,z)$ and $\boldsymbol{F} = 0$, we obtain the TM wave (transverse magnetic to the z-direction) with $H_z(x,y,z) = 0$. The explicit field expressions in rectangular and cylindrical coordinates are shown in Tables A.1 and A.2, respectively.

If the wavenumber in the z-direction is k, then $E_z = H_z = 0$. This type is referred to as the TEM wave.

Table A.1. Field representations in rectangular coordinates

Rectangular	TM wave	TE wave
E_x	$\dfrac{\mathrm{i}}{\omega\mu\epsilon}\dfrac{\partial^2 A_z}{\partial x \partial z}$	$-\dfrac{1}{\epsilon}\dfrac{\partial F_z}{\partial y}$
E_y	$\dfrac{\mathrm{i}}{\omega\mu\epsilon}\dfrac{\partial^2 A_z}{\partial y \partial z}$	$\dfrac{1}{\epsilon}\dfrac{\partial F_z}{\partial x}$
E_z	$\dfrac{\mathrm{i}}{\omega\mu\epsilon}\left(\dfrac{\partial^2}{\partial z^2}+k^2\right)A_z$	0
H_x	$\dfrac{1}{\mu}\dfrac{\partial A_z}{\partial y}$	$\dfrac{\mathrm{i}}{\omega\mu\epsilon}\dfrac{\partial^2 F_z}{\partial x \partial z}$
H_y	$-\dfrac{1}{\mu}\dfrac{\partial A_z}{\partial x}$	$\dfrac{\mathrm{i}}{\omega\mu\epsilon}\dfrac{\partial^2 F_z}{\partial y \partial z}$
H_z	0	$\dfrac{\mathrm{i}}{\omega\mu\epsilon}\left(\dfrac{\partial^2}{\partial z^2}+k^2\right)F_z$

Table A.2. Field representations in circular cylindrical coordinates

Cylindrical	TM wave	TE wave
E_r	$\dfrac{\mathrm{i}}{\omega\mu\epsilon}\dfrac{\partial^2 A_z}{\partial r \partial z}$	$-\dfrac{1}{\epsilon r}\dfrac{\partial F_z}{\partial \phi}$
E_ϕ	$\dfrac{\mathrm{i}}{\omega\mu\epsilon r}\dfrac{\partial^2 A_z}{\partial \phi \partial z}$	$\dfrac{1}{\epsilon}\dfrac{\partial F_z}{\partial r}$
E_z	$\dfrac{\mathrm{i}}{\omega\mu\epsilon}\left(\dfrac{\partial^2}{\partial z^2}+k^2\right)A_z$	0
H_r	$\dfrac{1}{\mu r}\dfrac{\partial A_z}{\partial \phi}$	$\dfrac{\mathrm{i}}{\omega\mu\epsilon}\dfrac{\partial^2 F_z}{\partial r \partial z}$
H_ϕ	$-\dfrac{1}{\mu}\dfrac{\partial A_z}{\partial r}$	$\dfrac{\mathrm{i}}{\omega\mu\epsilon r}\dfrac{\partial^2 F_z}{\partial \phi \partial z}$
H_z	0	$\dfrac{\mathrm{i}}{\omega\mu\epsilon}\left(\dfrac{\partial^2}{\partial z^2}+k^2\right)F_z$

Index

Printing: Mercedes-Druck, Berlin
Binding: Buchbinderei Lüderitz & Bauer, Berlin